Selected Titles in This Series

69 **C. T. C. Wall (A. A. Ranicki, Editor),** Surgery on compact manifolds, Second Edition, 1999

68 **David A. Cox and Sheldon Katz,** Mirror symmetry and algebraic geometry, 1999

67 **A. Borel and N. Wallach,** Continuous cohomology, discrete subgroups, and representations of reductive groups, Second Edition, 1999

66 **Yu. Ilyashenko and Weigu Li,** Nonlocal bifurcations, 1999

65 **Carl Faith,** Rings and things and a fine array of twentieth century associative algebra, 1999

64 **Rene A. Carmona and Boris Rozovskii, Editors,** Stochastic partial differential equations: Six perspectives, 1999

63 **Mark Hovey,** Model categories, 1999

62 **Vladimir I. Bogachev,** Gaussian measures, 1998

61 **W. Norrie Everitt and Lawrence Markus,** Boundary value problems and symplectic algebra for ordinary differential and quasi-differential operators, 1999

60 **Iain Raeburn and Dana P. Williams,** Morita equivalence and continuous-trace C^*-algebras, 1998

59 **Paul Howard and Jean E. Rubin,** Consequences of the axiom of choice, 1998

58 **Pavel I. Etingof, Igor B. Frenkel, and Alexander A. Kirillov, Jr.,** Lectures on representation theory and Knizhnik-Zamolodchikov equations, 1998

57 **Marc Levine,** Mixed motives, 1998

56 **Leonid I. Korogodski and Yan S. Soibelman,** Algebras of functions on quantum groups: Part I, 1998

55 **J. Scott Carter and Masahico Saito,** Knotted surfaces and their diagrams, 1998

54 **Casper Goffman, Togo Nishiura, and Daniel Waterman,** Homeomorphisms in analysis, 1997

53 **Andreas Kriegl and Peter W. Michor,** The convenient setting of global analysis, 1997

52 **V. A. Kozlov, V. G. Maz′ya, and J. Rossmann,** Elliptic boundary value problems in domains with point singularities, 1997

51 **Jan Malý and William P. Ziemer,** Fine regularity of solutions of elliptic partial differential equations, 1997

50 **Jon Aaronson,** An introduction to infinite ergodic theory, 1997

49 **R. E. Showalter,** Monotone operators in Banach space and nonlinear partial differential equations, 1997

48 **Paul-Jean Cahen and Jean-Luc Chabert,** Integer-valued polynomials, 1997

47 **A. D. Elmendorf, I. Kriz, M. A. Mandell, and J. P. May (with an appendix by M. Cole),** Rings, modules, and algebras in stable homotopy theory, 1997

46 **Stephen Lipscomb,** Symmetric inverse semigroups, 1996

45 **George M. Bergman and Adam O. Hausknecht,** Cogroups and co-rings in categories of associative rings, 1996

44 **J. Amorós, M. Burger, K. Corlette, D. Kotschick, and D. Toledo,** Fundamental groups of compact Kähler manifolds, 1996

43 **James E. Humphreys,** Conjugacy classes in semisimple algebraic groups, 1995

42 **Ralph Freese, Jaroslav Ježek, and J. B. Nation,** Free lattices, 1995

41 **Hal L. Smith,** Monotone dynamical systems: an introduction to the theory of competitive and cooperative systems, 1995

40.4 **Daniel Gorenstein, Richard Lyons, and Ronald Solomon,** The classification of the finite simple groups, number 4, 1999

40.3 **Daniel Gorenstein, Richard Lyons, and Ronald Solomon,** The classification of the finite simple groups, number 3, 1998

(*Continued in the back of this publication*)

Surgery on Compact Manifolds

Second Edition

Mathematical
Surveys
and
Monographs

Volume 69

Surgery on Compact Manifolds

Second Edition

C. T. C. Wall

Edited by A. A. Ranicki

American Mathematical Society

1991 *Mathematics Subject Classification.* Primary 57–02, 57R57;
Secondary 18F25, 19J25, 11E39.

ABSTRACT. This is a new edition of the classic 1970 book on the surgery theory of compact manifolds, which is the standard book on the subject. The original text has been supplemented by notes on subsequent developments, and the references have been updated. The book should appeal to any mathematician interested in the algebraic and geometric topology of manifolds.

Library of Congress Cataloging-in-Publication Data

Wall, C. T. C. (Charles Terence Clegg)
Surgery on compact manifolds / C. T. C. Wall.—2nd ed. / edited by A. A. Ranicki.
p. cm. — (Mathematical surveys and monographs, ISSN 0076-5376 ; vol. 69)
Includes bibliographical references and index.
ISBN 0-8218-0942-3 (alk. paper)
1. Manifolds (Mathematics) 2. Surgery (Topology) I. Ranicki, Andrew, 1948– . II. Title. III. Series: Mathematical surveys and monographs ; no. 69.
QA613.W35 1999
514′.3—dc21 99-12274
CIP

♾ The paper used in this book is acid-free and falls within the guidelines
established to ensure permanence and durability.
Visit the AMS home page at URL: http://www.ams.org/

10 9 8 7 6 5 4 3 2 1 04 03 02 01 00 99

Contents

Part 4: Postscript

Foreword to the first edition (1970)

This book is being published in the form in which it was originally planned and written. In some ways, this is not satisfactory: the demands made on the reader are rather heavy, though this is partly also due to a systematic attempt at completeness ('simplified' proofs have appeared of some of my results, but in most cases the simplification comes primarily from a loss of generality).

However, the partly historical presentation adopted here has its advantages: the reader can see (particularly in §5 and §6) how the basic problem of surgery leads to algebra, before meeting the abstract presentation in §9. Indeed, this relation of geometry to algebra is the main theme of the book. I have not in fact emphasised the algebraic aspects of the L-groups, though this is mentioned where necessary in the text: in particular, I have omitted the algebraic details of the calculations of the L-groups, since this is lengthy, and needs a different background. Though some rewriting is desirable (I would prefer to recast several results in the framework suggested in §17G; also, some rather basic results were discovered too late to be fully incorporated at the appropriate points – see the footnotes and Part 4) this would delay publication indefinitely, so it seemed better for the book to appear now, and in this form.

Chapters 0–9 were issued as duplicated notes from Liverpool University in Spring, 1967. They have been changed only by correcting minor errors, adding §1A (which originated as notes from Cambridge University in 1964), and correcting a mistake in the proof of (9.4). Part 2 was issued (in its present form) as duplicated notes from Liverpool University in May 1968. The rest of the material appears here for the first time.

Foreword to the second edition

It is gratifying to learn that there is still sufficient interest in this book for it to be worth producing a new edition. Although there is a case for substantially rewriting some sections, to attempt this would have delayed production indefinitely.

I am thus particularly pleased that Andrew Ranicki has supplemented the original text by notes which give hints to the reader, indicate relevant subsequent developments, and say where the reader can find accounts of such newer results. He is uniquely qualified to do this, and I am very happy with the result.

The first edition appeared before the days of TEX, so the entire manuscript had to be re-keyed. I am grateful to Iain Rendall for doing this efficiently and extremely accurately.

C. T. C. Wall, Liverpool, November 1998.

Editor's foreword to the second edition

The publication of this book in 1970 marked the culmination of a particularly exciting period in the history of the topology of manifolds. The world of high-dimensional manifolds had been opened up to the classification methods of algebraic topology by

- Thom's work on transversality and cobordism (1952)
- the signature theorem of Hirzebruch (1954)
- the discovery of exotic spheres by Milnor (1956).

In the 1960's there had been an explosive growth of interest in the surgery method of understanding the homotopy types of manifolds (initially in the differentiable category), including such results as

- the h-cobordism theorem of Smale (1960)
- the classification of exotic spheres by Kervaire and Milnor (1962)
- Browder's converse to the Hirzebruch signature theorem for the existence of a manifold in a simply connected homotopy type (1962)
- Novikov's classification of manifold structures within a simply connected homotopy type (1962)
- the s-cobordism theorem of Barden, Mazur and Stallings (1964)
- Novikov's proof of the topological invariance of the rational Pontrjagin classes of differentiable manifolds (1965)
- the fibering theorems of Browder and Levine (1966) and Farrell (1967)
- Sullivan's exact sequence for the set of manifold structures within a simply connected homotopy type (1966)
- Casson and Sullivan's disproof of the Hauptvermutung for piecewise linear manifolds (1967)
- Wall's classification of homotopy tori (1969)
- Kirby and Siebenmann's classification theory of topological manifolds (1970).

The book fulfilled five purposes, providing:

1. a coherent framework for relating the homotopy theory of manifolds to the algebraic theory of quadratic forms, unifying many of the previous results;
2. a surgery obstruction theory for manifolds with arbitrary fundamental group, including the exact sequence for the set of manifold structures within a homotopy type, and many computations;
3. the extension of surgery theory from the differentiable and piecewise linear categories to the topological category;
4. a survey of most of the activity in surgery up to 1970;
5. a setting for the subsequent development and applications of the surgery classification of manifolds.

However, despite the book's great influence it is not regarded as an 'easy read'. In this edition I have tried to lighten the heavy demands placed on the reader by suggesting that §§ 0, 7, 8, 9, 12 could be omitted the first time round – it is possible to take in a substantial proportion of the foundations of surgery theory in Parts 1 and 2 and the applications in Part 3 without these chapters.

Readers unfamiliar with surgery theory should have the papers of Milnor [M12], Kervaire and Milnor [K4] at hand, and see how the construction and classification of exotic spheres fits into the general theory. Also, the books of Browder [B24] and Novikov [N9] provide accounts of surgery from the vantage points of two pioneers of the field.

My own experience with reading this book was somewhat unusual. I was a first-year graduate student at Cambridge, working on Novikov's paper [N8], when the book reached the bookshops in early 1971*. When I finally acquired a copy, I was shocked to note that the very last reference in the book was to [N8], so that in effect I read the book backwards. The book accompanied me throughout my career as a graduate student (and beyond) – I always had it with me on my visits home, and once my mother asked me: 'Haven't you finished reading it yet?' My own research and books on surgery have been my response to this book, which I have still not finished reading.

Preparing the second edition of the book was an even more daunting experience than reading the first edition. It would be impossible to give a full account of all the major developments in surgery which followed the first edition without at least doubling the length of the book – the collections of papers [C7], [F10] include surveys of many areas of surgery theory. In particular, I have not even tried to do justice to the controlled and bounded theories (Quinn [Q6], Ferry and Pedersen [F9]), which are among the most important developments in surgery

*I have a vivid memory of telephoning the Foyles bookshop in London in search of a copy, and being directed to the medical department.

since 1970. But it is perhaps worth remarking on the large extent to which the formal structures of these theories are patterned on the methods of this book.

In preparing this edition I have added notes at the beginnings and ends of various chapters, and footnotes; I have also updated and renumbered the references. All my additions are set in italic type. However, I have not modified the text itself except to correct misprints and to occasionally bring the terminology into line with current usage.

A. A. Ranicki, Edinburgh, January 1999.

Introduction

This book represents an attempt to collect and systematise the methods and main applications of the method of surgery, insofar as compact (but not necessarily connected, simply connected or closed) manifolds are involved. I have attempted to give a reasonably thorough account of the theoretical part, but have confined my discussion of applications mostly to those not accessible by surgery on simply connected manifolds (which case is easier, and already adequately covered in the literature).

The plan of the book is as follows. Part 0 contains some necessary material (mostly from homotopy theory) and §1, intended as a general introduction to the technique of surgery. Part 1 consists of the statement and proof of our main result, namely that the possibility of successfully doing surgery depends on an obstruction in a certain abelian group, and that these 'surgery obstruction groups' depend only on the fundamental groups involved and on dimension modulo 4. Part 2 shows how to apply the result. §10 gives a rather detailed survey of the problem of classifying manifolds with a given simple homotopy type. In §11, we consider the analogous problem for submanifolds: it turns out that in codimension $\geqslant 3$ there are no surgery obstructions and in codimensions 1 and 2 the obstructions can be described by the preceding theory. Where alternative methods of studying these obstructions exist, we obtain calculations of surgery obstruction groups; two such are obtained in §12. In part 3, I begin by summarising all methods of calculating surgery obstructions, and then apply some of these results to homeomorphism classification problems: my results on homotopy tori were used by Kirby and Siebenmann in their spectacular work on topological manifolds. In Part 4 are collected mentions of several ideas, half-formed during the writing of the book, but which the author does not have time to develop, and discussions of some of the papers on the subject which have been written by other authors during the last two years.

The order of the chapters is not artificial, but readers who want to reach the main theorem as quickly as possible may find the following suggestions useful. Begin with §1, and read §4 next. Then glance at the statements in §3 and skip to §9 for the main part of the proof. Then read §10 and the first half of §11. Beyond this, it depends what you want: for the work on tori (§15), for example, you first need §12B, (13A.8) and (13B.8).

The technique of surgery was not invented by the author, and this book clearly owes much to previous work by many others, particularly Milnor, Novikov and Browder. I have tried to give references in the body of the book wherever a result or proof is substantially due to someone else.

Part 0

Preliminaries

Note on Conventions

Throughout this book we follow the convention (customary among topologists) of writing the operator before the operand, and thus writing compositions from right to left. Since the linear (and quadratic) algebra in this book is intimately related to the topology, we are forced to adopt the corresponding conventions there, with the following consequences.

Scalar multiplication in a module (which commutes with linear operators) is written on the *right*, thus we habitually study right modules. Given a linear map between free modules

$$\alpha : V \to W$$

where $\{e_i\}$ is a basis of V and $\{f_j\}$ a basis of W, we write

$$\alpha(e_i) \;=\; \sum_j f_j a_{ji}$$

and make α correspond to the matrix A with $(A)_{ji} = a_{ji}$. Frequently we denote an operator and its matrix by the same symbol, when the bases are understood. We also use matrix notation more generally for maps into or from a direct sum of modules, as e.g. in

$$\begin{pmatrix} a_1 \\ a_2 \end{pmatrix} : V \to W_1 \oplus W_2 \quad , \quad \begin{pmatrix} b_1 & b_2 \end{pmatrix} : W_1 \oplus W_2 \to X\,.$$

The composite map is to be evaluated by the usual rule for matrix products, not forgetting that we write composites from right to left. Thus, for example, the composite of the two maps above is

$$\begin{pmatrix} b_1 & b_2 \end{pmatrix}\begin{pmatrix} a_1 \\ a_2 \end{pmatrix} \;=\; b_1 a_1 + b_2 a_2 \;:\; V \to X\,.$$

For our sign conventions see the beginning of §2.

0. Basic Homotopy Notions

First-time readers may omit this chapter, proceeding directly to §1.

We will make much use of the standard notions of CW complex and CW pair (consisting of a complex and subcomplex [W44]). We also need more complicated arrangements of spaces. CW lattices in general are discussed in several papers by E. H. Spanier and J. H. C. Whitehead: see Vol. IV pp. 104–227 of the latter's collected works. We confine ourselves here to CW n-ads. A CW $(n+1)$-ad consists, by definition, of a CW complex and n subcomplexes thereof. In studying such an object, we are forced to consider the intersections of various families of subcomplexes: there are, of course, 2^n such. It is desirable to introduce a systematic notation. We must index all these complexes by reference to a standard model.

Consider an $(n+1)$-ad in general as a set (the 'total' set) with n preferred subsets. We can specify the intersections by a function S on the set of subsets of $\{1, 2, \ldots, n\}$, whose values are sets, and which preserves intersections (and hence, we note, S is compatible with inclusion relations). Then $S\{1, \ldots, n\}$ is the set, and the n preferred subsets are the values of S on the subsets of $\{1, \ldots, n\}$ obtained by deleting one of its members. We denote $S\{1, \ldots, n\}$ by $|S|$. If $|S|$ is a topological space, the subsets inherit topologies, and we speak of a topological $(n+1)$-ad. For a CW $(n+1)$-ad we require not merely that $|S|$ be a CW complex, but that the subsets be subcomplexes. We speak of a finite CW $(n+1)$-ad if $|S|$ is a finite complex. We can also regard the lattice of subsets of $\{1, \ldots, n\}$ as a category $\mathbf{2}^n$ (the morphisms are inclusion maps): S is then an intersection-preserving functor from $\mathbf{2}^n$ to the category of sets or spaces or CW complexes, and appropriate maps. We thus obtain categories of $(n+1)$-ads: in the CW case we permit any continuous maps here.

There are many operations on $(n+1)$-ads. The most natural ones arise as composition with an intersection-preserving functor $\mathbf{2}^m \to \mathbf{2}^n$: for example,

(1) Permutations (we introduce no special notation here).

(2) Given an injective map $f : \{1, \ldots, m\} \to \{1, \ldots, n\}$, take the induced map of subsets. This includes (1), but we are more interested in the maps $\partial_i : \mathbf{2}^{n-1} \to \mathbf{2}^n$ $(1 \leqslant i \leqslant n)$ induced by

$$j \mapsto j \quad (j < i) \qquad j \mapsto j+1 \quad (j \geqslant i).$$

The corresponding functor from $(n+1)$-ads to n-ads corresponds to taking number i of the n subspaces as total space, and using the intersections of the others with it as subsets.

(3) We define a functor $\delta_i : \mathbf{2}^{n-1} \to \mathbf{2}^n$ by $\delta_i(\alpha) = \partial_i(\alpha) \cup \{i\}$. This corresponds just to omitting the ith subset.

(4) Given f as in (2), we can take the inverse image of subsets. In particular, ∂_i defines $s_i : \mathbf{2}^n \to \mathbf{2}^{n-1}$ $(1 \leqslant i \leqslant n)$. This corresponds to introducing $|S|$ as new ith subset.

(5) In a category with initial object (the empty set or a base point), we can introduce this object as new ith subset. We will denote the corresponding operation on n-ads by σ_i: we can define σ_n simply by

$$\left.\begin{aligned} \sigma_n S(X) &= 0 \\ \sigma_n S(\alpha \cup \{n\}) &= S(\alpha) \end{aligned}\right\} \alpha \subset \{1, \ldots, n-1\}.$$

The operations ∂_i, δ_i, s_i, σ_i satisfy four analogues of the usual semi-simplicial identities, with minor changes.

(6) Given an $(m+1)$-ad S and an $(n+1)$-ad T we define an $(m+n+1)$-ad $S \times T$. If (for simplicity) we adjust notation so that T is defined on subsets of $\{m+1, \ldots, m+n\}$, then for $\alpha \subset \{1, \ldots, m\}$, $\beta \subset \{m+1, \ldots, m+n\}$, we define simply

$$S \times T(\alpha \cup \beta) \;=\; S(\alpha) \times T(\beta).$$

One can regard s_n and σ_n as the operations of multiplying by the 2-ads (pairs) $(\{P\}, \{P\})$ and $(\{P\}, \emptyset)$ respectively, where P is a point.

(7) If S is an $(n+1)$-ad, we can form an n-ad by (e.g.) amalgamating the last two subspaces. We will only use this construction in the case where these two subspaces are disjoint, and will denote the n-ad by cS.

There are many other operations, and many further relations between these: we will not attempt to list them here.

We next give a straightforward analogue of the usual mapping cylinder construction for converting maps into inclusion maps. Suppose X a functor from $\mathbf{2}^n$ to the category of topological spaces: we will define the mapping cube, $M(X)$, a topological $(n+1)$-ad. Begin with the disjoint union

$$\bigcup \{X(\alpha) \times I^{\alpha'} : \alpha \in \{1, 2, \ldots, n\}\},$$

where α' denotes the complement of α in $\{1, \ldots, n\}$. Now for $i : \alpha \subset \beta$, we have $\beta' \subset \alpha'$ and will identify each $(x, t) \in X(\alpha) \times I^{\beta'}$ with $\big(X(i)(x), t\big) \in X(\beta) \times I^{\beta'}$. Let $|M(X)|$ be the identification space. There is a well-defined projection

$$p_2 : |M(X)| \to I^{\{1,2,\ldots,n\}}.$$

For $t \in I^{\{1,2,\ldots,n\}}$, we set $t^{-1}\{0\} = \{i : 1 \leqslant i \leqslant n \text{ and } t(i) = 0\}$, and similarly for $t^{-1}\{1\}$. Then $p_2^{-1}(t)$ can be identified with $X(t^{-1}\{0\})$. Define $M(X)(\alpha) =$

$p_2^{-1}\{t \in I^{\{1,2,\ldots,n\}} : t^{-1}\{1\} \supset \alpha'\}$. Then $M(X)(\alpha) \cap M(X)(\beta)$ is defined by $t^{-1}\{1\} \supset \alpha' \cup \beta' = (\alpha \cap \beta)'$, so equals $M(X)(\alpha \cap \beta)$. Thus we have a topological $(n+1)$-ad. There is a canonical inclusion $X(\alpha) \subset M(X)(\alpha)$: we define the t-coordinate (uniquely) by $t(\alpha) = 0$, $t(\alpha') = 1$. Moreover, in the diagram valid for $i : \alpha \subset \beta$,

$$\begin{array}{ccc} X(\alpha) & \hookrightarrow & M(\alpha) \\ {\scriptstyle X(i)}\downarrow & & \downarrow \\ X(\beta) & \hookrightarrow & M(\beta) \end{array}$$

there is a canonical homotopy making the diagram commute: viz. leave the x-coordinate and the coordinates in $\alpha \cup \beta'$ fixed, but let each coordinate in $\beta - \alpha$ have value t at stage t. In the case where X takes values in CW complexes and cellular maps, it is easily verified by induction that $M(X)$ is a CW $(n+1)$-ad. Finally, $X(\alpha)$ is a deformation retract of $M(\alpha)$: we deform the coordinates in α linearly to 0. In the case of finite CW complexes, this is even a cellular collapse, so we have simple homotopy equivalences.

The above construction permits us to define homology, homotopy etc. of topological objects of type $\mathbf{2}^n$ (= functors from $\mathbf{2}^n$ to topological spaces) by considering only topological $(n+1)$-ads. Our definitions will be self-consistent, for if X is already a topological $(n+1)$-ad, we have a well-defined projection $p_1 : |M(X)| \to |X|$ with each inverse image a cube, and $M(X)(\alpha) \subset p_1^{-1}(X(\alpha))$ a homotopy equivalence. Thus in the CW case, p_1 is a homotopy equivalence of $(n+1)$-ads; in general, it is a singular homotopy equivalence.

We define the homology of a CW $(n+1)$-ad K by chain groups: we take the chains of $|K|$ modulo the union (denoted by $|\partial K|$) of all $K(\alpha)$ with α a proper subset of $\{1, \ldots, n\}$. For this we can use any coefficient module over the group ring of $\pi_1(|K|)$, or analogously if $|K|$ is not connected. The short exact sequences of chain complexes

$$0 \to C_*(\partial_i K) \to C_*(\delta_i K) \to C_*(K) \to 0$$

induce the usual homology exact sequences of the $(n+1)$-ad. Analogous observations apply to cohomology; for a topological $(n+1)$-ad we use singular chains.

Now let K be an $(n+1)$-ad in the category of based topological spaces. Define $F(K)$ as the function space of all maps $I^{\{1,\ldots,n\}} \to |K|$ such that $I^\alpha \times 0^{\alpha'} \to K(\alpha)$ for each $\alpha \subset \{1, \ldots, n\}$, and all proper faces with some coordinate 1 map to the base point. Of course, we give $F(K)$ the compact open topology, and base point the trivial map. Now for $r \geqslant 0$, define $\pi_{n+r}(K) = \pi_r(F(K))$: it is a group for $r \geqslant 1$, abelian for $r \geqslant 2$.

The face operator ∂_i induces (compose with I^{δ_i}) a projection $F(K) \to F(\partial_i K)$. This is a fibre map: the fibre is the subspace of maps sending the ith face to a point. This is just $F(\sigma_i \delta_i K)$. We observe that this is the loop space (with respect to the ith variable) of $F(\delta_i K)$. Hencc we have homotopy exact sequences

$$\cdots \to \pi_r(\partial_i K) \to \pi_r(\delta_i K) \to \pi_r(K) \to \pi_{r-1}(\partial_i K) \to \ldots$$

(indeed, we could map in any space, not just spheres).

We can also define bordism groups and obtain corresponding exact sequences. We adopt the philosophy of [W13]. We first define the boundary $(n+1)$-ad, ∂K, of K by

$$\partial K(\alpha) \;=\; \bigcup \{K(\beta) : \beta \subset \alpha, \beta \neq \alpha\}.$$

We call an $(n+1)$-ad M a *manifold* $(n+1)$-ad if for each $\alpha \subset \{1, 2, \ldots, n\}$, $M(\alpha)$ is a manifold with boundary $\partial M(\alpha)$. We obtain correspondingly PL manifold $(n+1)$-ads and smooth manifold $(n+1)$-ads: in the latter case, we usually assume that each $M(\alpha)$ and $M(\beta)$ meet transversely at $M(\alpha \cap \beta)$, so we have a 'variété à bord anguleux' in the sense of Cerf [C12] or Douady [D3].

Note that the contracted n-ad cM is again a manifold n-ad. In the case when $|\partial_{n-1} M|$ and $|\partial_n M|$ are disjoint, we regard either M or cM as a cobordism between the n-ads $\partial_{n-1} M$ and $\partial_n M$: the latter are effectively $(n-1)$-ads since $\partial_{n-1} M = \sigma_n \delta_n (\partial_{n-1} M)$; similarly for $\partial_n M$.

For cobordism purposes we will always assume $|M|$ (and hence each $M(\alpha)$) compact. One can study cobordism of manifold $(n+1)$-ads in general: this study is meaningful only if we impose various restrictions on the manifolds and cobordisms considered, with stronger restrictions on each $|\partial_i M|$ than on $|M|$. For example we have the plain bordism groups of an $(n+1)$-ad, K. Consider maps $\phi : M \to K$ with M a (smooth, compact) manifold $(n+1)$-ad, $\dim |M| = m$. Using disjoint unions of M gives the set of such (M, ϕ) the structure of an abelian monoid. We set $(M, \emptyset) \sim 0$ if there is a manifold $(n+2)$-ad N with $M = \partial_{n+1} N$, and an extension of ϕ to a map $\psi : N \to s_{n+1} K$ of $(n+2)$-ads. Then set $(M_1, \phi_1) \sim (M_2, \phi_2)$ if $(M_1, \phi_1) + (M_2, \phi_2) \sim 0$ (cf. the definition of cobordism above). The usual glueing argument (note the utility here of our corners) shows that this is an equivalence relation; it is evidently compatible with addition. We thus obtain an abelian group $\mathcal{N}_m(K)$. Clearly $\mathcal{N}_m(L) = \mathcal{N}_m(\sigma_i L)$: the inclusions $\partial_i K \subset \delta_i K$, $\sigma_i \delta_i K \subset K$, and restriction define sequences

$$\cdots \to \mathcal{N}_m(\partial_i K) \to \mathcal{N}_m(\delta_i K) \to \mathcal{N}_m(K) \to \mathcal{N}_{m-1}(\partial_i K) \to \ldots$$

which are easily seen to be exact. Since excision holds for unoriented bordism, it is easily seen that $\mathcal{N}_m(K) \cong \mathcal{N}_m(|K|, |\partial K|)$. This remark will not, however, apply to all the generalisations which we will need.

We introduce a convention for the oriented case which will be useful later on. Observe that for manifold $(n+1)$-ads in general, $\dim M(\alpha) - |\alpha|$ is independent of α (we ignore cases $M(\alpha) = \emptyset$ here). We denote this number by $\dim M\{\ \}$: of course if $M\{\ \}$ is empty, this is a convention. Suppose $|M|$ orientable: then so is $\partial |M| = \cup \{M(\alpha) : |\alpha| = n-1\}$; by downward induction on $|\alpha|$, we deduce that all $M(\alpha)$ are then orientable. More precisely, an orientation of $|M|$ induces one on each $M(\alpha)$ in the following way: Let $[M] \in H_m(|M|, \partial|M|)$ be the fundamental class. Let $\Lambda|M| = \cup\{M(\alpha) : |\alpha| \leqslant n-2\}$. Then the image of $[M]$ under

$$\begin{aligned} H_m(|M|, \partial|M|) &\xrightarrow{\partial_*} H_{m-1}(\partial|M|) \xrightarrow{j_*} H_{m-1}(\partial|M|, \Lambda|M|) \\ &\cong \bigoplus_{1 \leqslant i \leqslant n} H_{m-1}(|\partial_i M|, \partial|\partial_i M|) \end{aligned}$$

(where j_* is the inclusion map) shall be denoted by $\sum_{1 \leqslant i \leqslant n}(-1)^i[\partial_i M]$. Now by induction we obtain fundamental classes for each $M(\alpha)$; the usual combinatorial argument which shows that $\partial^2 = 0$ in a simplicial complex demonstrates that the class so obtained depends only on α (and not on any choice of construction).

1. Surgery Below the Middle Dimension

In order to decide if a space X is homotopy equivalent to a manifold it is convenient to start with a normal map $(\phi, F) : M \to X$ from a manifold M, and to consider the possibility of converting (ϕ, F) to a homotopy equivalence by a sequence of surgeries. By definition, a normal map (ϕ, F) is a map $\phi : M \to X$ together with a vector bundle ν over X and a stable trivialisation F of $\tau_M \oplus \phi^* \nu$. This chapter describes the basic procedure of surgery on a normal map : an element $\alpha \in \pi_{r+1}(\phi)$ can be killed by surgery on (ϕ, F) if α can be represented by an embedding $f : S^r \times D^{m-r} \to M^m$ such that (ϕ, F) extends to a normal bordism $(\psi, G) : (N; M, M_+) \to X$ with

$$\begin{aligned} M_+ &= (M - f(S^r \times D^{m-r})) \cup D^{r+1} \times S^{m-r-1} , \\ N &= M \times I \cup_f D^{r+1} \times D^{m-r} . \end{aligned}$$

The effect of the surgery is the bordant normal map $(\phi_+, F_+) : M_+ \to X$ given by the restriction of (ψ, G). The relative homotopy groups in dimensions $\leqslant r+1$ are such that

$$\pi_q(\psi) = \begin{cases} \pi_q(\phi) \\ \pi_{r+1}(\phi)/\langle\alpha\rangle \end{cases} \text{for } \begin{cases} q \leqslant r \\ q = r+1 \end{cases}$$

with $\langle\alpha\rangle \subseteq \pi_{r+1}(\phi)$ the $\mathbf{Z}[\pi_1(X)]$-submodule generated by α. Moreover, (ϕ, F) can be obtained from (ϕ_+, F_+) by a complementary surgery killing an element $\alpha_+ \in \pi_{m-r}(\phi_+)$. The main result of §1 is that every normal map $(\phi, F) : M \to X$ can be made highly connected by surgery below the middle dimension, i.e. is bordant to a normal map $(\phi', F') : M' \to X$ with

$$\pi_{r+1}(\phi') = 0 \text{ for } 2r < m .$$

In this chapter, in addition to obtaining some useful theorems, we try to describe what surgery is about. Let X be a topological space (usually a CW complex, and eventually a finite simplicial complex), M a manifold and $\phi : M \to X$ a map. Then the problem is to perform geometrical constructions on M (and ϕ) to make ϕ as near to a homotopy equivalence (eventually, simple) as possible.

Here we only discuss one construction, which has variously been called spherical modification [W36], χ-equivalence [M12], and Morse reconstruction [N4] by different authors: we follow Milnor [M10] in calling it surgery. Let $f : S^r \times D^{m-r} \to M^m$ be an embedding. We will replace M by the manifold M_+ formed by deleting the interior of the image of f, and glueing in its place $D^{r+1} \times S^{m-r-1}$. We must also do something with ϕ. The following seems to be a better description of the process. We form a new manifold from the union of $M \times I$ and $D^{r+1} \times D^{m-r}$ by attaching $S^r \times D^{m-r}$ to its image under $f \times 1$. Then M or rather $M \times 0$ appears at the 'lower' end of N, and M_+ at the 'top'. We describe this process as attaching an $(r+1)$-handle to $M \times I$.*

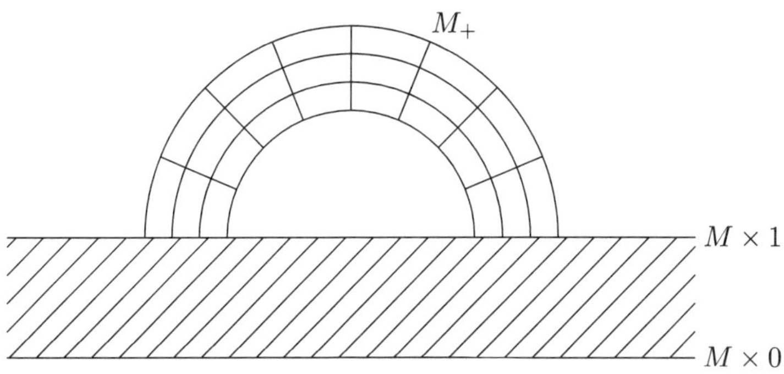

We will extend ϕ to a map ψ of all of N, not just the ends. This is (as is well-known) a homotopy problem, and up to homotopy, we may regard N as formed from M by attaching an $(r+1)$-cell, with attaching map $\overline{f} = f \,|\, S^r \times 0$. So up to homotopy the construction is defined by

(i) The map $\overline{f} : S^r \to M$

(ii) A nullhomotopy of $\phi \circ \overline{f}$.

But equivalence classes of such pairs define the relative homotopy group $\pi_{r+1}(\phi)$. Hence given $\alpha \in \pi_{r+1}(\phi)$, we may speak of surgery on the class α.

We now consider what conditions to impose in order to obtain results. Given α, we would like to be able to do surgery on it. This requires a technique for constructing embeddings f of $S^r \times D^{m-r}$ in M^m. At the time of writing, no such technique is known for a topological manifold:† we will assume, then, that M is either a smooth or a piecewise linear manifold. Even in these cases, we need a condition. Observe that $S^r \times D^{m-r}$ is parallelisable. Thus if the embedding f has the homotopy class of $\overline{f} : S^r \to M$, we must have $\overline{f}^* \tau_M$ trivial (recall that τ_M is the tangent bundle of M). As we are given $\overline{f}$ nullhomotopic in X, we can ensure this by requiring that τ_M be induced (by ϕ) from a bundle over X. In fact, we will require less: namely, that there exists a bundle ν over X (a vector bundle if M is smooth, a PL bundle if M is PL) such that $\phi^*\nu$ is (essentially) the stable normal bundle of M: a convenient way to express this (suggested by J. F. Adams) is to give a stable trivialisation F of $\tau_M \oplus \phi^*\nu$. We will want this condition not only at the outset but at each stage of the surgery;

* *Every cobordism $(W; M_1, M_2)$ has a handle decomposition, so that it is a union of elementary cobordisms of the type $(N; M, M_+)$ (Milnor [M12]).*

† Recent results of Kirby, Siebenmann and Lees have now (1969) provided such a technique. All our methods now extend to the topological case, with only trivial alteration. See [K9] [K10], [L11], [K11].

the convenient way to do this is to impose the additional requirement that F extend to a stable trivialisation of $\tau_N \oplus \psi^*(\nu)$ (where we identify $\tau_N \mid M$ with the direct sum of τ_M and the trivial line bundle).

This leads us to the formulation which we will adopt for the remainder of this book. Suppose given the space X, bundle ν, and an integer m. Consider the set of triples (M, ϕ, F): M a closed m-manifold (smooth or PL), $\phi : M \to X$ a map, F a stable trivialisation of $\tau_M \oplus \phi^*\nu$. Disjoint union gives an addition operation on this set, which is commutative, associative, and has a zero (M empty). Write $(M_1, \phi_1, F_1) \sim (M_2, \phi_2, F_2)$ if there exists a compact $(m+1)$-manifold N with $\partial N = M_1 \cup M_2$, a map $\psi : N \to X$ extending ϕ_1 and ϕ_2, and a stable trivialisation of $\tau_N \oplus \psi^*\nu$ extending F_1 and F_2. Some convention is necessary here to frame the normal bundles of M_1, M_2 in N: we use the inward normal along M_1 and the outward normal along M_2. It is now clear that this is an equivalence relation: we write $\Omega_m(X, \nu)$ for the set of equivalence classes. Clearly the above addition passes to equivalence classes; moreover, using $\partial(M \times I) = M \times 0 \cup M \times 1$ we see that (M, ϕ, F) has inverse $\big(M, \phi, F \oplus (-1)\big)$, so that $\Omega_m(X, \nu)$ is an abelian group.

The following terminology was introduced by W. Browder [B21], [B24]. A triple (ϕ, ν, F) defines a *normal map* from M to X; and (N, ψ, G) gives a *normal cobordism.*

The above equivalence is the same as equivalence by a sequence of surgeries. This follows at once from the fact (see e.g. [M12]) that N can be obtained from $M \times I$ by attaching a (finite) sequence of handles. We now give the theorem which shows how we can do surgery.

THEOREM 1.1. *Let M^m be a smooth or PL manifold (perhaps with boundary), $\phi : M \to X$ a continuous map, ν a vector bundle or PL bundle over X, and F a stable trivialisation of $\tau_M \oplus \phi^*\nu$. Then any $\alpha \in \pi_{r+1}(\phi)$, $r \leqslant m-2$, determines a regular homotopy class of immersions $S^r \times D^{m-r} \to M$. We can use the embedding $f : S^r \times D^{m-r} \to M$ to do surgery on α if and only if f is in this class.*

Proof First choose a representative for α, say

$$\begin{array}{ccc} S^r & \xrightarrow{\ i\ } & D^{r+1} \\ {\scriptstyle f_1}\downarrow & & \downarrow{\scriptstyle g_1} \\ M & \xrightarrow{\ \phi\ } & X \end{array}$$

The stable trivialisation F of $\tau_M \oplus \phi^*\nu$ induces by f_1 a stable trivialisation of $f_1^*\tau_M \oplus f_1^*\phi^*\nu = f_1^*\tau_M \oplus i^*g_1^*\nu$. But as D^{r+1} is contractible, we obtain a canonical trivialisation of $g_1^*\nu$, which induces a trivialisation of $i^*g_1^*\nu$. Putting these two together gives a stable trivialisation of $f_1^*\tau_M$, i.e. a stable isomorphism with it of the (trivial) tangent bundle of $S^r \times D^{m-r} \subset \mathbf{R}^m$.

We now appeal to the immersion classification theorems of Hirsch [H15] in the smooth case and Haefliger-Poenaru [H3] in the PL case; or rather to a later mild improvement by Haefliger [H6], as follows:†

PROPOSITION. *Let V^v and M^m be smooth or PL manifolds, $f : V \to M$ a map. Suppose V has a handle decomposition with no handle of dimension $> m-2$, and $v \leqslant m$. Then regular homotopy classes of immersions homotopic to f correspond bijectively (by the tangent map) to stable homotopy classes of stable bundle monomorphisms $\tau_V \to f^*\tau_M$.*

Thus if $r \leqslant m-2$, our stable trivialisation determines a regular homotopy class of immersions.

Now let f be an embedding. If we can use it to do surgery on α, it must belong to the homotopy class $\partial_*\alpha$: we suppose this. Then we can take f_1 above to be $f \mid S^r \times 0$. We can certainly construct the manifold N as described above, and use g_1 to extend ϕ over it. The only remaining problem is whether F extends to a stable trivialisation of $\tau_N \oplus \psi^*\nu$.

Certainly it induces such a trivialisation on $M \times I$: since the handle $D^{r+1} \times D^{m-r}$ is contractible, we have a unique trivialisation there also. We must see whether these agree on the intersection $S^r \times D^{m-r}$ – or equivalently, on S^r. But the first stable trivialisation is f^*F; the second is induced by a contraction of D^{r+1} and the isomorphism df of $f^*\tau_M$ and the trivial tangent bundle of $S^r \times D^{m-r}$. We saw above that these agree precisely when f lies in the specified regular homotopy class. This completes the proof of (1.1).

COROLLARY. *If $m > 2r$, we can do surgery on α.*

For then we may suppose by a general position argument that the immersion f defines an embedding of $S^r \times 0$. It must then also embed some neighbourhood, so we can obtain an embedding by shrinking the fibres.

This corollary resembles the original results of Milnor [M12] more closely than the theorem does, so it may seem that in spite of the apparent generality $r \leqslant m-2$ we have gained little. The fact is that in the case $m = 2r$ it is useful to have the class of immersions and even when $m = 2r+1$ we know that the embedding f can be varied by a regular homotopy (but no more). Also, we do not have to discuss structure groups of bundles and their Stiefel manifolds: this is all taken care of by the appeal to the Proposition above.

We can deduce a theorem, like [W18, (1.4)].

THEOREM 1.2. *Assume the hypotheses of* (1.1) *with M compact and X a finite simplicial complex. Then we can perform a finite sequence of surgeries on M (with handles of dimensions $\leqslant k$) to make ϕ k-connected provided that $m \geqslant 2k$. Thus every m-dimensional normal map is normally cobordant to a k-connected one.*

Proof Replace X by the mapping cylinder of ϕ, so that ϕ is an inclusion. We

†For the topological case see Lees [L11].

must first show by induction on i that if X_i is a sequence of subcomplexes of X, formed by attaching one at a time the simplices of $X - M$ of dimension $\leqslant k$ ($X_0 = M$), we can perform i surgeries on M to obtain a manifold N_i and a homotopy equivalence $\psi_i : N_i \to X_i$ ($N_0 = M \times I$). Assume this for $i-1$: set $\partial N_{i-1} = M \cup M_{i-1}$, $\psi_{i-1} \,|\, M_{i-1} = \phi_{i-1} : M_{i-1} \to X$. The ith simplex, of dimension $(r+1)$, say, determines an element α of $\pi_{r+1}(X, X_i)$. Now N_{i-1} is formed from M by attaching handles of dimension $\leqslant k$. Turning this upside down, it is formed from M_{i-1} by attaching handles of dimension $\geqslant (m+1-k) \geqslant k+1 \geqslant (r+2)$, so (N_{i-1}, M_{i-1}) is $(r+1)$-connected. Thus $\pi_{r+1}(\phi_{i-1}) \cong \pi_{r+1}(X, X_i)$: we choose α' mapping to α, and (by the corollary) perform surgery on α'. This completes the induction step. At the end of the induction we have $X_j = M \cup X^k$, so (X, X_j) is k-connected. By the argument above, (N_j, M_j) is k-connected and $N_j \simeq X_j$, so $\phi_j : M_j \to X$ is also k-connected. This proves (1.2).

We must now consider in more detail the case of manifolds with boundary: note however that both statement and proof of (1.1) and (1.2) are already valid for manifolds with boundary, where the boundary is left fixed by the surgeries. In fact a manifold and its boundary are in this sense (homotopy type up to just below half the dimension) independent. Thus we will prescribe a pair (Y, X) of spaces, and manifold N with boundary, and sharpen ϕ to a map of pairs

$$\phi : (N, \partial N) \to (Y, X).$$

The bundle ν over Y and stable trivialisation F of $\tau_N \oplus \phi^*\nu$ are given as before. We define a cobordism group as before (cf. [W13]). Set $(N_1, \phi_1, F_1) \sim (N_2, \phi_2, F_2)$ if there is a manifold Q with $\partial Q = N_1 \cup P \cup N_2$, $\partial P = \partial N_1 \cup \partial N_2$, an extension of $\phi_1 \cup \phi_2$ to $\psi : (Q, P) \to (Y, X)$, and an extension of F_1 and F_2 (with the same convention as before) to a stable framing of $\tau_Q \oplus \psi^*\nu$. This provides a group $\Omega_n(Y, X, \nu)$ (if we consider manifolds with $\dim N = n$). Also, by [W13, VA, 2.3], there is an exact sequence

$$\cdots \to \Omega_n(X, \nu) \to \Omega_n(Y, \nu) \to \Omega_n(Y, X, \nu) \to \Omega_{n-1}(X, \nu) \to \Omega_{n-1}(Y, \nu) \to \dots,$$

where the homomorphisms are the obvious ones. We note in passing that it is a simple matter to extend the definition to cobordism groups of $(n+1)$-ads X: we give the details in §3.

We must, however, consider the relation between cobordism of manifolds with boundary and surgery as described above. It turns out that we can perform a cobordism in two stages, doing surgery first on the boundary and then on the interior. Note that surgery on M produces eventually a cobordism P with $\partial P = M \cup M'$. If $\partial N = M$, we define a corresponding cobordism of N. Form V by attaching P to N along M (so $\partial V = M'$): then the cobordism is $V \times 1$, with the corner along $M' \times 0$ rounded and one introduced along $M \times 0$ instead (in the smooth case: in the PL case there is no need for this, but we must still specify in what way $V \times I$ is regarded as a cobordism; here, of $N \times 0$ to $V \times I$).

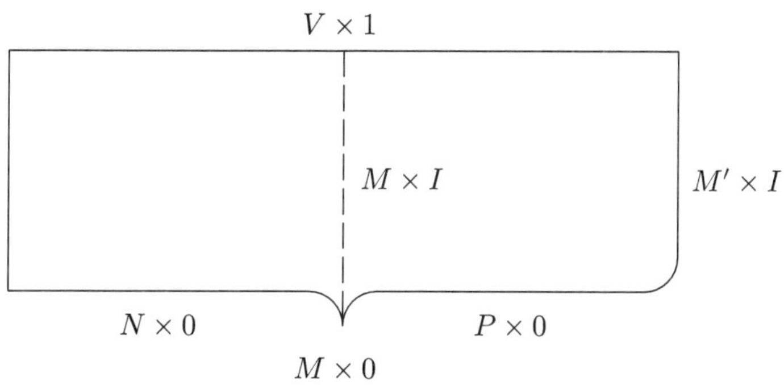

We now assert that any cobordism may be regarded as one of the type above, followed by a sequence of surgeries leaving the boundary fixed. For let Q be a cobordism of N to N', with $\partial N = M$, $\partial N' = M'$, $\partial Q = N \cup P \cup N'$ and $\partial P = M \cup M'$. We first construct V as above and the cobordism $V \times I$ of N to V. We can then (see figure) reinterpret Q as a cobordism of V to N' with the boundary (M') fixed. We do not write down the details (which would involve a collar neighbourhood of M'). Analogous remarks go for $(m+1)$-ads in general.

The above description is asymmetrical as regards the two ends of the cobordism. To restore this symmetry somewhat, we remark that the first step above, 'adding' P to N, has an inverse operation of a new type: 'subtracting' P from V (the cobordism is the same $V \times I$ as above, but now interpreted in the opposite sense). The processes of adding handles to the boundary of N and performing surgery on the interior are adequately described by (1.1). For subtraction, however, we need a little more.

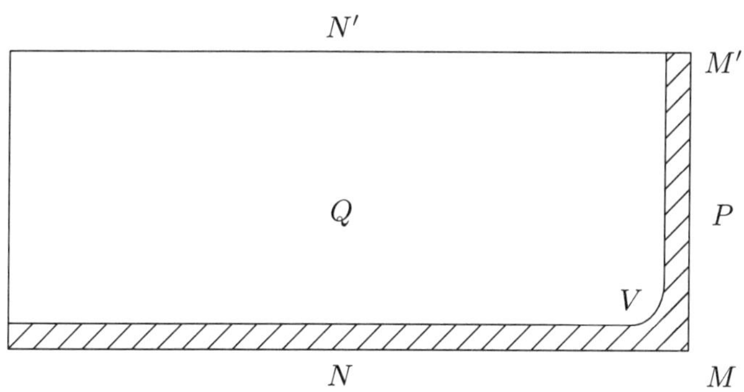

THEOREM 1.3. *Let (N, ϕ, F) define an element of $\Omega_m(Y, X, \nu)$. Write ψ for the quadruple*

$$\begin{array}{ccc} \partial N & \hookrightarrow & N \\ \downarrow & & \downarrow \\ X & \hookrightarrow & Y \end{array}$$

Then any $\alpha \in \pi_{r+1}(\psi)$, $r \leqslant m-2$, *determines a regular homotopy class of immersions*

$$(D^r_- \times D^{m-r}, S^{r-1} \times D^{m-r}) \to (N, \partial N).$$

We can use an embedding f *to do surgery on* α *if and only if* f *is in this class.*

Proof The argument which shows that α determines a class of immersions is the same as in (1.1) except that details are more complicated. Using f^*F and contractibility of the model $D^{r+1} \times D^{m-r}$ we can see that if $f_1 : (D^r_-, S^{r-1}) \to (N, \partial N)$ represents the appropriate $\partial_* \alpha$, we can define a stable trivialisation of $(f_1 \,|\, S^{r-1})^*(\tau_{\partial N})$ which extends to a stable trivialisation of $f_1^* \tau_N$. An appeal to the relative version of the immersion classification theorem now proves the first assertion.

For the second, the picture is a little different. Note that α provides a nullhomotopy

$$(D^{r+1} \times D^{m-r}, D^r_+ \times D^{m-r}) \to (Y, X)$$

of $\phi \circ f$. We regard this as a nullhomotopy of $\phi \,|\, \mathrm{Im}\, f$ (as map of pairs), and extend it to a homotopy of ϕ. Thus we can assume that $\phi(D^r_- \times D^{m-r}) = *$, and that the nullhomotopy is constant at $*$ (at the base point in X).

Form N_0 from N by deleting the interior of $f(D^r_- \times D^{m-r})$: then ϕ induces a map $\phi_0 : (N_0, \partial N_0) \to (Y, X)$, and N is obtained from N_0 by adding an $(m-r)$-handle. We can regard $N \times I$ (with corners adjusted) as a cobordism of (N, ϕ) to (N_0, ϕ_0): it remains, then, to check the stable framing. But as in (1.1), it is trivial that the given stable framing agrees on $D^r_- \times D^{m-r}$ with one induced by a contraction of the model if and only if the tangent map of f is stable as described above, which implies that f is in the class we have defined.

COROLLARY. *If* $m > 2r$, *we can do surgery on* α.

Again this follows by a simple general position argument.

We now obtain an analogue of (1.2) for bounded manifolds: the details are necessarily a little more complicated. One difficulty is that although it is reasonable to study only connected manifolds, we still wish to permit them to have disconnected boundaries. We can regard (1.2) as covering this case if we require a k-connected map $M \to X$ $(k \geqslant 1)$ to induce a bijection $\pi_0(M) \to \pi_0(X)$ of components and, in addition, a k-connected map of each corresponding pair of components. An equivalent condition is that ϕ be 0-connected (i.e. surjective on components) and also also k-connected in the ordinary homotopy sense with respect to each possible base point in M (one in each component is enough). One can make analogous definitions for maps of pairs, triads, etc., but unless the members of the pair have isomorphic fundamental groups, the notion does not seem to be a very useful extension of the one in the absolute case. If $\phi : (N, M) \to (Y, X)$ is a map of pairs, we will use instead the homological connectivity of ϕ (with coefficients $A = \mathbf{Z}[\pi_1(Y)]$).

THEOREM 1.4. *Let (Y, X) be a finite simplicial pair, ν a vector bundle or PL bundle over Y. Any element ξ of $\Omega_n(Y, X, \nu)$ has a representative (N, ϕ, F) such that:*

> *if $n = 2k$, ϕ induces a k-connected map $N \to Y$ and a $(k-1)$-connected map $\partial N \to X$, hence is (homologically) k-connected,*
> *if $n = 2k+1$, ϕ induces k-connected $N \to Y$ and $\partial N \to X$: moreover, ϕ is homologically $(k+1)$-connected.*

Proof Choose an arbitrary representative (N, ϕ, F) of ξ. Then $(\partial N, \phi \,|\, \partial N, F \,|\, \partial N)$ represents $\partial\xi$, and by (1.2) we can find a cobordism, (P, ψ, F_0) say, of this to a $(k-1)$-connected normal map (M, ϕ', F') according as $n = 2k$ or $2k+1$. By adding P to N we obtain a cobordism of N to (N', ϕ'', F''), where ϕ'' has the desired connectivity on $M = \partial N'$. Next we apply (1.2) to N' itself, keeping M fixed. This shows that we can do surgeries to make ϕ'' k-connected thus obtaining, say, (N''', ϕ''', F'''). This concludes the proof in the case $n = 2k$, and for $n = 2k + 1$ all is proved but the last clause, which is the interesting part, and which we obtained originally in [W18, (7.2)] by a different method.

Now the proof of (1.2) shows that, if ϕ' denotes the map $N \to Y$ induced by ϕ, then $\pi_{k+1}(\phi')$ is represented by a finite collection of cells, and so is a finitely generated Λ-module. We choose a finite set of generators; by Theorem 1, Corollary, we can represent each by a framed embedding of S^k. Connect each of these by a tube to ∂N (e.g. all to the same component), so that we have framed embeddings of D^k. Now perform handle subtraction as above: we claim that this achieves the desired result.

For let H denote the union of the handles, N_0 the constructed manifold, $\phi_0 : (N_0, \partial N_0) \to (Y, X)$ the resulting map, and note that $(N_0, \partial N_0) \to (N, H \cup \partial N_0)$ is an excision map. Thus we have an exact sequence (with coefficients Λ throughout)

$$H_k(H \cup \partial N_0, \partial N) \to H_{k+1}(\phi) \to H_{k+1}(\phi_0) \to H_{k-1}(H \cup \partial N_0, \partial N).$$

Clearly the lower relative homotopy (and homology) groups of ϕ_0 vanish as for ϕ. Also, $(H, H \cap \partial N) \to (H \cup \partial N_0, \partial N)$ is an excision map too, and $(H, H \cap \partial N)$ is a collection of copies of $(D^k \times D^{k+1}, S^{k-1} \times D^{k+1})$, so has vanishing homology except in dimension k (thus the final term in our sequence vanishes). In dimension k, we have the classes of the original k-discs, which we choose to represent images in $H_{k+1}(\phi)$ of a set of generators of $\pi_{k+1}(\phi') \cong H_{k+1}(\phi')$. But $H_{k+1}(\phi') \to H_{k+1}(\phi)$ is surjective (we have a k-connected map $\partial N \to X$), so the first map of our exact sequence is surjective too. It follows that $H_{k+1}(\phi_0)$ vanishes, as claimed.

Our proof yields the following strengthening of the asserted result, which we will have occasion to use.

ADDENDUM. *In the case $n = 2k+1$, if $\partial N \to X$ is already k-connected, future alterations of ∂N can be made in a prescribed (non-empty) open subset, and on ∂N have the effect of surgery on trivial framed embeddings of S^{k-1} in ∂N.*

1A. Appendix: Applications

Even the rather simple results of §1 have useful applications in differential topology. We first consider cobordism theory. Following Lashof [L2], given a space X and map $f : X \to BO$, we define an X-structure on a manifold M (with normal bundle ν_M) to be a homotopy class of factorisations through f of the classifying map of ν_M. We then investigate cobordism for manifolds with X-structure, say X-manifolds. This notion can be reformulated by noting that according to Milnor [M7] we can regard the loop space ΩX as a topological group, and $\Omega f : \Omega X \to O$ as a homomorphism. Then an X-structure is a reduction to ΩX of the structural group of the stable normal bundle.

We now suppose that X is a CW complex with finite skeletons. Then it follows from (1.2) that given an X-manifold M^m we can perform surgery to make the classifying map $M \to X$ k-connected, provided $m \geqslant 2k$; moreover, if M has boundary, the surgery can be chosen to leave the boundary fixed.

Our first application of this remark depends, really, only on the techniques of 1-connected surgery.

PROPOSITION 1.5. *In dimensions* $\geqslant 2$ *(resp.* 5*), cobordism groups defined by oriented* 1*-connected (resp.* 2*-connected, resp.* 3*-connected) manifolds map isomorphically to oriented (resp. spinor) cobordism groups.*

Proof Take $X = BSO$ (resp. BSpin) in the above. Then X is 1-connected (resp. 3-connected). Apply the above remark, noting that if $m \geqslant 2(i+1)$ we can make $M \to X$ $(i+1)$-connected, so that if X is i-connected, so is M. This shows that the map defined in the proposition is surjective in these dimensions: for injectivity, it suffices to do surgery similarly on cobordisms between them.

It remains to consider low dimensions. But the oriented cobordism group vanishes in dimensions 2 and 3, as does the spinor one in dimensions 5, 6 and 7. Also, the surgery argument above applies to cobordisms of i-connected $(2i+1)$-manifolds ($i = 1, 2, 3$). Finally, any 1-connected 2-manifold or 3-connected 5- or 6-manifold is already a sphere.

In still lower dimensions, the result breaks down: 2- (or 3-) connected cobordism of 4-manifolds gives h-cobordism of homotopy spheres; the group is zero. But the signature invariant is nonzero on the spinor cobordism group.

Our next application concerns characteristic numbers. Recall that for any manifold M^m with X-structure $g : M^m \to X$, and any class $\xi \in H^m(X)$, $\xi(M) = g^*(\xi)[M]$ is a cobordism invariant (here $[M]$ is an orientation of M for the homology theory H_*, presumed to be obtained from the X-structure).

THEOREM 1.6 *Let k be a field, $\xi \in H^i(X;k)$, M^m $(m \leqslant 2i-2)$ an X-manifold such that for all $\eta \in H^{m-i}(X;k)$, $\xi\eta(M) = 0$. Then there is an X-cobordism of M to a manifold M' with $g'^*\xi = 0$.*

Proof Perform surgery on $g : M \to X$ till we obtain an $(m-i+1)$-connected map $g' : M' \to X$. Suppose for simplicity X (hence also M') connected. Then $g'^*(\xi\eta) = 0$ for all $\eta \in H^{m-i}(X;k)$. But g'^* is an isomorphism in dimension $(m-i)$, so $(g'^*\xi).\zeta = 0$ for all $\zeta \in H^{m-i}(M;k)$. Now by Poincaré duality, $g'^*\xi = 0$.

This result may well extend to more general homology theories. More interesting would be to know exactly what happens in the next dimension $m = 2i-1$. Interesting special cases are obtained by taking $X = BO$, $k = \mathbf{Z}_2$ (when the problem of orientation does not arise) and $\xi = w^i$. The result holds here also when $i = 1$ [W4], and a full statement concerning the case $i = 2$ can be deduced from [B33]. In the unitary case, since BU is torsion free it is not necessary to take coefficients from a field: the result applies with $k = \mathbf{Z}$. Here the result is false for $\xi = c_1$ [B36].

We can also apply §1 to prove the following

THEOREM 1.7. *Let (K, L) be a CW-pair of dimension r, such that $K = L \cup_f e^2$, and K is contractible. Then if $m \geqslant 2r-1$, $m \geqslant 5$, there is a smooth embedding of S^{m-2} in S^m, with complement C, and an $(m-r)$-connected map $\psi : C \to L$.*

Proof In the notation of §1, set $M^m = S^1 \times D^{m-1}$, $X = L$, define ϕ by $\phi(x, y) = f(x)$, and let F be the obvious framing. Since $r \geqslant 2$ necessarily, we may suppose (altering (K, L) if necessary by a homotopy equivalence) that f identifies S^1 with a 1-cell of L which forms a loop. Now we need only attach the remaining cells of L to M to make ϕ a homotopy equivalence: since $m \geqslant 2r-1$, the Corollary to (1.1) asserts that we can do this. Let V be the manifold, and $\psi_V : V \to L$ the map so constructed; and let $N = \partial_+ V$.

Now $M^m \subset S^m = \partial D^{m+1}$, and – up to homotopy – D^{m+1} is formed from M^m by attaching a 2-cell. The definition of ϕ admits a natural extension which maps this to the 2-cell of K. We can regard the handles which form V as attached to ∂D^{m+1}; let the union be W. Then again $\psi_V : V \to L$ extends to $\psi_W : W \to K$; since ψ_V is a homotopy equivalence, and W and K are each formed by attaching a 2-cell, by maps which correspond, ψ_W is a homotopy equivalence. Thus W is contractible. Also, W is formed from D^{m+1} by attaching handles of dimensions $\leqslant r \leqslant m-2$, so ∂W is simply connected. Hence (Smale [S10]), W is diffeomorphic to D^{m+1}.

Now $S^m = (S^1 \times D^{m-1}) \cup (D^2 \times S^{m-2}) = M \cup (D^2 \times S^{m-2})$, where the boundaries are attached by the identity map. So we have $0 \times S^{m-2} \subset S^m$. This sphere also lies in ∂W, since all the handles were attached to its complement. Clearly, $C = \partial W - S^{m-2}$ has N as deformation retract. Now ψ_V is a homotopy equivalence, and V is formed from N by attaching handles of dimension $\geqslant m+1-r$. Hence $\psi_V | N$ is $(m-r)$-connected. Since $\partial W \cong S^m$, this completes the proof of the theorem.

Remark. The proof is valid throughout for the case $r = 2$, $m = 4$, except in the application of Smale's theorem. So the result holds in that case if S^m is replaced by an unknown homotopy sphere Σ^4.

COROLLARY 1.8. (*Kervaire* [K2]) *A group G can appear as fundamental group of some $C = S^m - S^{m-2}$ $(m \geqslant 5)$ if and only if G is finitely presented, $H_1(G) \cong \mathbf{Z}$, $H_2(G) = 0$, and G has an element α whose conjugates generate the whole group.*

Proof Since C has the homotopy type of a finite complex, G must be finitely presented; since C can be made contractible by adding a 2-cell, if α is the homotopy class of the attaching map then its conjugates generate G; and the spectral sequence of the universal cover of C gives $H_1(C) \cong H_1(G)$ and an exact sequence

$$\pi_2(C) \to H_2(C) \to H_2(G) \to 0$$

whence, since C is a homology circle, $H_1(G) \cong \mathbf{Z}$, $H_2(G) = 0$.

Conversely, let G have the properties stated. Use a finite presentation to construct a finite 2-complex P with fundamental group G. Then $H_1(P) \cong H_1(G) \cong \mathbf{Z}$, and we can attach a 2-cell (by α) to make P simply connected. This will kill $H_1(P)$, but leave $H_2(P)$ unaltered; hence $H_2(P)$ is free abelian. The exact sequence

$$\pi_2(P) \to H_2(P) \to H_2(G) \ = \ 0$$

shows that we can pick a base of $H_2(P)$, represent each element by a map of a 2-sphere, and attach corresponding 3-cells. This gives a complex L with $H_2(L) = H_3(L) = 0$ and still $\pi_1(L) \cong G$. If we now attach a 2-cell by α, we obtain a simply connected complex K with no homology; hence K is contractible.

Now apply the theorem with $r = 3$, $m \geqslant 5$; then ψ is at least 2-connected, so induces an isomorphism of fundamental groups. The result follows.

COROLLARY. *For $m \geqslant 5$, we can choose $\pi_1(C) \cong \mathbf{Z}$ and $\pi_2(C)$ non-finitely generated (as abelian group); for $m \geqslant 7$, we can have $\pi_1(C) \cong \mathbf{Z}$, $\pi_2(C) = 0$, and $\pi_3(C)$ non-finitely generated.*

Proof $\pi_2(S^1 \vee S^2)$ is the direct sum of a sequence of copies of $\pi_2(S^2) \cong \mathbf{Z}$ (with generator a), obtained by action of the fundamental group of the inclusion $S^2 \subset S^1 \vee S^2$. Attach a 3-cell by $2a^g - a$ (where g generates $\pi_1(S^1)$), forming L. Then $\pi_2(L)$ is isomorphic to the additive group of rationals with denominator a power of 2, and if we attach a 2-cell to span the S^1, the result is simply connected, and hence contractible. Now if $\psi : C \to L$ is $(m-3)$-connected, $\pi_1(C) \cong \pi_1(L)$, and $\pi_2(C)$ maps onto $\pi_2(L)$ (isomorphically, if $m \geqslant 6$), so the result follows from the theorem.

The second part is obtained similarly, by considering $S^1 \vee S^3$.

The last corollary gives an example which shows that some theorems of B. Mazur [M4] are incorrectly formulated.

We recall a definition of [M4]. Suppose $K \subset M$ is a subcomplex in some smooth triangulation of M. Then a submanifold U of M, containing K in its interior, is a simple neighbourhood of K in M if

(a) the inclusion of K in U is a simple homotopy equivalence;

(b) $\pi_1(U - K, \partial N) = 0 = \pi_2(U - K, \partial U)$.

Mazur's 'simple neighbourhood theorem' states that if U_1, U_2 are simple neighbourhoods of K in M, then there is a diffeotopy of M rel K which carries U_1 into U_2 provided $\dim M \geqslant 6$. We shall provide a counterexample to this by taking $M = S^m$, K a knotted S^{m-2} as constructed in Corollary 2.4.2, and finding two simple neighbourhoods U_1, U_2 such that $U_1 - K$ and $U_2 - K$ do not have the same homotopy type.

We take U_1 to be a tubular neighbourhood of S^{m-2}. Then $U_1 - S^{m-2}$ has the homotopy type of $S^{m-2} \times S^1$. Now let S^1 be a circle in L linking S^{m-2} once (for example, a fibre in the normal sphere bundle of S^{m-2} in S^m), let N be a tubular neighbourhood of S^1, and let U_2 be the closure of $S^m - N$. Then $U_2 - S^{m-2}$ is the closure in C of $C - N$, and has the homotopy type of $C - S^1$, hence the $(n-3)$-type of C. Thus if any $\pi_1(C)$ with $1 \leqslant i \leqslant n-3$ is nonzero, $U_1 - S^{n-2}$ and $U_2 - S^{n-2}$ have different homotopy types.

It remains to see whether U_2 is a simple neighbourhood of S^{m-2} (it is clear that U_1 is). Since N is a tubular neighbourhood of a (necessarily unknotted) S^1, U_2 is diffeomorphic to $S^{m-2} \times D^2$. Since S^{m-2} links S^1 once, the inclusion $S^{m-2} \subset U_2$ induces an isomorphism of $(m-2)$nd homology groups, from which it follows at once that this inclusion is a simple homotopy equivalence. In our case, condition (b) likewise reduces to the requirement that $\pi_2(U_2 - K) = 0$, and hence $(n \geqslant 5)$ that $\pi_2(C) = 0$. But according to Corollary 2.4.2 we may certainly have $\pi_2(C) = 0$, $\pi_3(C) \neq 0$, if $n \geqslant 7$.

Mazur's proof of his theorem appeared in [M4]. It is a straightforward deduction from his 'relative non-stable neighbourhood theorem'. We conclude that our example is a counterexample to this theorem also. The proof of this latter theorem has not been published: it is stated in [M6] that the proof is similar to that for the absolute case, given in [M5]. However, it is not difficult to see how to prove the relative non-stable neighbourhood theorem under the additional hypothesis that (in the notation of Mazur), the inclusion $Q \subset M$ induces an isomorphism of fundamental groups. Likewise, it is easy to prove the simple neighbourhood theorem provided the codimension of K is $\geqslant 3$, indeed, condition (b) in the definition of simple neighbourhoods can then be weakened to

(b′) the inclusion of ∂U in U induces an isomorphism of fundamental groups.

The simple neighbourhood theorem in this case is now a straightforward consequence of the s-cobordism theorem.

2. Simple Poincaré Complexes

The basic problem of surgery theory is to decide whether a given finite CW complex X is simple homotopy equivalent to a manifold. A first necessary condition is that X should be a simple Poincaré complex.

We have defined a Poincaré complex in [W18, §2] and [W21], essentially as a CW complex satisfying an appropriate form of the Poincaré duality theorem. In those papers, we permitted complexes dominated by a finite complex. The reader of [W18] may have wondered why we did not simply require our complexes to be finite. We can now justify our earlier hesitation on this point by giving a definition which is altogether more satisfactory for our purposes. We will not presuppose reading of these earlier papers.

Let X be a finite connected CW-complex (with base point $*$). let $w : \pi_1(X) \to \{\pm 1\}$ be a homomorphism. Loops having class α with $w(\alpha) = +1$ are to be thought of as orientation-preserving; if with $w(\alpha) = -1$, as orientation-reversing. Also, w is equivalent to a class in $H^1(X; \mathbf{Z}_2)$ (the first Stiefel-Whitney class). Write Λ for the integral group ring of $\pi_1(X)$: elements of Λ are finite formal integer linear combinations of elements g of $\pi_1(X)$, which we can write as $\sum n(g)g$. We define the symbol 'bar' by

$$\overline{\sum n(g)g} = \sum w(g)n(g)g^{-1};$$

it is simple to verify that bar is an involutory anti-automorphism (involution for short) of the ring Λ.

We now define the chain complex of X, $C_*(X)$. This is the complex of cellular chains (in the ordinary sense) of the universal covering space, $\widetilde{X}$, of X. We endow it with two additional structures. First, as $\pi_1(X)$ acts (on the right) cellularly on $\widetilde{X}$, it acts on the chain complex $C_*(X)$, which may thus be regarded as a complex of Λ-modules. Indeed, we have free Λ-modules, and bases can be obtained as follows: order the cells of X (of a given dimension r), orient each, and choose a lifting of each one to an r-cell of $\widetilde{X}$. Then we have a free Λ-basis of $C_r(X)$. This is unique up to order, sign, and multiplication of the basis elements by elements of $\pi_1(X)$.

We now define the homology and cohomology of X with respect to a (right) Λ-module B by

$$\begin{aligned} H^*(X; B) &= H\big(\mathrm{Hom}_\Lambda(C_*(X), B)\big) \\ H_*^t(X; B) &= H\big(C_*(X) \otimes_\Lambda B\big). \end{aligned}$$

In order to define the tensor product above, we must endow B with a left Λ-module structure: we do this using the involution bar – i.e. set $\lambda c = c\overline{\lambda}$. This differs from the usual conventions: the affix ^{t} indicates this fact. When we omit the coefficients, the module $B = \Lambda$ is always to be understood.

We will define duality with respect to a 'fundamental' class $[X] \in H_n^t(X; \mathbf{Z})$, for suitable n. Here, we give $\mathbf{Z}$ a Λ-module structure by making $\pi_1(X)$ act trivially. (Note: with other conventions – i.e. using the anti-automorphism $g \mapsto g^{-1}$ of Λ instead of the above – $H_n^t(X; \mathbf{Z})$ becomes $H_n(X; \mathbf{Z}^t)$, where $\mathbf{Z}^t$ is the group $\mathbf{Z}$ with a right Λ-module structure induced by w). We must define cap products: we will do this at the chain level with a representative cycle $\xi \in C_n(X) \otimes_\Lambda \mathbf{Z}$ for $[X]$, and so define

$$\xi \cap \; : \; C^*(X) \to C_*(X)$$

(in fact, $C^r(X) \to C_{n-r}(X)$), where $C^*(X) = \mathrm{Hom}_\Lambda\big(C_*(X), \Lambda\big)$. First suppose $\pi_1(X)$ finite: then $\widetilde{X}$ is a finite covering space of X. The transfer tr is then defined, and $tr\,\xi \in C_n(\widetilde{X})$ has homology class $tr\,[X] \in H_n^t(X; \Lambda) = H_n(\widetilde{X})$. Also in this situation, we can identify $C^*(X)$ and $C_*(X)$ respectively with $C^*(\widetilde{X})$ and $C_*(\widetilde{X})$. The above cap product is then defined as cap product with $tr\,\xi$ in the ordinary sense as chains on $\widetilde{X}$ (a finite CW complex). One could write down an explicit formula using values on cells, and a chain approximation to the diagonal map. If $\pi_1(X)$ is infinite, it can be verified without undue difficulty that the same formula again works: the justification is necessarily different, as (for example) $tr\,\xi$ is an infinite chain.

We have called $[X]$ a *fundamental class*, and X a *Poincaré complex*, with formal dimension n, when

$$[X] \cap \; : \; H^r(X) \to H_{n-r}^t(X)$$

is an isomorphism for all r. The map $\xi \cap$ is then a chain homotopy equivalence. Note that a change in our choices (of the representative cycle ξ and of the chain approximation to the diagonal map) will only affect $\xi \cap$ by a chain homotopy.

We now need the theory of Whitehead torsion. The most convenient account for our purposes is Milnor's survey article [M14] (other references will be found there). We refer specifically to paragraphs 2-5, but replacing $\overline{K}_1(A)$ by $Wh\big(\pi_1(X)\big)$ throughout. Suppose $f : C_* \to D_*$ a chain homotopy equivalence of chain complexes of free (finitely generated, left) Λ-modules, each with a preferred base. The algebraic mapping cone of f, $\mathscr{C}(f)$ has

$$\mathscr{C}_n(f) \;=\; C_{n-1} \oplus D_n$$

and boundary operator

$$C_n \oplus D_{n+1} \xrightarrow{\begin{pmatrix} d & 0 \\ (-1)^n f_n & d \end{pmatrix}} C_{n-1} \oplus D_n \,.$$

It is acyclic; its torsion $\tau\big(\mathscr{C}(f)\big)$ is thus defined, and is called the torsion of f. (cf. Milnor [M14, p. 382]). If it vanishes, we call f a *simple equivalence*.

We call X a *connected simple Poincaré complex** if it is a finite connected CW complex with a fundamental class $[X]$, represented by ξ, such that the chain homotopy equivalence

$$\xi\cap\ :\ C^*(X) \to C_*(X)$$

is simple. Here we refer to the basis of $C_*(X)$ given by cells and the dual basis of $C^*(X)$. Generally, a finite CW complex is a simple Poincaré complex if each component is, and the fundamental classes all have the same dimension.

Analogous considerations go for Poincaré pairs (Y, X). A representative cycle η for a class

$$[Y] \in H_n^t(Y, X; \mathbf{Z})$$

induces (by cap product) a chain map

$$\eta\cap\ :\ C^*(Y) \to C_*(Y, X)\,.$$

The understood group of coefficients is the integral group ring of $\pi_1(Y)$. We say the finite CW pair (with, in the first instance, Y connected) is a *simple Poincaré pair* if $\eta\cap$ is a simple equivalence and X is a simple Poincaré complex with fundamental class $\partial_*[Y]$. Note that this condition implies that the homomorphism w for X is the composite $\pi_1(X) \xrightarrow{i_*} \pi_1(Y) \xrightarrow{\omega} \{\pm 1\}$. If Y is not connected, we require this of each component. (Note that $X = \emptyset$ is permitted).

The following result shows why this notion will be important for our subsequent investigations. Note that a smooth manifold gives rise (by smooth triangulation) to an essentially unique PL manifold, and that a compact PL manifold has PL triangulations, any two of which admit a common subdivision.

THEOREM 2.1. *Let M^m be a compact triangulated homology manifold (resp. with boundary ∂M). Then M is a simple Poincaré complex (resp. $(M, \partial M)$ is a simple Poincaré pair) with formal dimension m.*

Proof The standard proof of Poincaré duality for M (see e.g. Lefschetz [L12, Chapter VI]) proceeds as follows. Let K be the finite simplicial complex (with simplices σ^r) which triangulates M, K' the derived complex. Then the vertices of K' are the barycentres $\widehat{\sigma}^r$ of the simplices of K, and its simplices have the form

$$\widehat{\sigma}^{i_0}\widehat{\sigma}^{i_1}\ldots\widehat{\sigma}^{i_s}$$

where, for each j, σ^{i_j} is a face of $\sigma^{i_{j+1}}$. Moreover, the simplex σ^r of K is the union of the simplices of K' which 'end' with $\widehat{\sigma}^r$. The dual complex K^* is now defined: its cells σ^{r*} correspond bijectively to the simplices σ^r of K, and σ^{r*} is the union of the simplices of K' which begin with $\widehat{\sigma}^r$. Since M is a homology manifold, the σ^{r*} are indeed homology cells; since K' is a common subdivision, we can use chains of K or of K^* as the chains of M (for a proof that such operations do not disturb torsion, see Milnor [M14, (5.2)]). Now evidently, $\dim(\sigma^{r*}) = m - r$, and σ^r and σ^{r*} intersect only in $\widehat{\sigma}^r$. More precisely, using the natural fundamental cycle ξ (which is unique up to sign), one can show that for

**In the first edition a simple Poincaré complex was called a finite Poincaré complex.*

a suitable chain approximation to the diagonal in $K \times k^*$, $\xi \cap$ takes the cochain on K^* dual to σ^{r*} to the chain σ^r on K. This argument lifts to the universal cover (in fact, it works for any non-compact manifold), so we see that

$$\xi \cap \; : \; C^*(K^*) \to C_*(K)$$

takes basis elements to basis elements. Hence it is a simple equivalence.

The argument in the bounded case is similar, using the usual proof (loc. cit.) of the Lefschetz duality theorem.

COROLLARY. *Suppose X a finite simplicial complex, M a closed PL manifold, $\phi : M \to X$ a simple homotopy equivalence. Then X is a simple Poincaré complex.*

For we have a (chain-homotopy-) commutative diagram

$$\begin{array}{ccc} C^*(M) & \xrightarrow{[M]\cap} & C_*(M) \\ \phi^* \uparrow & & \downarrow \phi_* \\ C^*(X) & \xrightarrow{\phi_*[M]\cap} & C_*(X) \end{array}$$

with sides and top simple equivalences. Since τ is additive for compositions [M14, (7.8)], the lower edge also is a simple equivalence.

This is a trivial corollary: a more detailed study, including a comparison of simple with general Poincaré complexes, is desirable. The above merely underlines that if we seek necessary and sufficient conditions that a (finite) simplicial complex be (simply) homotopy equivalent to a closed smooth or PL manifold, it is sensible to restrict attention at the outset to simple Poincaré complexes.

From time to time we need more general notions. If $(Y; X_-, X_+)$ is a finite CW-triad with $X_- \cap X_+ = W$, we call it a *simple Poincaré triad* if (X_-, W), (X_+, W), and $(Y, X_- \cup X_+)$ are simple Poincaré pairs with

$$j_*\partial_*[Y] \;=\; [X_+] - [X_-] \;,$$

where

$$H_n(Y, X_- \cup X_+) \xrightarrow{\partial_*} H_{n-1}(X_- \cup X_+) \xrightarrow{j_*} H_{n-1}(X_- \cup X_+, W) \cong H_{n-1}(X_+, W) \oplus H_{n-1}(X_-, W)\,.$$

Again this last condition implies that the homomorphisms w for (the components of) X_-, X_+, $X_- \cup X_+$ and W are induced from w for Y.

More generally, we have the following. A (finite) CW $(n+1)$-ad X is a (simple) Poincaré $(n+1)$-ad if

(a) for each $\alpha \subset \{1, 2, \ldots, n\}$, $(X(\alpha), \partial X(\alpha))$ is a (simple) Poincaré pair, with fundamental class $[X(\alpha)]$,

(b) for each $\alpha = \{i_1, \ldots, i_s\}$ with $1 \leqslant i_1 < i_2 < \cdots < i_s \leqslant n$, we have

$$j_* \partial_* [X(\alpha)] = \sum_{t=1}^{s} (-1)^t [X(\alpha - \{i_t\})],$$

where j is the appropriate inclusion map.

Clearly this includes manifold $(n+1)$-ads; condition (b) is here not merely a sign convention, but also again ensures that the various maps w are all induced from w for $|X|$. Again, $\dim[X(\alpha)] - |\alpha|$ is independent of α.

This completes our list of definitions: we now turn to a number of properties of maps of degree 1 which will be needed below. A map is called *of degree* 1 if it preserves fundamental homology classes: by (b) above, if $[|X|]$ is preserved, so are all of the $[X(\alpha)]$.

LEMMA 2.2. *Let M, X be connected Poincaré complexes, $\phi : M \to X$ a map of degree 1, B a $\mathbf{Z}[\pi_1(X)]$-module, $\phi^* B$ the module over $\mathbf{Z}[\pi_1(M)]$ induced by $\phi_* : \pi_1(M) \to \pi_1(X)$. Then the diagram*

$$\begin{array}{ccc} H^r(M; \phi^* B) & \xleftarrow{\phi^*} & H^r(X; B) \\ \downarrow{\scriptstyle [M]\cap} & & \downarrow{\scriptstyle [X]\cap} \\ H^t_{m-r}(M; \phi^* B) & \xrightarrow{\phi_*} & H^t_{m-r}(X; B) \end{array}$$

is commutative, so $[M]\cap$ induces an isomorphism of the cokernel $K^r(M; \phi^* B)$ of ϕ^* on the kernel $K^t_{m-r}(M; \phi_* B)$ of ϕ_*. Thus if ϕ is k-connected, ϕ_* and ϕ^* are isomorphisms for $r < k$ and for $r > m - k$.

Similarly let $\phi : (N, M) \to (Y, X)$ be a map of degree 1 of Poincaré pairs. Then ϕ_* gives split surjections of homology groups $M \to X$, $N \to Y$ and $(N, M) \to (Y, X)$ with kernels K_*, and split injections of cohomology with cokernels K^*. The duality map $[N]\cap$ induces isomorphisms

$$K^*(N) \xrightarrow{\cong} K_*(N, M),\ K^*(N, M) \xrightarrow{\cong} K_*(N).$$

The homology (cohomology) exact sequence of (N, M) is isomorphic to the direct sum of the sequence for (Y, X) and a sequence of groups K_* (K^*).

Analogous results hold for maps of degree 1 of Poincaré n-ads.

Proof (cf. [W18, (2.1) and (2.2)]). Commutativity of the first diagram follows from naturality of cap products. The other assertions in the first paragraph are immediate consequences. The same goes for the second paragraph up to the remark about exact sequences: this is justified by noting that ϕ_* and ϕ^* are morphisms of exact sequences, so the sequence operations preserve K_* (or K^*).

For n-ads, we observe that if $\phi_*[|M|] = [|X|]$, then it follows by downward induction on $|S|$ (using (b)) that for all $S \subset \{1, 2, \ldots, n\}$, $\phi_*[M(\alpha)] = [X(\alpha)]$. We can then apply the preceding to each Poincaré pair $(X(\alpha), \partial X(\alpha))$.

LEMMA 2.3. *Let $\phi : (N, M) \to (Y, X)$ be a map of CW pairs with Y connected (not necessarily M and X, which may also be empty). Suppose, with $\Lambda = \mathbf{Z}[\pi(Y)]$ as coefficients, that $H_i(\phi) = 0$ for $i < r$. Then*

(a) *If $H^{r+1}(\phi; B) = 0$ for every Λ-module B, then $H_r(\phi)$ is a projective Λ-module.*

(b) *If N and Y are finite, $H_r(\phi)$ is finitely generated.*

(c) *If, in addition to* (a) *and* (b), *we suppose $H_i(\phi) = 0$ for $i \neq r$, then $H_r(\phi)$ is stably free, and has a preferred equivalence class of s-bases.*

(For the terms 'stably free' and s-base see Milnor [M14, §4]).

Proof By replacing (Y, X) (if necessary) by the mapping cylinder of ϕ, we may suppose that ϕ is an inclusion, and that $M = N \cap X$. The homology groups of ϕ are then calculated from the chain complex $C_*(\phi) = C_*(Y, N \cap X)$. The lemma is thus reduced to a proposition about chain complexes. We abbreviate $C_i = C_i(\phi)$; let Z_i be the submodule of cycles, B_i of boundaries.

Now for $i \leqslant r$ we have a short exact sequence

$$0 \to Z_i \to C_i \to Z_{i-1} \to 0\,.$$

Also $Z_0 = C_0$. By induction on i, Z_i is projective, and the above sequence splits. Thus C_* is chain homotopy equivalent to the complex obtained by replacing the terms of degree $\leqslant r$ by Z_r,

$$\cdots \to C_{r+2} \to C_{r+1} \to Z_r \to 0 \qquad (1).$$

Now (a) follows as in [W14] or [W15] by deducing from the hypothesis that the $(r+1)$-cocycle $C_{r+1} \to B_r$ is a coboundary, so that the inclusion $B_r \subset Z_r$ splits. For (b), note that Z_r is a direct summand of the module C_r, which is now finitely generated, and $H_r(\phi)$ is the quotient module Z_r/B_r.

Finally for (c) we observe that $Z_r \cong B_r \oplus H_r$, and the complex (of projective modules)

$$\cdots \to C_{r+2} \to C_{r+1} \to B_r \to 0 \qquad (2)$$

has zero homology, so is contractible, and (1) is chain homotopy equivalent to the complex $0 \to H_r \to 0$, being its direct sum with the contractible complexes (2) and (3)

$$0 \to Z_{r-1} \to C_{r-1} \to \cdots \to C_0 \to 0 \qquad (3).$$

But for a contractible complex A_* of projective modules, we have $\bigoplus A_{2i} \cong \bigoplus A_{2i+1}$. Applying this to the direct sum of (2) and (3), we have

$$B_r \oplus Z_{r-1} \oplus \bigoplus_{i \neq 0} C_{r+2i} \cong \bigoplus_i C_{r+2i+1}$$

or, adding H_r,

$$\bigoplus_i C_{r+2i} \cong H_r \oplus \bigoplus_i C_{r+2i+1} .$$

Thus H_r is stably free. An explicit isomorphism here would give an s-base for H_r, but the treatment above is not careful enough to prove uniqueness. Choose an s-base for H_r. Then [M14, §3 and §4], $\tau(C)$ is defined. If we change the s-base $\{a\}$ of H_r to another, $\{b\}$ such that the identity map of $(H_r, \{a\})$ to $(H_r, \{b\})$ has torsion τ, then by [M14, (3.2)] the torsion of C becomes $\tau(C) \pm \tau$ (the sign depends only on r). Now it is clear that $\{b\}$ can be chosen with τ arbitrary, thus if we stipulate that $\tau(C) = 0$, the s-base of H_r is determined up to equivalence. This completes the proof of the lemma.

COROLLARY. *If N and Y are finite the same applies to cohomology.*

For in this case we also have $C_*(X) = \mathrm{Hom}_\Lambda\big(C^*(X), \Lambda\big)$, etc., and the whole argument dualises. In this case, the hypothesis $H^{r+1}(\phi; B) = 0$ becomes $H_{r-1}(\phi; B) = 0$.

We will sometimes use the following terminology. If P is an s-based Λ-module, then another s-base is called *preferred* if the identity map of P, regarded as a morphism between these s-based modules, is a simple isomorphism. We can always replace the given s-base by a preferred one. A short exact sequence

$$0 \to P \to Q \to R \to 0$$

of s-based Λ-modules will be called *a based short exact sequence* if, regarded as a chain complex, it has zero torsion. An equivalent condition is that if $\{\alpha_i\}$ is the s-base of P, and the s-elements $\{\beta_j\}$ of Q lift to the s-base $\{\beta_j'\}$ of R, then $\{\alpha_i, \beta_j\}$ is a preferred s-base of Q. A short exact sequence of (s-) based chain complexes over Λ is based if it is so in each dimension.

LEMMA 2.4. *Let $0 \to C_*' \to C_* \to C_*'' \to 0$ be a based short exact sequence of free based chain complexes over Λ, each with finite total rank. Assume that $H_i(C) = 0$ for $i \neq r$, $H_i(C'') = 0$ for $i \neq r+1$ and $H^{r+1}(C; B)$, $H^{r+2}(C''; B)$ vanish for any Λ-module B. Then C' satisfies the same condition as C, so by (2.3) $H_{r+1}(C'')$, $H_r(C')$ and $H_r(C)$ each have a preferred class of s-bases. Moreover, the exact homology sequence*

$$0 \to H_{r+1}(C'') \to H_r(C') \to H_r(C) \to 0$$

is based.

Proof The assertions about C' are immediate consequences of the exact homology and cohomology sequences of the given exact sequence of chain complexes. Since the s-bases of the homology modules are defined so as to make $\tau(C'') = \tau(C') = \tau(C) = 0$, the last assertion follows from Milnor [M14, Theorem 3.2].

Of course we could prove similar results if the exact homology sequence collapsed – for example – to

$$0 \to H_r(C') \to H_r(C) \to H_r(C'') \to 0 ,$$

but the above is the case we will need.

COROLLARY. *Let $\phi : (N, M) \to (Y, X)$ be a map of finite CW-complexes with Y connected, $\Lambda = \mathbf{Z}[\pi_1(Y)]$ be the coefficient module and suppose that $H_i(\phi) = 0$ for $i \neq r+1$; if $\phi' : N \to Y$ and $\phi'' : M \to X$ are induced by ϕ, that $H_i(\phi') = 0$ for $i \neq r$, and that $H^{r+2}(\phi; B)$, $H^{r+1}(\phi'; B)$ vanish for all Λ-modules B. Then the following is a based short exact sequence:*

$$0 \to H_{r+1}(\phi) \to H_r(\phi'') \to H_r(\phi') \to 0\,.$$

For we may replace ϕ by an inclusion map, and apply (2.4) to the sequence

$$0 \to C_*(X, M) \to C_*(Y, N) \to C_*(Y, N \cup X) \to 0\,.$$

LEMMA 2.5. *Suppose given a commutative diagram of (s-)based chain complexes each of finite total rank,*

$$\begin{array}{ccccccccc} 0 & \longrightarrow & C'_* & \longrightarrow & C_* & \longrightarrow & C''_* & \longrightarrow & 0 \\ & & \downarrow & & \downarrow & & \downarrow & & \\ 0 & \longrightarrow & D'_* & \longrightarrow & D_* & \longrightarrow & D''_* & \longrightarrow & 0 \end{array}$$

whose rows are based short exact sequences. If two of the vertical maps are simple equivalences so is the third.

Proof Taking the algebraic mapping cones of the vertical maps, we get a based short exact sequence of acyclic complexes, say

$$0 \to E'_* \to E_* \to E''_* \to 0\,.$$

The hypothesis implies that two of these have zero torsion: [M14, (2.3)] implies that the third also does. Hence the third vertical map in the above diagram is a simple equivalence (that it is an equivalence follows, of course, from the exact homology sequence of the E's).

LEMMA 2.6. *Let $\phi : (N, M) \to (Y, X)$ be a map of degree 1 of simple Poincaré pairs which satisfies the assumptions of (2.3 (c)). Then the induced map $\psi : N \to Y$ also satisfies those assumptions for cohomology, with r replaced by $s = \dim[Y] - r + 1$. Moreover, the duality map*

$$[N] \cap \; : \; H^s(\psi) \; = \; K^s(N) \to K_{r-1}(N, M) \; = \; H_r(\phi)$$

is a simple isomorphism with respect to preferred s-bases. Similarly if ϕ and ψ are interchanged.

Proof As usual, we may suppose ϕ an inclusion. Then we have the commutative diagram of based short exact sequences of based chain complexes of finite total rank

$$\begin{array}{ccccccccc} 0 \longleftarrow & C^*(\psi) & \longleftarrow & C^*(N) & \longleftarrow & C^*(Y) & \longleftarrow 0 & & \\ & & & \downarrow {\scriptstyle [N] \cap -} & & \downarrow {\scriptstyle [Y] \cap -} & & & \\ & 0 & \longrightarrow & C_*(N, M) & \longrightarrow & C_*(Y, X) & \longrightarrow & C_*(\phi) & \longrightarrow 0 \end{array}$$

in which, by hypothesis, the vertical maps are simple equivalences. Now replace $C_*(N, M)$ by a suitable equivalent complex such that $C'_*(N, M) \to C_*(Y, X)$ is surjective, with kernel $D_*(\phi)$, say (this is the chain complex with homology groups $K_*(N, M)$). Then there is a simple equivalence of degree 1, $C_*(\phi) \to D_*(\phi)$, and the diagram shows that $[N]\cap$ induces another simple equivalence (unique up to chain homotopy) $C^*(\psi) \to D_*(\phi)$. The desired result now follows as in (2.5).

We conclude this chapter by recapitulating some geometrical facts concerning Poincaré complexes and pairs from [W21, Chapter 2]. These will be needed especially in §9. Small changes need to be made, since we are interested here only in simple Poincaré complexes.

PROPOSITION 2.7. *Let $(Z; Y, Y')$ be a CW-triad, with $Y \cap Y' = X$, Z connected, and $w : \pi_1(Z) \to \{\pm 1\}$ defining twistings for all four. Let $[Z] \in H^t_m(Z; \mathbf{Z})$ have image $[Y] + [Y']$ in $H^t_m(Y, X; \mathbf{Z}) \oplus H^t_m(Y, X'; \mathbf{Z})$.*

(i) *If (Y, X) and (Y', X) are simple Poincaré pairs with fundamental classes $[Y]$ and $[Y']$, Z is a simple Poincaré complex with fundamental class $[Z]$.*

(ii) *If Z is a simple Poincaré complex with fundamental class $[Z]$, and (Y', X) a simple Poincaré pair with fundamental class $[Y']$, and if for each component Y_i of Y $\mathbf{Z}[\pi_1(Y_i)]$ can be regarded as a $\pi_1(Z)$-module inducing the natural left action of $\pi_1(Y_i)$ then (Y, X) is a simple Poincaré pair with fundamental class $[Y]$.*

Proof Apply (2.5) to the diagram

$$\begin{array}{ccccccccc}
0 & \longrightarrow & C^*(Y, X) & \longrightarrow & C^*(Z) & \longrightarrow & C^*(Y') & \longrightarrow & 0 \\
 & & \Big\downarrow [Y]\cap & & \Big\downarrow [Z]\cap & & \Big\downarrow [Y']\cap & & \\
0 & \longrightarrow & C_*(Y) & \longrightarrow & C_*(Z) & \longrightarrow & C_*(Y', X) & \longrightarrow & 0
\end{array}$$

where the coefficients are $\mathbf{Z}[\pi_1(Z)]$ for (i), and the various $\mathbf{Z}[\pi_1(Y_i)]$ (taken in turn) for (ii). The idea of this proof comes from [B16].

As in [W21] the above has various extensions proved by the same method, e.g.

2.7 ADDENDUM. *Let N, N' be simple Poincaré $(n+1)$-ads with $\partial_{n-1}N = \emptyset$, $\partial_n N = \partial_{n-1}N'$, and $\partial_{n-1}N' \cap \partial_n N' = \emptyset$: suppose $|N| \cap |N'| = |\partial_n N|$. Define N'' by $\partial_{n-1}N'' = \partial_{n-1}N$, $\partial_n N'' = \partial_n N'$, and for $\{n-1, n\} \subset \alpha \subset \{1, 2, \ldots, n\}$, $N''(\alpha) = N(\alpha) \cup N'(\alpha)$. Then N'' is a simple Poincaré $(n+1)$-ad with $\partial_{n-1}N'' \cap \partial_n N'' = \emptyset$.*

The next result is a re-statement of [W21, (2.3.2) and (2.3.3)]: in the case of finite complexes, the proof given in [W21] does give a simple homotopy equivalence.

LEMMA 2.8. *Let X be a simple Poincaré complex, $n = \dim[X] \geqslant 4$. Then X is simply homotopy equivalent to a complex X' obtained from a subcomplex Z of dimension $\leqslant n-2$ (or, in case $n=4$, dominated by a 2-complex) by attaching along ∂H a smooth manifold H which is formed from D^n by attaching 1-handles. Inclusion induces a surjective map $\pi_1(H) \to \pi_1(X)$.*

We will not repeat the proof: we observe, however, that a similar statement and proof are valid for a Poincaré pair (Y, X): that we can replace Y by $Y' = Y_0 \cup_{\partial H} H$, where now Y_0 has X as subcomplex and $\dim(Y_0 - X) \leqslant (n-2)$.

The proof is modified mainly by noting that the bottom cells of Y are dual to the top cells of Y mod X.

The following is essentially [W21, (2.4)].

LEMMA 2.9. *Let X be a simple Poincaré complex, $n = \dim[X] \geqslant 3$. Then X is simply homotopy equivalent to a complex $K \cup_f e^n$ with $\dim K \leqslant n-1$ (or, in case $n=3$, K dominated by a 2-complex). The pair (K, f) is unique up to homotopy type of K, and homotopy and orientation of f.*

LEMMA 2.10. *The product of a (simple) Poincaré $(m+1)$-ad and a (simple) Poincaré $(n+1)$-ad is a (simple) Poincaré $(m+n+1)$-ad.*

Proof We must verify that various sub-pairs are simple Poincaré pairs: each of these has the form $(Y_1 \times Y_2, Y_1 \times X_2 \cup X \times Y_2)$ where (Y_1, X_1) and (Y_2, X_2) are simple Poincaré pairs. We may suppose Y_1, Y_2 connected; then

$$\begin{aligned}
\pi_1(Y_1 \times Y_2) &= \pi_1(Y_1) \times \pi_1(Y_2)\,, \\
C^*(Y_1 \times Y_2) &= C^*(Y_1) \otimes_{\mathbf{Z}} C^*(Y_2)\,, \\
C_*(X \times Y_2, Y_1 \times X_2 \cup X_1 \times Y_2) &= C_*(Y_1, X_1) \otimes_{\mathbf{Z}} C_*(Y_2, X_2)\,.
\end{aligned}$$

Since the tensor product of two simple equivalences is another (an easy result, generalised in [K17]), the result follows (cf. [W21, (2.5)]).

Part 1

The Main Theorem

3. Statement of Results

We now describe the problem to which our work is directed. We seek, in fact, a characterisation of the simple homotopy types of compact manifolds. As shown in §1, our methods necessitate simultaneous description of the (stable) tangent bundle. Thus let X be a finite simplicial complex and ν a bundle over X. We seek normal maps (M, ϕ, F) with M a manifold, $\phi : M \to X$ a map, and F a stable trivialisation of $\tau_M \oplus \phi^* \nu$, such that ϕ is a simple homotopy equivalence. Again, we explained in §1 that it is convenient to classify normal maps (M, ϕ, F) (not necessarily satisfying the last condition) into bordism classes, and then study one such class at a time.

A first necessary condition (on X) that there should exist a normal map (M, ϕ, F) with ϕ a simple homotopy equivalence was found in (2.1): that X be a simple Poincaré complex. A second such condition concerns ν: if the induced bundle over M agrees (stably) with the normal bundle of an embedding in Euclidean space, then it follows by a standard application of the Thom-Pontrjagin construction (shrinking the complement of a tubular neighbourhood to a point) that the Thom space of ν is (stably) reducible. Thus ν must be stably fibre homotopy equivalent to the 'Spivak normal fibration' of the Poincaré complex X, as described in [S5] and [W4]. One simple and useful deduction from this is that the first Stiefel-Whitney class w^1 of ν must coincide with $w^1(X)$ – the cohomology class corresponding to the homomorphism $w : \pi_1(X) \to \{\pm 1\}$. For the Thom isomorphism corresponding to ν must take $[X]$ to a spherical (and hence untwisted) class.

Our third necessary condition restricts the bordism class. For if $\phi : M \to X$ is a simple homotopy equivalence, then (if $[X]$ is chosen with appropriate sign) it has degree 1. Now consider any bordism $\psi : (N; M_-, M_+) \to X$. We have $\partial_*[N] = [M_+] - [M_-]$, and applying ψ_*, it follows that $\psi_*[M_+]$ and $\psi_*[M_-]$ give the same homology class in X. Thus if one map in the bordism class has degree 1, so have they all. We thus restrict consideration to those classes which consist of maps of degree 1. We will see in §10, when we give a fuller discussion, that for a given (X, ν) such classes exist if and only if ν is stably fibre homotopy equivalent to the Spivak normal fibration of X.

We have defined bordism groups $\Omega_m(X, \nu)$. The above shows for (M, ϕ, F) in a class $\alpha \in \Omega_m(X, \nu)$ that $\phi_*[M] \in H^t_m(X; \mathbf{Z})$ depends only on α (where the twisting is by $w^1(\nu)$). Thus taking fundamental classes defines an augmentation – clearly a homomorphism –

$$\varepsilon : \Omega_m(X, \nu) \to H^t_m(X; \mathbf{Z}) .$$

Our classes are those in $\varepsilon^{-1}([X])$: they do not form a subgroup, but a coset of one, so the set has a natural affine structure.

We now relativise the above: the most interesting case is that of pairs, but it is no harder to discuss $(n+1)$-ads. Let X be a simple Poincaré $(n+1)$-ad, ν a (vector or PL) bundle over $|X|$, and consider normal maps (M, ϕ, F) where M is a compact (smooth or PL) manifold $(n+1)$-ad, $\phi : M \to X$ a map of $(n+1)$-ads of degree 1, and F a stable trivialisation of $\tau_{|M|} \oplus \phi^*\nu$. We arrange these in bordism classes as follows: a bordism of normal maps (M, ϕ, F) to (M', ϕ', F') is a normal map (N, ψ, G) with N a compact manifold $(n+2)$-ad with $\partial_{n+1}N = M + (-M')$ (the disjoint union with the sign of $[M']$ changed), $\psi : N \to s_{n+1}X$ a map of degree 1 of $(n+2)$-ads extending ϕ and ϕ', and G a stable framing of $\tau_{|N|} \oplus \psi^*\nu$ extending F and F' in the sense that one extra stabilising vector is identified with the inward (outward) normal to N along $M(M')$. With these conventions, bordism is an equivalence relation; as discussed in §1, it can be generated by 'surgeries'.

Various modifications of the above will be needed from time to time; we may, for example, restrict ϕ by requiring it to induce a simple homotopy equivalence of n-ads $\partial_i M \to \partial_i X$ for some i. One would require correspondingly that in a bordism, ψ induced a simple homotopy equivalence of $(n+1)$-ads $\partial_i N \to \partial_i s_{n+1} X = s_n \partial_i X$. Or alternatively, some $\partial_j X$ may be already a manifold; then we would require ϕ to induce a (smooth or PL) homeomorphism of n-ads $\partial_j M \to \partial_j X$ and correspondingly ψ a homeomorphism $\partial_j N \to \partial_j X \times (I, \partial I)$. We may of course require conditions of both types simultaneously (with $i \neq j$).

The problem, then, is: given a bordism class of normal maps satisfying the above conditions, to classify those of its members for which ϕ is a simple homotopy equivalence. We will confine ourselves to the problem of existence of such a number. We can then consider uniqueness by taking a cobordism N with $\partial_n N$ a union of two members each mapped by a simple homotopy equivalence to X, taking a product map $N \to X \times (I, \partial I)$, and attempting surgery on this, with $\partial_n N$ fixed: this was one possibility referred to above. If such surgery can be performed, N will be an 's-cobordism', and the s-cobordism theorem [K3] will give us uniqueness up to diffeomorphism. Of course, it may happen that N is not surgerable even though its 'ends' are diffeomorphic, thus care is needed to obtain a complete numerical result in any particular case.

Another apparently reasonable extension of our problem would be, to suppose that $\phi : M \to X$ induces a simple homotopy equivalence of certain subcomplexes of M and X, and to consider surgery leaving fixed the subcomplex of M. We do not achieve quite this. However, in the case when X is a manifold also, and ϕ transverse regular on the subcomplex in question, we have an induced simple homotopy equivalence of a regular neighbourhood of it, which is a manifold, and of the corresponding boundaries. Thus if the subcomplex has codimension $\geqslant 3$ (so that troubles about the fundamental group of its complement do not occur) we are reduced to a problem already discussed. We can treat similarly the case of a map of triads

$$\phi : (M\{12\}, M\{1\}, M\{2\}, M\{\ \}) \to (X\{12\}, X\{1\}, X\{2\}, X\{\ \})$$

in which the induced map $M\{\ \} \to X\{\ \}$ is a simple homotopy equivalence and we wish to do surgery leaving $M\{\ \}$ fixed. Here, we write $L = M\{\ \}$, choose a product neighbourhood $L \times I$ of L in $M\{1\}$ (which we may suppose ϕ maps to $X\{\ \} = W$, say), and then replace $|\partial X|$ by the simple homotopy equivalent complex $|\partial X'| = X\{1\} \times 0 \cup W \times I \cup X\{2\} \times 1$ and $(|X|, |\partial X|)$ by a corresponding pair $(|X'|, |\partial X'|)$. Set $M''\{1\} = M\{1\} - L \times [0, 1)$; then $\phi||\partial M|$ and the projection of $L \times I$ on I combine to define a map $|\partial M| \to |\partial X'|$ which takes $M''\{1\}$ to $X\{1\} \times 0$, $L \times I$ to $W \times I$, and $M\{2\}$ to $X\{2\} \times 1$: use homotopy extension to extend to $M\{12\} \to |X'|$. But this defines a map of triads $\phi' : M' \to X'$ where $|M'| = |M|$ and

$$\begin{aligned} M'\{1\} &= M''\{1\} \cup M\{2\}, \ M'\{2\} = L \times I, \ M'\{\ \} = L \times \partial I \\ X'\{1\} &= X\{1\} \times 0 \cup X\{2\} \times 1, \ X'\{2\} = W \times I, \ X'\{\ \} = W \times \partial I; \end{aligned}$$

moreover, ϕ' gives a simple homotopy equivalence of pairs $\partial_2 M' \to \partial_2 X'$. Adjusting tangential data appropriately, we are again back to a problem already discussed. We leave to the reader further extensions of this trick, which shows that our level of generality is useful to work at.

So much for the problem: now we describe the results. Consider first the simplest case, of closed connected manifolds and Poincaré complexes. Then a first statement of our results is as follows. First, an algebraic assertion.

There exist abelian groups $L_m(\pi_1(X), w)$ depending only on the group $\pi_1(X)$, the homomorphism $w : \pi_1(X) \to \{\pm 1\}$, and the value of m modulo 4.

Next the geometrical result. Let X be a simple Poincaré complex, ν a bundle over X, M a closed manifold, $\phi : M \to X$ a map of degree 1, and F a stable trivialisation of $\tau_M \oplus \phi^*\nu$; let x be the bordism class of the normal map (M, ϕ, F), and $m = \dim M$.

There is an obstruction $\theta(x) \in L_m(\pi_1(X), w)$ defined by (M, ϕ, F) but depending only on x, which vanishes if x contains a simple homotopy equivalence. Conversely, if $\theta(x) = 0$ and $m \geqslant 5$, x has a representative (M_0, ϕ_0, F_0) with ϕ_0 a simple homotopy equivalence.

Such a result is of course somewhat unsatisfactory, as it is necessary to know the groups L_m for the purposes of any computations. The problem of calculating L_m is of considerable difficulty, but we will be able to give some results: we postpone these till after the proof of the main theorem (§13A).

The result in the general case is of a similar nature, but naturally the details are more complicated. First of all we drop the assumption that X be connected: as it is a finite complex, it will only have a finite number of components, and if we choose a base point in each we find a finite number of fundamental groups, each finitely presented. It seems neater to consider instead the fundamental groupoid $\pi(X)$ of X: this, we note, is a category (with all morphisms equivalences). The components of $\pi(X)$ (as groupoid) correspond to the (path-) components of the space X, and the vertex groups [H6] of $\pi(X)$ are the fundamental groups of X based at the points of X. Picking one point from each component of X

determines an equivalent (skeletal) subcategory. We call a groupoid of *finite type* if it has a finite number of components and each vertex group is finitely presented.

We now introduce the homomorphism w; we regard this as defining an element ± 1 for each *closed* loop in X (to do more would involve too many arbitrary choices). Clearly, for closed loops α_1, α_2 with the same base point, we have $w(\alpha_1\alpha_2) = w(\alpha_1)w(\alpha_2)$, and if β is any path with $\alpha\beta$ defined (where α is a loop), $w(\beta^{-1}\alpha\beta) = w(\alpha)$. The first identity expresses that w is a homomorphism of each vertex group; the second gives the change of base point formula, so there are no other identities. We define a category $\mathcal{G}pd$ as follows. An object is a groupoid of finite type, together with a function w as above. A morphism $F : \pi \to \pi'$ is a functor such that for any 'closed loop' α in π (i.e. a morphism whose domain and codomain coincide) we have $w(F\alpha) = w(\alpha)$. This can be regarded, if the reader prefers, as a collection of homomorphisms between vertex groups.

If X is a Poincaré complex, we have the fundamental groupoid $\pi(X)$: we associate with it the w which is part of the definition, and so regard $\pi(X)$ as an object of $\mathcal{G}pd$. Similarly if (Y, X) is a Poincaré pair, $\pi(X)$ and $\pi(Y)$ are both objects of this category, and (by definition) the induced morphism of groupoids $\pi(X) \to \pi(Y)$ is a morphism in the category. More generally, let X be a Poincaré $(n+1)$-ad. Then for any $\alpha \subset \{1, 2, \ldots, n\}$, $\big(X(\alpha), \partial X(\alpha)\big)$ is a Poincaré pair, and in particular $\pi\big(X(\alpha)\big)$ is an object of $\mathcal{G}pd$. If $\beta \subset \alpha$, we have an inclusion-induced functor $\pi\big(X(\beta)\big) \to \pi\big(X(\alpha)\big)$: that these are compatible with w has been noted as a consequence of our definitions. Thus X determines an object $\pi(X)$ of type $\mathbf{2}^n$ in the category $\mathcal{G}pd$.

Our main algebraic result now states that we have functors L_n from the category $F(\mathbf{2}^n, \mathcal{G}pd)$ to the category $\mathcal{A}b$ of abelian groups, and with these the exact sequences commonly associated (see §0) with objects of type $\mathbf{2}^n$. A precise statement must take into account the interplay here between different values of n.

THEOREM 3.1. *There are symmetric functors* $L_m : F(\mathbf{2}^n, \mathcal{G}pd) \to \mathcal{A}b$, *natural transformations* $\partial_* : L_m(\pi) \to L_{m-1}(\partial_n\pi)$, *and natural equivalences* $\alpha : L_m(\pi) \to L_m(\sigma_{n+1}\pi)$ *such that if* π *is a groupoid of type* $\mathbf{2}^n$, *the inclusions* $\partial_n\pi \xrightarrow{i} \partial_n\pi$, $\sigma_n\delta_n\pi \xrightarrow{j} \pi$ *induce an exact sequence*

$$\cdots \to L_m(\partial_n\pi) \xrightarrow{L_m(i)} L_m(\delta_n\pi) \xrightarrow{L_m(j)\circ\alpha} L_m(\pi) \overset{\partial_m}{\to} L_{n-1}(\partial_n\pi) \to \ldots$$

Moreover, all are periodic in m *with period* 4.

A word of explanation is in order here. The functors σ_n use the initial object in $\mathcal{G}pd$: the empty groupoid. We call a functor on $F(\mathbf{2}^n, \mathcal{C})$ *symmetric* if it is invariant under permutations of the factors of $\mathbf{2}^n$. This permits us to discuss the exact sequence associated with the nth suffix and deduce the others by symmetry.

The above result is fairly powerful, and we will spell out one or two of its more elementary consequences, since they are perhaps obscured by the generality of the above statement.

COROLLARY 3.1.1. *If $f : \pi \to \pi'$ is an equivalence in $\mathcal{G}pd$, then for each m, $L_m(f) = 0$.*

For in the exact sequence of f,

$$\cdots \to L_m(\pi) \xrightarrow{L_m(f)} L_m(\pi') \xrightarrow{L_m(j)\circ\alpha} L_m(f) \xrightarrow{\partial_m} L_{m-1}(\pi) \xrightarrow{L_{m-1}(f)} L_{m-1}(\pi'),$$

the morphisms $L_m(f)$, $L_{m-1}(f)$ are equivalences since f is. The result follows by exactness.

Analogous deductions can be made if f has a left or a right inverse; since they do not generally imply a vanishing theorem, they are less interesting:

COROLLARY 3.1.2. *Given an object P of type $\mathbf{2}^n$ in $\mathcal{G}pd$,*

$$\begin{array}{ccc} \pi_1 & \xrightarrow{f_1} & \pi_2 \\ {\scriptstyle g_1}\downarrow & & \downarrow{\scriptstyle g_2} \\ \pi_3 & \xrightarrow{f_2} & \pi_4 \end{array}$$

if f_1 is an equivalence $L_m(P) \cong L_m(f_2)$, and if f_2 is an equivalence $L_m(P) \cong L_{m-1}(f_1)$. If both are, $L_m(P) = 0$.

This follows from the preceding corollary and the exact sequence

$$L_m(f_1) \to L_m(f_2) \to L_m(P) \to L_{m-1}(f_1) \to L_{m-1}(f_2).$$

COROLLARY 3.1.3. *Given two composable morphisms $\pi \xrightarrow{f} \pi' \xrightarrow{f'} \pi''$ in $\mathcal{G}pd$, we have an exact sequence*

$$\cdots \to L_m(f) \to L_m(f'f) \to L_m(f') \to L_{m-1}(f) \to \cdots.$$

Indeed, using the preceding corollary, we may obtain this as the (horizontal) exact sequence of either of the two objects

$$\begin{array}{ccc} \pi & \xrightarrow{1} & \pi \\ {\scriptstyle f}\downarrow & & \downarrow{\scriptstyle f'f} \\ \pi' & \xrightarrow{f'} & \pi'' \end{array} \qquad \text{or} \qquad \begin{array}{ccc} \pi & \xrightarrow{f} & \pi \\ {\scriptstyle f'f}\downarrow & & \downarrow{\scriptstyle f'} \\ \pi'' & \xrightarrow{1} & \pi'' \end{array}$$

These remarks show that we have (in particular) an Eckmann-Hilton functor in the sense of Pressman [P6]. However, his methods are of no use in relativising our result: the functors L_m do not appear to be effaceable.

We now state the geometrical part of our results.

THEOREM 3.2. *Let X be a simple Poincaré $(n+1)$-ad, M a compact manifold $(n+1)$-ad, $\phi : M \to X$ a map of degree 1; ν a bundle over $|X|$, F a stable trivialisation of $\tau_{|M|} \oplus \phi^*\nu$. Then there is an obstruction θ in $L_{\dim|M|}(\pi(X))$ which depends only on the bordism class of (M, ϕ, F) and vanishes if that class contains a simple homotopy equivalence; conversely, if $\dim M\{\ \} > 4$, its vanishing is sufficient.*

If we suppose also that ϕ induces a simple homotopy equivalence $\partial_j M \to \partial_j X$, and keep $\partial_j M$ fixed in the surgeries, then the appropriate obstruction lies in $L_{\dim|M|}(\delta_j \pi(X))$, and its vanishing suffices if $\dim M\{\ \} \geqslant 4$.

In each case, the surgery obstruction for the induced map $\partial_n M \to \partial_n X$ is $\partial_* \theta$. Moreover, θ is natural for inclusion maps.

We now formulate what we mean by the naturality for inclusion maps. Let X, ν, M, ϕ, F be as in the theorem, with $\partial_n M \to \partial_n X$ a simple homotopy equivalence of n-ads; let N be a manifold $(n+2)$-ad with $\partial_n N = \sigma_{n+1}\partial_n M$ (and thus $|\partial_n N| \cap |\partial_{n+1} N|$ empty). We form a new manifold $(n+1)$-ad M' by $\partial_n M' = \delta_n \partial_{n+1} N$ and

$$M'(\alpha) \;=\; M(\alpha) \cup N(\alpha \cup \{n+1\}) \quad \text{for} \quad n \in \alpha \subset \{1, 2, \ldots, n\}.$$

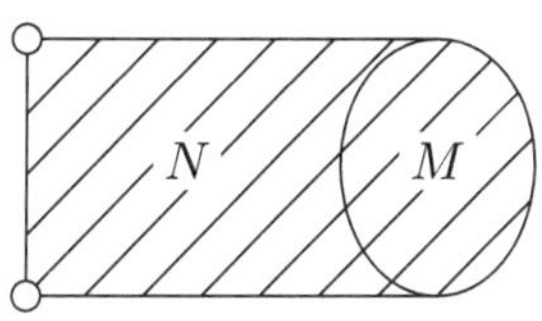

(glue along common face).

Similarly form X' by glueing N to X via the simple homotopy equivalence $\partial_n M \to \partial_n X$. Thus ϕ induces a map (of degree 1) of $(n+1)$-ads $M' \to X'$; the usual tangential conditions are easily verified, and the induced map $\partial_n M' \to \partial_n X'$ is a simple homotopy equivalence. Clearly if $M \to X$ is a simple homotopy equivalence, so is $M' \to X'$. Our naturality for inclusions says generally that the surgery obstruction for $M \to X$, which lies in $L_m(\delta_n \pi(X))$ maps, under the map induced by inclusion $\delta_n X \subset \delta_n X'$, to the surgery obstruction in $L_m(\delta_n \pi(X'))$ for the map $M' \to X'$.

The above does not exhaust the list of properties of surgery obstructions that we possess, but it does at least make it clear in what sense they are obstructions, and the proof of the theorems will clear the way for a more detailed study, of which some aspects will be presented in the later parts of this book.

One corollary of (3.2) (following (3.1.1)) is particularly noteworthy: in some cases we can say that surgery is certainly possible.

THEOREM 3.3 *Let X be a simple Poincaré pair, with* $\dim|X| \geqslant 6$, *such that inclusion induces an isomorphism* $\pi(X\{\ \}) \to \pi(X\{1\})$. *Then given a bundle* ν *over* X, *a compact manifold pair* M, *a map* $\phi : M \to X$ *of degree* 1, *and a stable framing of* $\tau_{|M|} \oplus \phi^*\nu$, *we can perform surgery to make* ϕ *a simple homotopy equivalence of pairs. Similarly if* X *is a simple Poincaré triad, we have the isomorphism* $\pi(X\{2\}) \to \pi(X\{12\})$, *and we restrict to maps* $\phi : M \to X$ *inducing simple homotopy equivalences* $\partial_2 M \to \partial_2 X$.

Follows at once from (3.2) and (3.1.1). In logistic order, its place is less simple; as it seems to be the most important case of our result, we give a direct proof of it in §4.

COROLLARY 3.3.1. *With the assumptions of* (3.3), *the constructed* (M_0, ϕ_0, F_0) *with* ϕ_0 *a simple homotopy equivalence is unique (in its bordism class) up to diffeomorphism.*

An argument above shows that we must do surgery with X replaced by $X \times I$, and $X \times \partial I$ already covered by a simple homotopy equivalence. But such surgery is possible by the relativised (3.3), and the result follows by the s-cobordism theorem [K3].

The proof of our main result is organised as follows. In §4 we prove (3.3). In the next two chapters we give direct proofs of the results in the 'absolute case' (i.e. where all of $\partial|M|$ is to be kept fixed by surgery), together with an explicit expression for the corresponding groups L_m in algebraic terms. Then §7 and §8 attempt to do the same in the relative case. Our results here cover only surgery of pairs, and even so, §8 is inconclusive; these two chapters are not necessary for the main theorem, but do (we hope) give extra insight into the algebra. Finally in §9 we give an argument based on general principles in which all our problems are collected in one big cobordism group, and the main result deduced by using §4 and also a little surgery on Poincaré complexes. The more detailed results of §5 and §6 are needed to establish the periodicity in m.

4. An Important Special Case

The important special case is the 'π-π theorem' (3.3): for $m \geqslant 6$ a normal map $(\phi, F) : (N, M) \to (Y, X)$ from an m-dimensional manifold with boundary to a simple m-dimensional Poincaré pair with $\pi_1(X) \cong \pi_1(Y)$ is normal bordant to a simple homotopy equivalence of pairs. The general case (including a rel∂ version) is considered in §§ 5, 6, 7, 8 below – for any morphism of groups $\pi \to \rho$ there are defined algebraic L-groups $L_*(\pi \to \rho)$; a normal map $(\phi, F) : (N, M) \to (Y, X)$ determines a surgery obstruction element $\theta(\phi, F) \in L_m\big(\pi_1(X) \to \pi_1(Y)\big)$ such that $\theta(\phi, F) = 0$ if (and for $m \geqslant 6$ only if) (ϕ, F) is normal bordant to a simple homotopy equivalence of pairs. From this point of view, the special case follows from the algebraic result that $L_*(\pi \to \rho) = 0$ for an isomorphism $\pi \cong \rho$. However, the special case is technically easier, since no obstruction to surgery is encountered.

This chapter is devoted to the proof of Theorem 3.3. Since $\partial_2 M$ and $\partial_2 X$ are fixed throughout, they will play no rôle in the proof. So we will only discuss the map $\delta_2 M \to \delta_2 X$. It is then notationally simpler to denote this map of pairs by

$$\phi : (N, M) \to (Y, X).$$

Our hypothesis is that $\pi_1(X) \cong \pi_1(Y)$, and that $\dim[Y] \geqslant 6$. The argument is somewhat different according as $\dim[Y]$ is even or odd.

Proof when $\dim[Y] = 2k$.

The argument here was anticipated in [W18, (7.1)]: we follow the earlier version closely, with modifications to cover the simple homotopy condition.

By (1.4), we can perform surgery on ϕ to make the induced map $M \to X$ $(k-1)$-connected and $N \to Y$ k-connected. Since $k \geqslant 3$, all four fundamental groups are now isomorphic. By (2.3 (c)) $K_k(N, M)$ is stably free and s-based. Now perform surgery on a trivial $(k-1)$-sphere in M. This has the effect of replacing N by its boundary-connected sum with a copy of $S^k \times D^k$, and thus of adding to $K_k(N, M)$ a free module of rank 1, with the natural basis. Iterating this construction, we may suppose $K_k(N, M)$ free and based: let $\{e_i\}$ be a preferred base.

As the fundamental groups are isomorphic, we can apply a theorem of Namioka [N1] to the universal cover of the quadruple ϕ to deduce that the Hurewicz map defines an isomorphism

$$\pi_{k+1}(\phi) \cong \pi_{k+1}(\widetilde{\phi}) \to H_{k+1}(\widetilde{\phi}) \;=\; H_{k+1}(\phi; \Lambda) \;=\; K_k(N, M).$$

Thus the e_i determine classes in $\pi_{k+1}(\phi)$; by Theorem 1.3, these determine immersions

$$f_i : (D^k \times D^k, \partial D^k \times D^k) \to (N, M)$$

and we can perform surgery on any embeddings regularly homotopic to these.

In fact the f_i are regularly homotopic to disjoint embeddings. It is enough to show this for the restricted immersions $\overline{f}_i : (D^k, \partial D^k) \to (N^{2k}, M^{2k-1})$, for we can then use small enough neighbourhoods of the resulting discs. Our assertion follows from a standard sort of 'piping' argument (cf. [Z1, lemma 48], [M13, p. 71]), so we give only an indication here. Put the maps $\overline{f}_i$ in mutual general position. The only intersections and self-intersections are then isolated points P in the interior of N, at each of which two branches of the same disc or different discs meet transversely. Choose arcs α, α' from P along these branches to M, meeting no other singularities on the way. Then $\alpha \cup \alpha'$ is an arc in N with both ends on M; since (N, M) is 1-connected, we can find a singular triangle δ in N with edges α, α' and α'', where α'' lies in M. Put δ in general position: since $k \geqslant 3$, it is then embedded, disjointly from all the discs $\overline{f}_i(D^k)$, except along α and α'. One can now (see especially [M13, pp. 73–84]) construct a regular homotopy of the $\overline{f}_i$ which leaves fixed all except a neighbourhood of the arc α, and 'pulls' this across δ to the 'other side' of α', thus getting rid of the intersection point P and introducing no new undesirable points. Proceeding in this manner, we convert the $\overline{f}_i$ (and the f_i) into disjoint embeddings.

Now perform handle subtraction: let N_0 be obtained from N by deleting the interiors of the $f_i(D^k \times D^k)$: let U be the union of the images of the f_i: let M_0 be obtained from M. By construction, we have simple isomorphisms

$$H_{k+1}(U, U \cap M) \cong H_{k+1}(U \cup M, M) \to K_{k+1}(N, M)$$

and as $C_*(U, U \cap M)$ and $D_*(\phi)$ have zero homology in other dimensions it follows (by the definition of the s-base on $K_{k+1}(N, M)$ and using the natural cells for $(U, U \cap M)$) that $C_*(U, U \cap M) \to D_*(\phi)$ is a simple equivalence. Here, $D_*(\phi)$ is the natural chain complex of ϕ (regarded e.g. as an inclusion), re-graded by $+1$ (since $H_{r+1}(\phi) = K_r(N, M)$). Up to chain homotopy equivalence, we now have a commutative diagram, with rows based short exact sequences,

$$\begin{array}{ccccccccc}
0 & \longrightarrow & C_*(U \cup M, M) & \longrightarrow & C_*(N, M) & \longrightarrow & C_*(N, U \cup M) & \longrightarrow & 0 \\
 & & \downarrow & & \| & & \downarrow & & \\
0 & \longrightarrow & D_*(\phi) & \longrightarrow & C_*(N, M) & \longrightarrow & C_*(Y, X) & \longrightarrow & 0.
\end{array}$$

The first two maps are simple equivalences; by (2.5), so is the third. By excision, $C_*(N_0, M_0) = C_*(N, U \cup M)$. So ϕ induces a simple equivalence $C_*(N_0, M_0) \to C_*(Y, X)$.

Now the diagram

$$\begin{array}{ccc}
C^*(N_0) & \xleftarrow{\ \phi^*\ } & C^*(Y) \\
{\scriptstyle [N_0]\cap}\downarrow & & \downarrow{\scriptstyle [Y]\cap} \\
C_*(N_0, M_0) & \xrightarrow{\ \phi_*\ } & C_*(Y, X)
\end{array}$$

is (chain homotopy-) commutative, since $\phi_*[N_0] = [Y]$. Three maps are simple equivalences (the vertical ones since we have simple Poincaré pairs), hence so is the lower ϕ^*. Taking duals, we see that $\phi_* : C_*(N_0) \to C_*(Y)$ is a simple equivalence too, hence $N_0 \to Y$ is a simple homotopy equivalence.

Finally, applying (2.5) to the diagram

$$\begin{array}{ccccccccc} 0 & \longrightarrow & C_*(M_0) & \longrightarrow & C_*(N_0) & \longrightarrow & C_*(N_0, M_0) & \longrightarrow & 0 \\ & & \downarrow & & \downarrow & & \downarrow & & \\ 0 & \longrightarrow & C_*(X) & \longrightarrow & C_*(Y) & \longrightarrow & C_*(Y, X) & \longrightarrow & 0\,, \end{array}$$

we deduce that $C_*(M_0) \to C_*(X)$ is a simple equivalence; since the fundamental groups are all the same (so the group ring of $\pi_1(X)$ is the same as the coefficients used throughout), the map $M_0 \to X$ is a simple homotopy equivalence as well.

Proof when $\dim[Y] = 2k+1$.

This time, (1.4) permits us to suppose that ϕ induces k-connected maps $M \to X$ and $N \to Y$, and moreover that $K_k(N, M)$ is zero. By the Corollary to (2.4), we now have a based short exact sequence of s-based modules

$$0 \to K_{k+1}(N, M) \to K_k(M) \to K_k(N) \to 0\,.$$

As before, we can perform surgery on trivial $(k-1)$-spheres in M, thus taking the boundary-connected sum with copies of $S^k \times D^{k+1}$; again, this allows us to convert all the s-bases into actual bases.

The reader may perhaps now expect us to represent the basis elements of $K_{k+1}(N, M)$ by disjoint embeddings and then proceed as above. But we need (N, M) to be 2-connected for the desired embedding theorem (at least, with the present state of knowledge), and I do not wish to assume this. We can, however, again say that $\pi_{k+2}(\phi) \cong K_{k+1}(N, M)$, and apply (1.4) to represent the basis elements by framed immersions $f_i : (D^{k+1}, \partial D^{k+1}) \to (N, M)$.

We assert, however, that we can at least modify the f_i by regular homotopies so that their boundaries define disjoint embeddings $S^k \to M$. For again, put the f_i in general position and consider the intersections and self-intersections. These are 1-dimensional, and along each of them two branches meet transversely (see [A5], [W45], or – preferably – [R17] for the PL case). Thus they form certain circles (which do not concern us here) and arcs α with both ends on M. Now we can find in each branch at α a disc δ_i whose other side α_i lies in M. Moreover, the loop $\alpha_1 \cup \alpha_2$ spans the disc $\delta_1 \cup \delta_2$ in N, so is nullhomotopic in N, hence also $\big(\text{as } \pi_1(M) = \pi_1(N)\big)$ in M. We can thus span the loop by a disc δ in M: now since $k \geqslant 3$ (this is the crucial use of this hypothesis), we may suppose that δ is embedded, and meets the images of the f_i only in $\alpha_1 \cup \alpha_2$. Now by [M13, p. 71] we can deform a neighbourhood of α_1 across δ so as to get rid of the intersections at the two ends. Proceeding in this way, we get rid of all the intersections and self-intersections on M.

The reader who returns to this point after §5 will recognise $K_{k+1}(N,M)$ as a lagrangian (in fact, this follows from (5.7), short-circuiting the preceding paragraph). A quicker conclusion to the proof is then found by choosing a complementary lagrangian, and attaching handles corresponding to a preferred base, to kill $K_k(N)$. The proof may then be completed as in the first case. We prefer, however, to derive the result independently of the algebra of the next chapter.

For the rest of the proof, all homology groups will be free and based in such a way that the Whitehead torsion of chain complexes with respect to these homology bases is zero. As the chain complex is in these cases simply equivalent to the complex formed by the homology groups (with zero differential), we may safely replace it by this. The condition holds with the assumptions above; moreover, by (2.6), we have a simple isomorphism

$$[N]\cap \,:\, K^k(N) \to K_{k+1}(N,M)\,,$$

where $K^k(N)$ is dual to $K_k(N)$, with the dual base.

We have represented a preferred base of $K_{k+1}(N,M)$ by disjoint framed embeddings $S^k \to M$. Attach corresponding $(k+1)$-handles to N, thus performing surgery. Let U be the union of the added handles, and the pair (N,M) be replaced by (N',M'). Since our spheres are nullhomotopic in N, $K_k(N)$ is unaltered: in fact N is replaced (up to simple homotopy) by its bouquet with corresponding $(k+1)$-spheres, so we acquire a free module $K_{k+1}(N')$.

Dually, the exact sequence of the triple $M' \subset M' \cup U \subset N'$ reduces (using excision) to

$$0 \to K_{k+1}(N',M') \to K_{k+1}(N,M) \to K_k(U, U\cap M') \to K_k(N',M') \to 0\,.$$

The module $K_k(U, U\cap M')$ is free, with one basis element corresponding to each handle (represented by the core of the dual handle). The map to it of $K_{k+1}(N,M)$ is dual to the map $K_{k+1}(U, U\cap M) \to K_k(N)$ representing the attaching maps, and so is zero. Thus $K_{k+1}(N,M)$ is unaltered and we acquire a free (based) module $K_k(N',M')$ dual to $K_{k+1}(N')$.

The attached handles correspond (by construction) to a preferred base of $K_{k+1}(N,M)$, hence the map

$$K_{k+1}(N') \to K_{k+1}(N',M')$$

is a simple isomorphism. It follows that the map of the dual modules $K_k(N') \to K_k(N',M')$ also is, and hence that $K_k(M')$ vanishes (by the exact sequence). In fact (as we will see below) the trivial base for $0 = K_k(M')$ is preferred, so that the map $M' \to X$ is already a simple homotopy equivalence.

Now choose a preferred base of $K_k(N')$, and (using (1.1), Corollary) perform surgery on the elements of $\pi_{k+1}(\phi')$ corresponding to the base elements under the Hurewicz-Namioka isomorphism. Write P for the cobordism so obtained of N' to N'', say, and consider the induced map of degree 1 of Poincaré triads (in the relative case, there will be tetrads)

$$(P; N' \cup (\partial N' \times I), N'') \to (Y \times I; Y \times 0 \cup X \times I, Y \times I):$$

we will identify $N' \cup (\partial N' \times I)$ with N'. In the exact sequence

$$0 \to K_{k+1}(N') \to K_{k+1}(P) \to K_{k+1}(P, N') \xrightarrow{d} K_k(N') \to K_k(P) \to 0,$$

the map d is, by construction, a simple isomorphism, so $K_k(P)$ vanishes (with trivial preferred base) and $K_{k+1}(N') \to K_{k+1}(P)$ is a simple isomorphism. Now in the sequence

$$0 \to K_{k+1}(N'') \to K_{k+1}(P) \to K_{k+1}(P, N'') \to K_k(N'') \to 0$$

the middle map is dual to d, so is a simple isomorphism, hence all $K_i(N'')$ vanish, with trivial preferred bases.

So we have a simple homotopy equivalence $N'' \to Y$. Arguing as in the first case, it follows that we have a simple homotopy equivalence of pairs $(N'', M') \to (Y, X)$. This completes the proof of the theorem.

5. The Even-dimensional Case

The even-dimensional surgery obstruction groups $L_{2k}(\Lambda)$ of a ring with involution Λ are Witt groups of stable isomorphism classes of $(-1)^k$-hermitian forms over Λ, where stability is with respect to hyperbolic forms. These L-groups are constructed by analogy with the projective class group $K_0(\Lambda)$, but using forms instead of modules. A k-connected $2k$-dimensional normal map $(\phi, F) : M^{2k} \to X$ determines a $(-1)^k$-hermitian intersection form (λ, μ) on the kernel $\mathbf{Z}[\pi_1(X)]$-module $K_k(M) = \pi_{k+1}(\phi)$. The surgery obstruction is the class of the intersection form

$$\theta(M, \phi, F) \;=\; (K_k(M), \lambda, \mu) \in L_{2k}\big(\mathbf{Z}[\pi_1(X)]\big)\,.$$

For $k \geqslant 3$ the normal map (ϕ, F) is bordant to a homotopy equivalence if and only if $\theta(M, \phi, F) = 0$, i.e. if and only if the form $(K_k(M), \lambda, \mu)$ is stably hyperbolic. The surgery obstruction of an arbitrary $2k$-dimensional normal map (M, ϕ, F) is obtained by first applying (1.2) to make ϕ k-connected by surgery below the middle dimension. See the notes at the end of section 17G for the chain complex method of Ranicki [R5], which obtains $\theta(M, \phi, F)$ directly from the normal map, without preliminary surgeries below the middle dimension.

We commence our proof of the main theorem by considering the case of surgery on an even-dimensional manifold M, leaving the boundary fixed: this was essentially solved in our paper [W18], but we repeat the details here for the reader's convenience, and because we wish to introduce some terminology which will be needed in later chapters.

Suppose given a Poincaré complex X of formal dimension $2k \geqslant 4$, a bundle over X, and a normal map (M, ϕ, F) such that $\phi : M \to X$ has degree 1. By (1.2), we may suppose ϕ k-connected. Then (2.3) and (2.6) show that

LEMMA 5.1. *$G = K_k(M) = \pi_{k+1}(\phi)$ is a stably free, stably based Λ-module where $\Lambda = \mathbf{Z}[\pi_1(X)]$. So is $G^* = K^k(M)$, and Poincaré duality induces a simple isomorphism of G and G^*.*

We can identify G^* with $\mathrm{Hom}_\Lambda(G, \Lambda)$ since ϕ is k-connected. Thus the map of G to G^* transposes to give a map $\lambda : G \times G \to \Lambda$. This is induced by intersection numbers. We next study intersections and self-intersections geometrically. For this we do not need such strong hypotheses.

Let M^{2k} ($k \geqslant 2$) be a connected smooth or PL manifold with fundamental group π, E the total space of the bundle associated to the tangent bundle of M with fibre the Stiefel manifold $V_{k,k}$ (in the PL case we will use the stable Stiefel manifold V_k instead: see [H6] and [H3]). Then regular homotopy classes of immersions $S^k \to M$ correspond bijectively to elements of $\pi_k(E)$ – this follows from the Proposition in §1. We have an exact sequence

$$\pi_k(V_{k,k}) \to \pi_k(E) \to \pi_k(M) \to \{1\};$$

the first term is cyclic of order ∞ or 2 according as k is even or odd. Projection induces $\pi_1(E) \cong \pi_1(M) = \pi$: thus π operates (on the right) on $\pi_k(E)$. We represent elements of $\pi_k(E)$ as immersions $f : S^k \to M$ which do not necessarily preserve the base point, hence some convention is necessary. Let $\widetilde{M}$ be the universal covering space of M: we can either specify a lifting $\widetilde{f} : S^k \to \widetilde{M}$, or (equivalently) a (homotopy) class of paths in M joining the base point $*$ to (1): we shall sometimes use a lifting, sometimes a path. Note that π operates via its operation on $\widetilde{M}$, or equivalently by composing the path with a loop on $*$. Addition of two immersed spheres S^k in $\widetilde{M}$ is represented by joining by an arc, which is then thickened to a copy of $D^k \times I$ (with ends $D^k \times \partial I$ on the two spheres), and we use $\partial D^k \times I$ to form the connected sum of the spheres.

Write G for any Λ-module provided with a Λ-homomorphism $G \to \pi_k(E)$, so that elements of G are represented by immersed spheres in M.

THEOREM 5.2. *Intersections define a map $\lambda : G \times G \to \Lambda$ such that*

(i) *for $x \in G$ fixed, $y \mapsto \lambda(x, y)$ is a Λ-homomorphism $G \to \Lambda$.*

(ii) $\lambda(y, x) = (-1)^k \overline{\lambda(x, y)} \quad (x, y \in G)$.

Write Q_k for the quotient group $\Lambda/\{\nu - (-1)^k \overline{\nu} : \nu \in \Lambda\}$. Then self-intersections define a map $\mu : G \to Q_k$ such that

(iii) $\lambda(x, x) = \mu(x) + (-1)^k \overline{\mu(x)} + \chi_N(x) \quad (x \in G)$,

(iv) $\mu(x + y) - \mu(x) - \mu(y) = \lambda(x, y) \quad (x, y \in G)$,

(v)* $\mu(xa) = \bar{a}\mu(x)a \quad (x \in G, a \in \Lambda)$.

If $k \geqslant 3$, x is represented by an embedding if and only if $\mu(x) = 0$†.

* *Peter Teichner has pointed out that the formula $\mu(xa) = \bar{a}\mu(x)a$ is only correct in general if $a = ng \in \Lambda$ ($n \in \mathbf{Z}, g \in \pi_1(X)$), or if $\chi_N(x) = 0$. In fact, for an arbitrary $a = \sum_i n_i g_i \in \Lambda$ there is a correction term involving the Euler number $\chi_N(x) \in \mathbf{Z}$*

$$\mu(xa) \;=\; \bar{a}\mu(x)a + \Big(\sum_{i<j} n_i n_j w(g_i) g_j (g_i)^{-1}\Big)\chi_N(x) \in Q_k\,.$$

† *Thus for $k \geqslant 3$ it is possible to kill $x \in G$ by surgery on (ϕ, F) if and only if $\mu(x) = 0$. The effect of the surgery is a bordant normal map $(\phi', F') : M' \to X$ with kernel Λ-modules*

$$K_k(M') \;=\; \langle x\rangle^{\perp}/\langle x\rangle\;,\;\; K_{k-1}(M') \;=\; \mathrm{Coker}(\lambda(x, -) : G \to \Lambda)$$

where

$$\langle x\rangle = \{xa : a \in \Lambda\} \subseteq \langle x\rangle^{\perp} = \mathrm{Ker}(\lambda(x, -) : G \to \Lambda) = \{y \in G : \lambda(x, y) = 0 \in \Lambda\} \subset G\,.$$

The normal map (ϕ', F') is k-connected if and only if x generates a direct summand $\langle x\rangle \subset G$.

Notes. In (iii), although $\mu(x) \in Q_k$, $\mu(x) + (-1)^k\overline{\mu(x)}$ is a well-defined element of Λ. We use $\chi_N(x)$ for the normal Euler number of an immersion representing x. In (iv), we should really replace $\lambda(x, y)$ by the element of Q_k it determines. As to (v), although the group Q_k is not a Λ-module, the symbol $\bar{a}ba$ is well-defined for $a \in \Lambda$, $b \in Q_k$.

Proof Let S_1, S_2 be two immersed k-spheres in M; we may suppose that they meet in general position, i.e. transversely in a finite set of points P. To each such P we assign an element $g_P \in \pi$, and a sign $\varepsilon_P = \pm 1$ as follows. In terms of paths, g_P is the class of the loop at $*$ which starts along the path to the base point of S_2, round S_2 (avoiding other singularities) to P, round S_1 to its base point, and back along the given path to $*$. In terms of prescribed lifts to $\widetilde{M}$, if $\widetilde{P}$ is the point of $\widetilde{S}_1$ lying over P, g_P is that element of π such that $\widetilde{S}_2 g_P^{-1}$ passes through $\widetilde{P}$.

To define ε_P we must orient M at $*$ (it may be globally nonorientable), and transport the orientation to P by the path for S_1. Then ε_P is the sign of the intersection of S_1 and S_2 with respect to this orientation at P. Equivalently, the orientation at $*$ orients $\widetilde{M}$, and ε_P is the sign of the intersection of $\widetilde{S}_1$ and $\widetilde{S}_2 g_P^{-1}$ at $\widetilde{P}$.

We now define $\lambda(S_1, S_2) = \sum_P \varepsilon_P n_P$ over all intersection points P. If we write the sum as

$$\sum_{g \in \pi} n(g)g$$

(note, however, that it is finite), then $n(g)$ is the intersection number of $\widetilde{S}_1$ and $\widetilde{S}_2 g^{-1}$ in $\widetilde{M}$, so depends only on the homotopy classes; thus it is certainly well-defined for elements of G, so $\lambda : G \times G \to \Lambda$ is defined. The fact that λ is linear over $\mathbf{Z}$ is evident (we can ignore the connecting tubes); also that

$$\lambda(S_1, S_2 g) = \lambda(S_1, S_2)g$$

(for we must replace each g_P by $g_P g$). Thus (i) is established. As to (ii), we have the same intersection points, but must recompute ε_P and g_P. But π acts on $\widetilde{M}$ by homeomorphisms, so if $\widetilde{S}_1$ meets $\widetilde{S}_2 g_P^{-1}$, then $\widetilde{S}_1 g_P$ meets $\widetilde{S}_2 : g'_P = g_P^{-1}$. The sign of the intersection changes by $(-1)^k$ on interchanging the order of the two spheres; there is a further change if g_P changes the orientation of $\widetilde{M}$. Hence $\varepsilon'_P = (-1)^k w(g_P)\varepsilon_P$, which proves (ii).

This completes the properties of λ; we come now to μ, which is more complicated, and will be a constant source of extra snags for the rest of the book. Let S_1 be an immersed sphere in general position, so that it has only a finite set of self-intersections, all transverse. At each such point P, two branches of S_1 cross. If we impose an order on these branches, we can compute ε_P and g_P as before. Moreover, by the argument above, if we interchange the order $\varepsilon_P g_P$ is replaced by $(-1)^k w(g_P)\varepsilon_P g_P^{-1} = (-1)^k \varepsilon_P \bar{g}_P$. Consider $\sum_P \varepsilon_P g_P$, where at each self-intersection point P we choose arbitrarily an ordering of the two branches. Then if $\mu(S_1)$ is the element of Q_k defined by $\sum_P \varepsilon_P g_P$, alteration of our choices will not affect $\mu(S_1)$. Let us now change S_1 by a regular homotopy. If this is generic (see Cerf [C13]), the self-intersections, and hence μ, vary continuously except at a finite set of points where two self-intersections appear (or disappear) together. At such a birth or death point, the two self-intersections determine the same g_P and opposite values of ε_P; thus μ is constant. So we obtain a map $\mu : G \to Q_k$.

Now (iv) is immediate, since the self-intersections of the connected sum of two spheres S_1 and S_2 are just the self-intersections of S_1, and those of S_2, together with the intersections of S_1 with S_2. For (iii), note that $\lambda(x, x)$ is the mutual intersection of two different spheres S_1, S_1' representing x. We choose S_1 as above; then take a tubular neighbourhood, and take S_1' as a cross-section of the normal bundle of S_1 (with fibre D^k). This will intersect S_1 (the zero cross-section) with intersection number $\chi_N(x)$. In addition, each self-intersection P of S_1 gives rise to two intersections of S_1 with S_1': for one, the two branches at P correspond to S_1 and S_1' respectively, for the other they correspond to S_1' and S_1. This proves (iii).

In view of (iv), it is sufficient to prove (v) when $a = g$ is an element of π, and in this case it is immediate (see (i) above). Thus it remains to show only that if $k \geqslant 3$ and $\mu(x) = 0$ we can find an embedding representing x. Note that (for all $k \geqslant 2$) $\mu(x) = 0$ is clearly a necessary condition for x to be representable by an embedding; also that for $k = 2$ it is known [K4] to be insufficient. Now if $\mu(x) = 0$, we can put the self-intersections of a sphere S representing x in pairs (P_i, Q_i) such that (with appropriate choices of order of the two branches at each intersection) $g(P_i) = g(Q_i)$ and $\varepsilon(P_i) = -\varepsilon(Q_i) = 1$. Join P_i to Q_i by an arc α_i along one branch and an arc β_i along the other; then the loop defined by α_i and β_i is nullhomotopic, and the two intersections P_i, Q_i have opposite signs on it. The result now follows from [M13, 6.6].

REMARK 5.2.1. The fact that $\widetilde{M}$ is the universal covering of M is not essentially used in the above. We can consider the more general case when $M \subset N$, N connected (but M need no longer be), and we use the covering of M induced from the universal covering of N. Since this is not connected, we must permit an element of G to be represented by a finite union of spheres in $\widetilde{M}$, and permit these to be varied by regular homotopies, and by replacing two spheres in the same component by their connected sum. The above proof shows that λ and μ are unchanged by such operations. All assertions of the Theorem then remain valid except, of course, our criterion for embeddings (which assumes $\widetilde{M}$ connected and simply connected).

In the situation of (5.1), we can apply (5.2). The map λ is induced by intersection numbers, so is the same (i.e. up to sign-convention) as the map which corresponds to Poincaré duality. We define $A\lambda : G \to \mathrm{Hom}_\Lambda(G, \Lambda)$ by $A\lambda(x)(y) = \lambda(x, y)$: with our conventions, this is a map of Λ-modules; indeed, a simple isomorphism of stably free and stably based Λ-modules. When G is free and based, we call (G, λ, μ) a *simple hermitian form** (or, more precisely,

**In the first edition a simple hermitian form was called a special hermitian form.*

$(-1)^k$-hermitian). Note that as our immersions are now always framed, (5.2 (iii)) simplifies as $\chi_N \equiv 0$. Correspondingly, we can apply (5.2.1) with an appropriate Λ and again obtain a simple hermitian form.

Suppose given a free Λ-module G with base $\{e_i\}$. Choose arbitrarily elements $b_i \in Q_k$, $a_{ij} \in \Lambda\,(i < j)$. Then there is a unique (G, λ, μ), satisfying (5.2 (i)–(v)), with $\lambda(e_i, e_j) = a_{ij}$ for $i < j$ and $\mu(e_i) = b_i$. However, $A\lambda$ will in general not be a simple isomorphism. It is in the case when the base has two elements $\{e, f\}$, and we set

$$\mu(e) \;=\; \mu(f) \;=\; 0 \qquad \lambda(e, f) \;=\; 1\,.$$

The resulting form we call the *standard plane*, and a direct sum of copies of it – or any isomorph – a *hyperbolic form**. (Observe that given simple hermitian forms (G_1, λ_1, μ_1) and (G_2, λ_2, μ_2) we can form their orthogonal direct sum $(G_1 \oplus G_2, \lambda_1 \oplus \lambda_2, \mu_1 \oplus \mu_2)$).

LEMMA 5.3. *A simple hermitian form (G, λ, μ) is hyperbolic if and only if G has a free based submodule H, with a preferred base extending to one of G, and so defining a preferred class of bases of G/H, such that*

$$\lambda(H \times H) \;=\; 0\;,\;\; \mu(H) \;=\; 0\;,$$

and the map $G/H \to \mathrm{Hom}_\Lambda(H, \Lambda)$ induced by λ is a simple isomorphism.

Such a submodule H we will henceforth call a *lagrangian*†.

Proof Let $\{e_i\}$ be a preferred base of H. There is a dual base of $\mathrm{Hom}_\Lambda(H, \Lambda)$, which induces (by the above isomorphism) a base of G/H: we choose representative elements f_i' in G. By hypothesis, $\{e_i, f_i'\}$ is a preferred base of G, and we have

$$\mu(e_i) \;=\; 0 \qquad \lambda(e_i, e_j) \;=\; 0 \qquad \lambda(e_i, f_j') \;=\; \delta_{ij}\,.$$

Now choose $\mu_i \in \mu(f_i')$ and make the elementary basis change

$$f_j \;=\; f_j' + (-1)^{k-1}\big(e_j\mu_j + \sum_{i<j} e_i\lambda(f_i', f_j')\big)\,.$$

We obtain

$$\mu(f_i) \;=\; 0 \qquad \lambda(e_i, f_j) \;=\; 0 \quad \text{and} \quad \lambda(f_i, f_j) \;=\; 0$$

thus G is a hyperbolic form, as asserted: the base $\{e_i, f_i\}$ provides an isomorphism of G with a direct sum of standard planes. Since $\{e_i\}$ could be any preferred base of the lagrangian H, our argument shows

* *In the first edition a hyperbolic form was called a kernel, being a generalisation of a Witt kernel. The kernel terminology was introduced in [W9].*

† *In the first edition a lagrangian was called a subkernel. The lagrangian terminology was introduced by Novikov [N8], who related surgery obstruction theory to the formalism of hamiltonian physics.*

COROLLARY 5.3.1. *Suppose for $i = 1, 2$, H_i is a lagrangian in the simple hermitian form (G_i, λ_i, μ_i). Then any simple isomorphism $H_1 \to H_2$ extends to a simple isomorphism of (G_1, λ_1, μ_1) on (G_2, λ_2, μ_2).*

We call two lagrangians H_1, H_2 in (G, λ, μ) *complementary* if they are complementary submodules, i.e.

$$H_1 \cap H_2 = \{0\} , \quad H_1 + H_2 = G .$$

There is then an obvious simple (Noether) isomorphism of H_2 on G/H_1, and in (5.3) above we can choose the f_1 (lifting a base of G/H_1) to lie in H_2: no further adjustment is then necessary. So any two complementary lagrangians are isomorphic to the pair described above (with bases $\{e_i\}$, $\{f_i\}$ respectively, $1 \leqslant i \leqslant r$, for suitable r).

LEMMA 5.4. *If (G, λ, μ) is a simple hermitian form, then $(G, \lambda, \mu) \oplus (G, -\lambda, -\mu)$ is a hyperbolic form.*

Proof Let $\{e_i\}$ be a preferred base of G: write e'_i, e''_i for the corresponding elements of the two summands. We have

$$\begin{aligned} \lambda(e'_i + e''_i, e'_j + e''_j) &= \lambda(e'_i, e'_j) + \lambda(e''_i, e''_j) \\ &= \lambda(e_i, e_j) - \lambda(e_i, e_j) = 0 \end{aligned}$$

and

$$\begin{aligned} \mu(e'_i + e''_i) &= \mu(e'_i) + \mu(e''_i) \\ &= \mu(e_i) - \mu(e_i) = 0 . \end{aligned}$$

To show that the submodule H of $G \oplus G$ freely generated by the $e'_i + e''_i$ is a lagrangian, it remains only to verify that the map of $(G \oplus G)/H$ to $\mathrm{Hom}_\Lambda(H, \Lambda)$ induced by λ is a simple isomorphism. Now the classes of the e'_i give a basis of the former module; and for the latter we use the dual basis of $\{e'_j + e''_j\}$. Thus our map has matrix (a_{ij}), where

$$a_{ij} = \lambda(e'_i, e'_j + e''_j) = \lambda(e_i, e_j) .$$

But this is also the matrix of the map of G to $\mathrm{Hom}_\Lambda(G, \Lambda)$ which is, by hypothesis, a simple isomorphism. The result follows from 5.3.

We now define a group $L_m(\pi)$ when $m = 2k$. Consider the semigroup under $\oplus$ of simple hermitian forms. Write $X \sim X'$ if there are hyperbolic forms K, K' such that $X \oplus K$ and $X' \oplus K'$ are isomorphic: since the sum of hyperbolic forms is a hyperbolic form, this is an equivalence relation, and it is clearly compatible with addition, so we have a quotient semigroup. But by (5.4) $(G, -\lambda, -\mu)$ is inverse to (G, λ, μ) modulo the equivalence relation, so our quotient semigroup is a group: we call it $L_m(\pi)$.

If we relax the conditions that G be free and based to require merely an s-base, we appear to get a different group. But given such a (G, λ, μ), we can add a hyperbolic form of large dimension $2r$: if r is big enough, this makes G free, and if r is even larger, the resulting s-base will be equivalent to an actual base. Thus if we define an equivalence relation as above, we get the same equivalence classes as before.

The reason why hyperbolic forms are ignored in our equivalence relation is explained by

LEMMA 5.5. *Assume the hypotheses preceding* (5.1), *and perform surgery on a further* $(k-1)$*-sphere. Then the effect on* (G, λ, μ) *is to add a standard plane.*

Proof Since $\pi_k(\phi) = 0$, we are doing surgery on the zero element of $\pi_k(\phi)$. Thus our $(k-1)$-sphere is regularly homotopic, and so (by general position) isotopic to an unknotted one inside a disc $D^m \subset M$, with the standard framing. Surgery now replaces M by the connected sum $M\#(S^k \times S^k)$: G is correspondingly replaced by a direct sum – evidently an orthogonal one – and if e, f are the classes of $S^k \times 1$ and $1 \times S^k$ respectively, we have a preferred base, and λ and μ are as for a standard plane.

We can make $L_m(\pi)$ into a functor. For if $r : \pi \to \pi'$ is a homomorphism compatible with w, it induces $r : \Lambda \to \Lambda'$ compatible with bar, and any simple hermitian form (G, Λ, μ) over Λ induces by tensoring a simple hermitian form (with carrier $G \otimes_\Lambda \Lambda'$) over Λ': indeed, we can map the matrix elements of λ by r, and analogously for μ. This preserves direct sums, so induces a homomorphism $L_m(\pi) \to L_m(\pi')$.

THEOREM 5.6. *Let* (X, W) *be a connected simple Poincaré pair with formal dimension* $m = 2k$, ν *a bundle over* X, M *a compact manifold with boundary,* $\phi : (M, \partial M) \to (X, W)$ *a map of degree* 1 *inducing a simple homotopy equivalence* $\partial M \to W$, F *a stable trivialisation of* $\tau_M \oplus \phi^*\nu$. *If* $k \geqslant 2$ *we define* $\theta \in L_m(\pi_1(X))$ *by performing surgery rel* ∂M *till* ϕ *is* k*-connected and then taking the class of* $K_k(M)$. *Then* θ *depends only on the bordism class of* (M, ϕ, F) *relative to* ∂M*; if* $k \geqslant 3$*, the class has a representative with* ϕ *a simple homotopy equivalence if and only if* $\theta = 0$. *The obstruction* θ *is natural for inclusion maps.*

Note that this is stronger than the result announced in §3, for we have a precise definition of θ which is not beyond the reach of computation. We will proceed analogously for m odd and for pairs, though the algebraic definitions of the L_m get progressively more complicated.

Proof We must show that θ depends only on the bordism class. So suppose given bordant normal maps (M_-, ϕ_-, F_-) and (M_+, ϕ_+, F_+), with ϕ_- and ϕ_+ each k-connected: let (N, ψ, F) be the cobordism. We can write $\partial M_- = \partial M_+$ and take N to be a manifold triad: we regard ψ as a map of triads,

$$\psi : (N; M_-, M_+) \to (Y; X \times 0 \cup W \times I, X \times I).$$

By (1.2) we can do surgery rel ∂N to make the map $N \to Y$ k-connected.

We wish to have $K_k(\psi) = 0$. Theorem 1.4 states that we can ensure this by performing suitable handle subtractions, and the addendum shows that (e.g.)

M_- can be kept fixed. By (5.5), the only effect on $K_k(M_+)$ is to add a hyperbolic form [which does not affect class in $L_{2k}(\pi_1(X))$].

At this stage we know that all groups K_i of M_-, M_+ vanish except with $i = k$; likewise for ∂N. In fact all groups K_i of ∂N, N and $(N, \partial N)$ vanish except (cf. (1.4)) those in the exact sequence

$$0 \to K_{k+1}(N, \partial N) \to K_k(\partial N) \to K_k(N) \to 0\,.$$

Now we can make $K_k(M_-)$ and $K_k(M_+)$ free and based by adding hyperbolic forms of large enough rank: by (5.5) again, we can achieve this by adding enough k-handles to N along M_- or M_+. Also, each such addition adds a free module (with its natural basis) to $K_{k+1}(N, \partial N)$ and $K_k(N)$, so we may suppose each of these free and based. By (2.4), Corollary, the exact sequence is based (i.e. as chain complex, it has zero torsion).

Since $j_*\partial_*[N] = [M_+] - [M_-]$, and $K_k(\partial N) = K_k(M_+) \oplus K_k(M_-)$, (with coefficients $\Lambda = \mathbf{Z}[\pi_1(X)]$ understood: note that $\pi_1(X) \neq \pi_1(\partial N)$ here), the simple hermitian form defined on $K_k(\partial N)$ by (5.2.1) is the sum of a form representing the surgery obstruction for M_+ and the negative of a corresponding form for M_-. So to prove these are equal, it suffices to show that $K_k(\partial N)$ is a hyperbolic form. By (5.3), it will be enough to show that $K_{k+1}(N, \partial N)$ is a lagrangian. This result is important enough for us to be worth stating as a separate lemma.

LEMMA 5.7. *Let $\phi : (N, M) \to (Y, X)$ be a map of degree 1, where N is a compact manifold with boundary M and (Y, X) is a simple Poincaré pair with Y connected; suppose ν and F as usual. Let $\dim N = 2k+1 \geqslant 5$, and suppose ϕ induces k-connected maps $M \to X$ and $N \to Y$ and that $K_k(N, M) = 0$ (with coefficients $\Lambda = \mathbf{Z}[\pi_1(Y)]$ understood). Assume that the stable base of $K_{k+1}(N, M)$ given by (2.3) is equivalent to an actual base. Then $K_{k+1}(N, M)$ is a lagrangian of $K_k(M)$.*

We emphasise again that the coefficients are $\mathbf{Z}[\pi_1(Y)]$ and not $\mathbf{Z}[\pi_1(X)]$. We will conclude the proof of (5.6) using this result, and then return to prove (5.7).

We have seen that (5.7) implies that θ depends on the bordism class; clearly it vanishes for a simple homotopy equivalence. Suppose, conversely, that θ vanishes and that $k \geqslant 3$. Using (5.5) as above, we see that we may suppose $K_k(M)$ a hyperbolic form, with standard base, say $\{e_i, f_i : 1 \leqslant i \leqslant r\}$. As $\mu(e_r) = 0$, the class $e_r \in K_k(M) \cong \pi_{k+1}(\phi : M \to X)$ is represented by an embedded sphere S_r in M, by (5.2) (here we use $k \geqslant 3$). By (1.1), we can now do surgery on M using this sphere. Let N be the support of the surgery: up to homotopy, $N \simeq M \cup e^{k+1}$, and if M_+ is the resulting manifold, $N \simeq M_+ \cup e^k$. The homomorphism

$$K_k(M) \to H_k(N) \to H_k(N, M_+) \cong \Lambda$$

has an immediate geometrical interpretation by intersection numbers with e_r; since $\lambda(e_r, f_r) = 1$, it is surjective. So ϕ_+ is still k-connected, the surgery on M_+ to return to M is made on a trivial $(k-1)$-sphere, and (cf. (5.5)) $M \cong$

$M_+\#(S^k \times S^k)$. We claim (cf. [W18, (3.3)]) that $K_k(M_+)$ may be identified with the hyperbolic form with base $\{e_i, f_i : 1 \leqslant i \leqslant r-1\}$: the desired result then follows by induction on r. But this can be seen from the commutative exact diagram

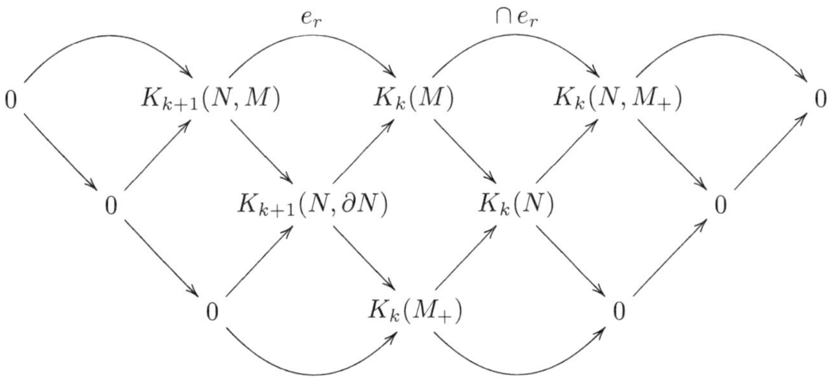

in which $K_{k+1}(N, M)$ and $K_k(N, M_+)$ can be identified with Λ: we can identify $K_{k+1}(N, \partial N)$ with the submodule of $K_k(M)$ generated by all e_i and f_i except f_r, and $K_k(M_+)$ with the quotient of this by e_r. Moreover, these identifications preserve preferred classes of bases, as is evident since the isomorphisms with Λ correspond geometrically to cells: in fact this construction is simply the reverse of the preceding (i.e. of taking connected sums with $S^k \times S^k$).

To complete the proof of (5.6) it remains to show naturality. Write $\partial M = L$; let $\partial M' = L \cup L'$, and suppose given a simple homotopy equivalence ϕ' : $(M'; L, L') \to (X'; W, W')$ of triads, with associated ν and F, extending $\phi \,|\, L$ etc. Then we must consider the combined map

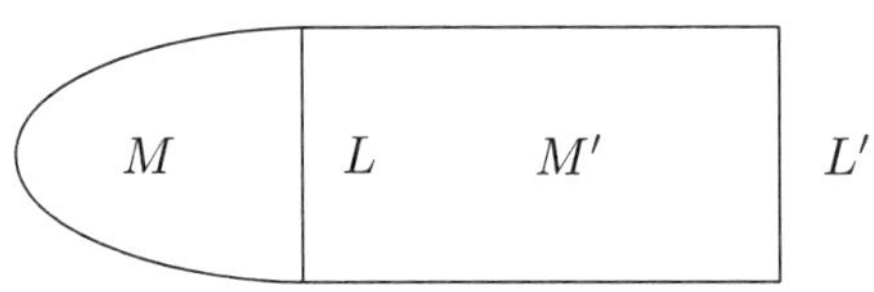

$$\phi'' : (M \cup M', L') \to (X \cup X', W')\,.$$

But since ϕ' is a simple homotopy equivalence we have $K_k(M \cup M') = K_k(M)$ with, however, different coefficients. Since we can use the same spheres to calculate λ and μ, we must get the same results, composed with the map $\mathbf{Z}[\pi_1(X)] \to \mathbf{Z}[\pi_1(X \cup X')]$ induced by inclusion. Hence we obtain the image of the simple hermitian form ϕ under the associated map, so θ too behaves naturally.

Proof of (5.7). (cf. [W18, (7.3)]). It follows from our assumptions that the only non-vanishing K-groups are those in the sequence

$$0 \to K_{k+1}(N, M) \to K_k(M) \to K_k(N) \to 0\,.$$

By (2.4), Corollary, this is a based short exact sequence of s-based modules. By (2.6), duality induces a simple isomorphism of $K_{k+1}(N, M)$ on $K^k(N) \cong \mathrm{Hom}_\Lambda\big(K_k(N), \Lambda\big)$. Since, by hypothesis, the s-base of $K_{k+1}(N, M)$ is equivalent to a base, the same holds for $K^k(N)$, and hence also for the dual module $K_k(N)$. To check the conditions (5.3) that $K_{k+1}(N, M)$ be a lagrangian, it remains only to show that λ and μ vanish identically on it. This can easily be shown algebraically in the case of λ, but we give a geometrical argument which covers μ also.

Since the restricted map of $\phi : M \to X$ is k-connected, it follows that although $\Lambda\ \big(= \mathbf{Z}[\pi_1(Y)]\big)$ is not closely related to the fundamental groups of the components of X, nevertheless $K_k(M)$ is generated as Λ-module by classes represented by maps of spheres. (One can regard M as a subcomplex of X containing the k-skeleton; the relative $(k+1)$st homology module is then generated by the classes of the $(k+1)$-cells of $X - M$). Indeed, we have used spheres to define (5.2.1) the simple hermitian form on $K_k(M)$. Now let $x \in K_{k+1}(N, M)$. We represent $\partial x \in K_k(M)$ as a sum of maps of spheres: these maps may be taken to be framed immersions (e.g. just one in each component of $\widetilde{M}$ where ∂x has a nonzero summand). These spheres have classes in $\pi_k(\widetilde{N})$. I say that the sum of these classes is zero. For the image in $K_k(N)$ is zero by the homology exact sequence, and $\pi_{k+1}(\widetilde{\phi}) \cong H_{k+1}(\widetilde{\phi}) = K_k(N)$ maps to $\pi_k(N)$. Hence there is a map into $\widetilde{N}$ of a $(k+1)$-sphere, with discs removed, whose boundary spheres are mapped by our framed immersions. More precisely it can be seen (arguing as in (1.3)) that the framed immersions in $\widetilde{M}$ of the spheres S^k extend to a framed immersion in $\widetilde{N}$ of the punctured $(k+1)$-sphere, representing x.

Let T be the immersed punctured $(k+1)$-sphere representing $x \in K_{k+1}(N, M)$; let T' similarly represent x'. We may suppose that T and T' meet transversely: i.e. in a finite set of circles (which do not concern us) and arcs with both ends representing intersections of ∂x and $\partial x'$. Hence (homologically) all such intersections cancel in pairs, so $\lambda(\partial x, \partial x') = 0$. Similarly consider the self-intersections of T: along each arc, we choose an order of the two branches of T meeting there, and note that the self-intersections of ∂T at the two ends determine the same $g \in \pi_1(Y)$, but opposite signs: so the self-intersections cancel in pairs too, and $\mu(\partial x) = 0$. This proves Lemma 5.7.

We have now proved the main theorem in the special case considered in this paragraph, but before we leave it, we wish to show that the groups $L_m(\pi)$ defined above are not too large, in that all elements do occur as obstructions to surgery problems.

THEOREM 5.8. *Let $m = 2k \geqslant 6$; let X^{m-1} be a connected compact smooth or PL manifold with fundamental group π and stable normal bundle ν. Then we can find a compact manifold triad $(M; \partial_- M, \partial_+ M)$, a map*

$$\phi : (M; \partial_- M, \partial_+ M) \to (X \times I; X \times 0 \cup \partial X \times I, X \times I)$$

of degree 1, and a stable framing F of $\tau_M \oplus \phi^* \nu$, such that

(i) *$\phi \,|\, \partial_- M$ is an identity map, $\partial_- M \to X \times 0 \cup \partial X \times I$,*

(ii) *$\phi \,|\, \partial_+ M$ is a simple homotopy equivalence and*

(iii) *the surgery obstruction for ϕ keeping ∂M fixed is a prescribed element of $L_m(\pi)$. More precisely the bordism set for all (M, ϕ, F) satisfying* (i) *and* (ii) *is mapped bijectively to $L_m(\pi)$ by θ.*

Proof By definition, any element of $L_m(\pi)$ is represented by a simple hermitian form: let $\{e_i\}$ be a preferred base, $1 \leqslant i \leqslant r$. Choose r disjoint discs $D_i^{2k-1} \subset \operatorname{Int} X$; let $f^0 : S^{k-1} \times D^k \to D^{2k-1}$ be the standard embedding, so by composition we obtain r disjoint embeddings $f_i^0 : S^{k-1} \times D^k \to \operatorname{Int} X$, and choose lifts to $\widetilde{X}$ (such embeddings are called unknotted and unlinked). We now subject the f_i^0 to simultaneous regular homotopies η_i, to new disjoint embeddings f_i^1.

We can regard η_i as a framed immersion $S^{k-1} \times I \to X \times I$ with boundary embedded. We can thus count intersections and self-intersections exactly as in (5.2). I say (cf. [W18, p. 247]) that we can choose the self-intersections of the η_i, and the intersections of η_i with η_j $(i < j)$, arbitrarily and independently. Since all is additive under composition of regular homotopies, it is sufficient to be able to introduce a single (self-) intersection with invariant $\pm g$, $g \in \pi$. To do this, we join $\widetilde{f}_i^0(S^{k-1})g$ to $\widetilde{f}_j^0(S^{k-1})$ by an arc in $\widetilde{X}$, projecting to an arc in X in a prescribed homotopy class, move part of the latter sphere along the arc till it is close to the former, and then deform across a disc D^k meeting $\widetilde{f}_i^0(S^{k-1})g$ transversely. If care is taken with orientation (in fact, we have room for manoeuvre along the arc), one sees that either sign can be achieved. The argument is valid for $i \leqslant j$.

Now attach k-handles to $X \times I$ with attaching maps $f_i^1 \times 1$. Since we chose trivial spheres to start with, we can regard this as a surgery on the identity map $X \to X$. Let M be the resulting manifold. All assertions of the Theorem up to (i) are clear. To obtain (ii) and (iii) we use our simple hermitian form; in fact, we choose

$$\text{self-intersection of } \eta_i = \mu(e_i)$$
$$\text{intersection of } \eta_i \text{ with } \eta_j = \lambda(e_i, e_j) \quad (i < j):$$

it follows as for (5.2) (ii) and (iii) that the second line then holds for all i and j. Clearly $K_k(M) = K_k(M, \partial_- M)$ has as preferred basis the classes of the cores of the attached bundles. We complete these to spheres S_i by adjoining the images in $X \times I$ of the η_i, and discs in the D_i^{2k-1} spanning the images of the f_i^0, and (in the smooth case) rounding the resulting corners.

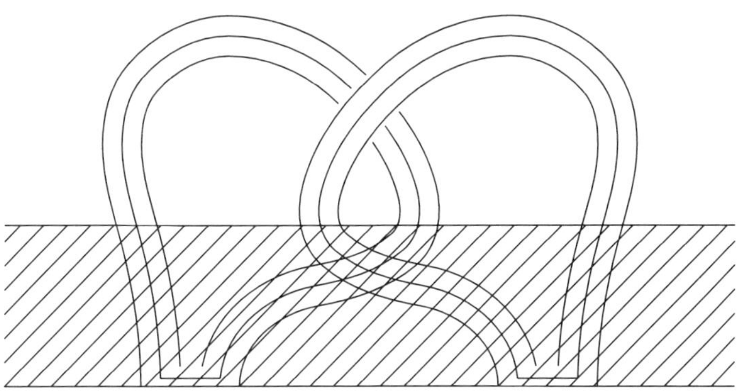

Then $S_i \cap S_j = \eta_i \cap \eta_j$, and similarly for self-intersections, so we recover the given simple hermitian form. Its associated simple isomorphism $G \to \mathrm{Hom}_\Lambda(G, \Lambda)$ can be identified with the natural map

$$K_k(M) \to K_k(M, \partial_+ M)\,,$$

so (by the exact sequence) all $K_i(\partial_+ M)$ vanish. M is formed from $\partial_+ M$ by attaching k-handles, so $\partial_+ M$ has fundamental group π, too (here is the only place in the proof where the hypothesis $k > 2$ is used. The result is known to be false if $k = 2$). Hence ϕ induces a homotopy equivalence $\partial_+ M \to X$; using (2.4) Corollary we see that this is simple. We have already seen that the surgery obstruction is represented by the prescribed simple hermitian form.

As to the last part, suppose (M, ϕ, F) and (M', ϕ', F') correspond to the same surgery obstruction. We glue M to M' along $X \times 0$, correspondingly for the two copies of $X \times I$ (so change one of them the $X \times [0, -1]$), and modify ϕ, F appropriately. In the result, we still have a simple homotopy equivalence on the boundary. The surgery obstruction is zero, since the two original ones were equal (note that the sign of $[M']$ was changed for the glueing). We thus find a manifold N, an extended $\psi : N \to X \times [-1, 1] \times I$ and stable framing G; ψ is a simple homotopy equivalence on all faces of the boundary except $X \times [-1, 1] \times 0$. But we can regard (N, ψ, G) as the required bordism, by reinterpreting $[-1, 1] \times I$ so that $[-1, 0] \times 0$ becomes the 'lower end', $[0, 1] \times 0$ becomes the 'top' and the rest of $\partial([-1, 1] \times I)$ becomes a 'side'. In the smooth case, this requires some adjustment of corners.

Note that in particular the bordism provides an s-cobordism of M_+ to M'_+ (viz. the 'side' mentioned above). By the s-cobordism theorem, we actually have a (smooth or PL) homeomorphism between these two. We make no assertion about the nonexistence of such a homeomorphism if the surgery obstructions for M and M' differ.

Plumbing. The construction in (5.8) *of a k-connected $2k$-dimensional normal map $(\phi, F) : M \to X$ realising a prescribed element of $L_{2k}(\pi_1(X))$ is a non-simply connected generalisation of the plumbing construction of $(k-1)$-connected $2k$-dimensional manifolds with prescribed intersection form due to Milnor [M10]. In particular, the generator $(\mathbf{Z}^8, E_8) \in L_8(\{1\}) = \mathbf{Z}$ (13A.1) is realised in the differentiable category as the rel ∂ surgery obstruction of a 4-connected 8-dimensional degree 1 normal map $(\phi, F) : (W^8, \Sigma^7) \to (D^8, S^7)$ with W^8 the 3-connected 8-dimensional manifold of signature 8 obtained by the E_8-plumbing of 8 copies of τ_{S^4}, with boundary an exotic sphere Σ^7, and $\Sigma^7 \to S^7$ a homotopy equivalence.*

6. The Odd-dimensional Case

The odd-dimensional surgery obstruction groups $L_{2k+1}(\Lambda)$ *of a ring with involution* Λ *are stable unitary groups of automorphisms of hyperbolic* $(-1)^k$*-hermitian forms over* Λ*. The construction of these* L*-groups is motivated by the Whitehead torsion group* $K_1(\Lambda)$ *in algebra, but using forms instead of modules, and by the duality properties of handlebody splittings of odd-dimensional manifolds in topology.*

Every odd-dimensional manifold M *is a twisted double*, i.e.* $M = W \cup_h W$ *for a manifold with boundary* $(W, \partial W)$ *and an automorphism* $h : \partial W \to \partial W$ *of the even-dimensional boundary. The model for this is a Heegaard splitting of a connected 3-dimensional manifold*

$$M^3 = (\#_r S^1 \times D^2) \cup_h (\#_r S^1 \times D^2)$$

expressing M *as a twisted double of an* r*-fold connected sum of a solid torus* $S^1 \times D^2$*, with*

$$h : \partial(\#_r S^1 \times D^2) = \#_r S^1 \times S^1 \to \#_r S^1 \times S^1$$

an automorphism of the boundary surface of genus r*. A Heegaard decomposition is not unique, and* h *can be changed by elementary moves of the following type without changing* M *:*

(H1) *h can be replaced by* $(\partial g)^{-1} h (\partial f)$ *for any automorphisms*

$$f, g : \#_r S^1 \times D^2 \to \#_r S^1 \times D^2 ,$$

(H2) *h can be replaced by the stabilisation*

$$h \# \sigma : \#_{r+1} S^1 \times S^1 \to \#_{r+1} S^1 \times S^1$$

with

$$\sigma = \begin{pmatrix} 0 & 1 \\ 1 & 0 \end{pmatrix} : S^1 \times S^1 \to S^1 \times S^1 \; ; \; (x, y) \mapsto (y, x)$$

such that

$$S^3 = (S^1 \times D^2) \cup_\sigma (S^1 \times D^2) .$$

The odd-dimensional surgery obstruction theory developed here is based on an algebraic analogue of a Heegaard decomposition, obtained by viewing an odd-dimensional normal map as a twisted double and considering the automorphism

* *This is one of the many consequences of the surgery obstruction theory in this book, see Ranicki [R13, Chapter 30].*

of a hyperbolic form induced by the twisting. Given a k-connected (2k + 1)-dimensional normal map $(\phi, F) : M^{2k+1} \to X$ *it is possible to realise every choice of* $\mathbf{Z}[\pi_1(X)]$*-module generators* $\{x_1, x_2, \ldots, x_r\} \subset K_k(M) = \pi_{k+1}(\phi)$ *by an embedding* $U = \cup_r S^k \times D^{k+1} \subset M^{2k+1}$ *such that*

$$(\phi, F) \;=\; (\phi_0, F_0) \cup (\phi_1, F_1) \;:\; M \;=\; M_0 \cup U \longrightarrow X \;=\; X_0 \cup D^{2k+1}$$

with (ϕ_0, F_0), (ϕ_1, F_1) *null-bordisms of a k-connected 2k-dimensional normal map* $(\phi, F)| : \partial U \to S^{2k}$*, corresponding to an automorphism* $\alpha : H \to H$ *of the hyperbolic* $(-1)^k$*-hermitian form on the kernel* $\mathbf{Z}[\pi_1(X)]$*-module* $K_k(\partial U) = H$*. The surgery obstruction*

$$\theta(M, \phi, F) \;=\; [\alpha] \in L_{2k+1}\big(\mathbf{Z}[\pi_1(X)]\big)$$

is the equivalence class of the α*'s which arise from all the choices of generators* $\{x_1, x_2, \ldots, x_r\} \subset K_k(M)$*.*

The surgery obstruction $\theta(M, \phi, F)$ *of an arbitrary* $(2k+1)$*-dimensional normal map* $(\phi, F) : M \to X$ *is obtained by first applying* (1.2) *to make* ϕ *k-connected by surgery below the middle dimension. See the note at the end of §17G for the chain complex formulation of the odd-dimensional surgery theory, which does not require preliminary surgeries below the middle dimension.*

In §5, we were able to give a comparatively minor reformulation of the results of [W18, §3]. In the odd-dimensional case, however, a completely fresh approach is needed. Again we will simplify the exposition by omitting mention of the boundary, which can be mapped by a simple homotopy equivalence throughout and will not affect the argument.

Suppose then X a connected Poincaré complex of formal dimension $m = (2k + 1) \geqslant 5$, ν a bundle over X, M a compact manifold, $\phi : M \to X$ of degree 1, and F a stable framing of $\tau_M \oplus \phi^*\nu$, so that (M, ϕ, F) represents an element of degree 1 of $\Omega_m(X, \nu)$. By (1.2), we may suppose ϕ k-connected. These notations will be fixed for the subsequent discussion.

Choose a set of generators of $\pi_{k+1}(\phi) = K_k(M; \Lambda)$ with $\Lambda = \mathbf{Z}[\pi_1(X)]$: by the usual general position argument, we can represent them by disjoint framed embeddings $f_i : S^k \times D^{k+1} \to M$, each joined by a path to the base point: or equivalently, by $\widetilde{f}_i : S^k \times D^{k+1} \to \widetilde{M}$. Let U be the union of the images of the f_i, $M_0 = M - \operatorname{Int} U$. Since the f_i are trivial in X with given nullhomotopies, we can replace ϕ by a homotopic map and so suppose $\phi(U) = *$, the base point in X (and the nullhomotopies constant).

Now by (2.9) we may suppose that $\dim X = m$ and X has only one m-cell, so that by (2.7) we have a simple Poincaré pair (X_0, S^{m-1}), and $X = X_0 \cup D^m$. Using a cellular approximation of $\phi|M_0$, we may suppose after a further homotopy that ϕ is a map of degree 1 of Poincaré triads,

$$\phi \;:\; (M; M_0, U) \to (X; X_0, D^m)\,.$$

We combine the exact sequences (2.2) of groups K_i (with coefficients $\Lambda =$

$\mathbf{Z}[\pi_1(X)]$ throughout) in the diagram*

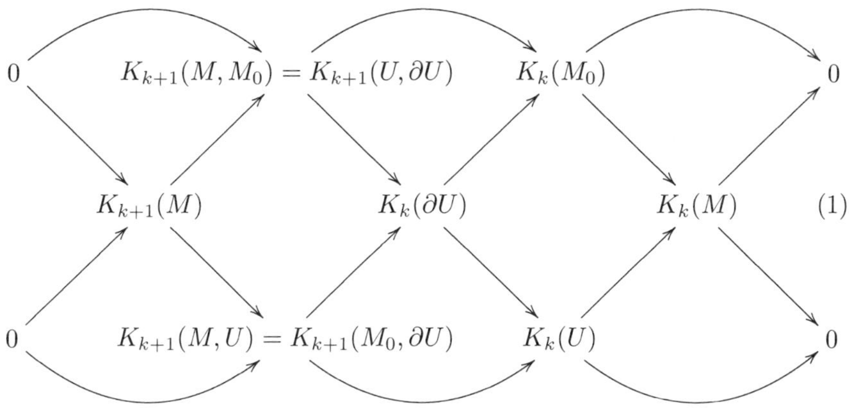

(1)

We can say something about the groups in this diagram. Since ϕ maps $(U, \partial U)$ to (D^m, S^{m-1}), which has zero absolute and relative homology in the middle dimensions, we can replace the groups $K_{k+1}(U, \partial U)$, $K_k(\partial U)$ and $K_k(U)$ by the straight homology groups H_i (with coefficients Λ). Also, by (2.3 (c)) the modules $K_{k+1}(M_0, \partial U)$ and $K_k(M_0)$ are stably free and s-based, and by (2.4), Corollary, the short exact sequences through $H = K_k(\partial U)$ are based. In fact, (5.7) shows that $K_k(\partial U)$ is a hyperbolic form, and that $K_{k+1}(U, \partial U)$ and $K_{k+1}(M_0, \partial U)$ are lagrangians: in the case of $K_{k+1}(U, \partial U)$, this is obvious *a priori*, as U is a disjoint union of copies of $S^k \times S^k$, and we take the classes of the $S^k \times 1$ and $1 \times S^k$ as basis. The above assumes that $K_{k+1}(M_0, \partial U)$ is free and based: but since we can always adjoin an extra f_i, and this clearly replaces $K_{k+1}(M_0, \partial U)$ by its direct sum with Λ, this assumption is not inconveniently restrictive.

We use the base of $K_k(\partial U) = H$ above to identify H with a standard hyperbolic form. By (5.3.1) any (simple) isomorphism of the lagrangian $K_{k+1}(U, \partial U)$ onto $K_{k+1}(M_0, \partial U)$ extends to an automorphism of H: let α be the corresponding automorphism of the standard hyperbolic form. Our plan is to use α to find our surgery obstruction. We will have to check the effect on α of all the choices made hitherto. First, however, we need some more algebraic notation.

Recall that the standard hyperbolic form has basis $\{e_i, f_i : 1 \leqslant i \leqslant r\}$; that μ vanishes on all these basis elements, and that λ vanishes on all pairs of them except for $\lambda(e_i, f_i) = (-1)^k \lambda(f_i, e_i) = 1\,(1 \leqslant i \leqslant r)$. We write $SU_r(\Lambda)$ for the group of automorphisms of this standard hyperbolic form: that is, module automorphisms preserving λ, μ, and the preferred class of bases, hence simple. Write $TU_r(\Lambda)$ for the subgroup leaving the lagrangian with base $\{e_i\}$ invariant, and inducing a simple automorphism of it; $UU_r(\Lambda)$ for the subgroup leaving each

* *The kernels are the homology Λ-modules $K_*(M) = H_*(C)$ of the based f. g. free Λ-module chain complex*

$$C \ : \ \cdots \to 0 \to K_{k+1}(M, U) \to K_k(U) \to 0 \to \ldots$$

which is simple chain equivalent to its $(2k+1)$-dual $C^{2k+1-*} = \mathrm{Hom}_\Lambda(C, \Lambda)^{2k+1-*}$.

element of this lagrangian fixed. Denote the group of simple automorphisms of the lagrangian by $SL_r(\Lambda)$ (note: our notation conflicts with that of [B5] in the commutative case): then inclusion and restriction homomorphisms clearly give an exact sequence

$$1 \to UU_r(\Lambda) \to TU_r(\Lambda) \to SL_r(\Lambda) :$$

moreover, the last map is surjective, as we can define a splitting homomorphism $SL_r(\Lambda) \xrightarrow{H} TU_r(\Lambda)$. For if $A = (a_{ij})$ is the matrix of an element of $SL_r(\Lambda)$, $A^* = (\overline{a}_{ji})$ the conjugate matrix, also (clearly) giving a simple automorphism, and $B = (A^*)^{-1}$ we use

$$H(A) \ : \ e_i \mapsto e_j a_{ji} \qquad f_i \mapsto f_j b_{ji} .$$

Note that $*$ is an involutory anti-automorphism of $M_r(\Lambda)$, hence also of $SL_r(\Lambda)$. We will also need a matrix description of $UU_r(\Lambda)$: an element here induces the identity on the submodule of the $\{e_i\}$, hence also on the dual module, which is the same as the quotient module, so has the form

$$e_i \mapsto e_i \qquad f_i \mapsto f_i + e_j c_{ji} .$$

This transformation preserves λ only if we have $C + (-1)^k C^* = 0$: for it to preserve μ we need further that C have the form $D - (-1)^k D^*$: note that this further restriction affects only the diagonal elements.

Now given the embeddings f_i (or rather, $\widetilde{f_i}$), the lagrangians are already determined. Thus β satisfies the same condition as α if and only if $\beta\alpha^{-1}$ preserves the lagrangian $\{e_i\}$ corresponding to $K_{k+1}(U, \partial U)$, and preserves the preferred class of bases on it, hence lies in $TU_r(\Lambda)$. So the sequence $\{\widetilde{f_i}\}$ determines uniquely a right coset $TU_r(\Lambda)\alpha$ of $TU_r(\Lambda)$ in $SU_r(\Lambda)$.

We next investigate the effect of changing the embeddings $\widetilde{f_i}$ in their regular homotopy classes. Now a regular homotopy of S^k in M^{2k+1} is an immersion of $S^k \times I$ in $M^{2k+1} \times I$. As, moreover, the ends are embedded, we can calculate the self-intersection invariant (in Q_{k+1}) of a regular homotopy or the mutual intersection (in Λ) of two such, as in §5. Denote the regular homotopy by $\{\eta_i\}$: let the self-intersection of η_i be ν_i, and the intersection of η_i with η_j be ρ_{ij}. Then $\rho_{ii} = \nu_i + (-1)^{k+1}\overline{\nu_i}$, and $\rho_{ij} = (-1)^{k+1}\overline{\rho_{ij}}$ for all $1 \leqslant i, j \leqslant r$. Thus the ρ_{ij} form a general matrix P of the form $D - (-1)^k D^*$. By the above, this determines an element γ of $UU_r(\Lambda)$. We claim that α is replaced by $\alpha\gamma$ as a result of the regular homotopy.

Consider the diagram (1). The regular homotopy evidently leaves unchanged the exact sequences of (M, U) and of $(U, \partial U)$, and hence all the modules in the diagram are replaced by isomorphic copies: in fact the exact sequence of (M, M_0) is isomorphic by duality with the cohomology sequence of (M, U). The maps in the exact sequence of $(M_0, \partial U)$, however, will change. In fact, given an element of $K_{k+1}(M_0, \partial U)$, its boundary in ∂U will alter, although of course its image in $K_k(U)$ will not. The spheres $1 \times S^k$ in ∂U bound discs in U, and are evidently not essentially changed. The ith sphere $S^k \times 1$, however, is subjected to the regular homotopy η_i (translated along a normal vector). This has intersection number ρ_{ij} with the core of the jth handle, as does $\sum \rho_{ij}(1 \times D^{k+1})$: the difference of these two chains lies in M_0. Thus if $x \in K_{k+1}(M_0, \partial U)$, then $\partial x \in K_k(\partial U)$, expressed as a linear combination of the standard basis elements, will change by altering the spheres $(S^k \times 1)$ by the formula above. This amounts to composing with γ, as asserted.

So far we have proved that a sequence of r elements of $G = \pi_{k+1}(\phi) = K_k(M; \Lambda)$, which generate G, determine the double coset

$$TU_r(\Lambda)\alpha\, UU_r(\Lambda) \subset SU_r(\Lambda)\,,$$

and that α can be replaced by any element of the double coset.

Now suppose given a sequence of s elements which together generate G: suppose our two sequences are $\{x_1, \ldots, x_r\}$ and $\{y_1, \ldots, y_s\}$. We will go from the first to the second by a sequence of operations. Write $y_i = \sum x_j \lambda_{ji}$ (possible as the x's generate G). Then

$$\begin{aligned}\{x_1, \ldots, x_r\} &\to \{x_1, \ldots, x_r, 0\} \to \{x_1, \ldots, x_r, y_1\}\\ &\to \{x_1, \ldots, x_r, y_1, 0\} \to \{x_1, \ldots, x_r, y_1, y_2\} \to \ldots\\ &\to \{x_1, \ldots, x_r, y_1, \ldots, y_s\} \to \{y_1, \ldots, y_s, x_1, \ldots, x_r\}\\ &\to \cdots \to \{y_1, \ldots, y_s\}\,.\end{aligned}$$

Each of these operations has one of the following types:

(T1) Adjoin (or delete) a zero.

(T2) Permute the elements.

(T3″) Add to the last element a linear combination of the others.

We can simplify this last type of operation by adding the terms in turn:

(T3′) Replace the last element by its sum with $\pm$ an element of $\pi_1(X)$ times one of the other elements.

A further simplification using (T2) reduces this to two operations

(T3) Replace the first element by its product with $\pm$ an element of $\pi_1(X)$.

(T4) Replace the first element by the sum of the first two.

It remains to investigate the effect of each of (T1)–(T4) on α.

(T1) Geometrically, this means adjoining an embedding of $S^k \times D^{k+1}$ whose image lies in a disc D^{2k+1} disjoint from the remaining tubes. This takes the direct sum of (1) with the diagram

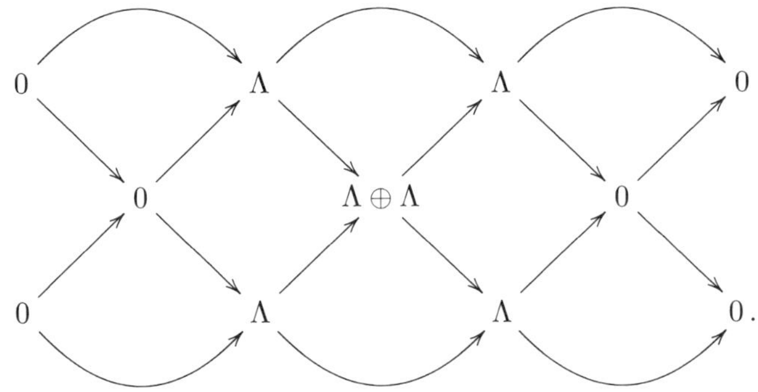

The maps are easy to compute (e.g. the maps $\Lambda \to \Lambda$ are identity maps), and we conclude:

The effect of (T1) *is to form the direct sum of* α *with the matrix*

$$\sigma = \begin{pmatrix} 0 & 1 \\ (-1)^k & 0 \end{pmatrix}.$$

(T2) *This has the effect only of altering the defining basis of* H, *and thus of conjugating* α, *by a permutation of the* Λ^2. Clear.

(T3) *This also only alters the basis of* H, *and thus conjugates* α, *by the direct sum of* $\begin{pmatrix} \pm g & 0 \\ 0 & \pm w(g)g \end{pmatrix}$ *and the identity.*

(T4) First join the two copies of $S^k \times D^{k+1}$ in M by an arc (in the right homotopy class: we really join them in $\widetilde{M}$): and thicken the arc, so as to obtain an embedding of the trivial handlebody of type $\mathscr{H}(2k+1, 2, k)$ [W7]. We effect our change by performing an appropriate diffeomorphism of the handlebody; in fact the one constructed in [W8, p. 272]. Adjoining the thickened arc to U does not affect (1), and now the diffeomorphism will change the preferred basis of H, and hence conjugate α, by the automorphism

$$\begin{aligned} e_1' &= e_1 + e_2 & \qquad f_1' &= f_1 \\ e_2' &= e_2 & \qquad f_2' &= f_2 - f_1 . \end{aligned}$$

We can combine the effects of (T2), (T3) and (T4) as follows. They represent (formally) all the elementary basis changes of $r\Lambda$, and thus generate the subgroup $E_r(\Lambda)$ [B5]. Thus for any $\xi \in E_r(\Lambda)$ we can conjugate α by $H(\xi) \in TU_r(\Lambda)$. As we have already shown that α is equivalent to any member of the double coset $TU_r(\Lambda)\alpha\, UU_r(\Lambda)$, we can deduce already if $E_r(\Lambda) = SL_r(\Lambda)$ that the effect of all our operations excluding (T1) is to replace α by an arbitrary element of the double coset $TU_r(\Lambda)\alpha\, TU_r(\Lambda) \subset SU_r(\Lambda)$. Thus here we have as invariant just the isomorphism class of a pair of lagrangians, which recalls the main problem treated in [W9]*.

* *See the note on forms and formations at the end of this chapter.*

The effect of (T1) is to stabilise, but in a somewhat unexpected way: $\alpha \mapsto \alpha \oplus \sigma$ instead of $\alpha \oplus 1$. At any rate, the distinction between E_r and SL_r disappears stably. In terms of lagrangians, we can say: the k-connected map $\phi : M \to X$ of degree 1 (plus ν, F) determines a pair of lagrangians up to stable isomorphism, where we stabilise by taking orthogonal direct sums with a system of two complementary lagrangians. Note for later reference that this is an invariant of the map, but is not a surgery invariant. For group theory, we have natural inclusions

$$SU_r(\Lambda) \subset SU_{r+1}(\Lambda) \subset \ldots$$

We denote the limit by $SU(\Lambda)$: the (infinite) simple unitary group* over Λ. We have different injective maps $SU_r(\Lambda) \to SU_{r+1}(\Lambda)$ by taking direct sums with σ. This is *not* a group homomorphism, but it is compatible with the natural left and right actions of $SU_r(\Lambda)$, with respect to each of which it is a principal homogeneous space. Our invariant is the equivalence class of $\alpha \in S'U(\Lambda)$ under the product action of $TU(\Lambda) \times TU(\Lambda)$. The set $S'U(\Lambda)$ has a natural base point Σ: the direct sum of copies of σ.

We now define $RU(\Lambda)$ to be the subgroup of $SU(\Lambda)$ generated by $TU(\Lambda)$ and the element $\sigma \in SU_1(\Lambda)$.

LEMMA 6.1. *Surgery can be performed if and only if α is equivalent to Σ under the 2-sided action of $RU(\Lambda)$.*

Proof We first investigate the effect on α of a single surgery. We can choose the framed embeddings of $S^k \times D^{k+1}$ on which surgery is to be performed as the first of those used in calculating α. It follows that the effect of surgery is to leave the exact sequence

$$0 \to K_{k+1}(M_0, \partial U) \to K_k(\partial U) \to K_k(M_0) \to 0$$

unchanged, but the basis of $K_k(\partial U)$ must be reinterpreted, as the rôles of the first $S^k \times 1$ and $1 \times S^k$ are interchanged. The effect on α is to replace it by $\alpha\sigma$. It follows that the class of α under $RU(\Lambda)$ is invariant under surgeries on k-spheres. To prove it a surgery invariant, we now show that if, in fact, $\phi : M \to X$ and $\phi' : M' \to X$ (with F, F') are cobordant k-connected maps, we can go from one to the other by surgeries on k-spheres. For let (N, ψ, G) be a cobordism. By (1.2), we can suppose ψ $(k+1)$-connected. Hence the pairs (N, M) and (N, M') are k-connected. By [W13, IV, 5.5], N has a handle decomposition based on M with no handles of dimension $\leqslant k$, and another with no handles of dimension $> k+1$. We must prove that both conditions can be satisfied simultaneously. Now by the argument of (2.3), $K_{k+1}(N, M)$ is stably free and s-based. Performing surgery on a trivial k-sphere in N changes N to $N\#(S^{k+1} \times S^{k+1})$, and adds to this module a free module of rank 2. Do this enough times to make $K_{k+1}(N, M)$ free and based, and represent the basis elements by handles attached to M. The rest of N is an s-cobordism of the result to M', and hence diffeomorphic to $M' \times I$. Our assertion is proved.

In the first edition the simple unitary group was called the special unitary group.

Necessity of the stated condition now follows from the observation that if ϕ is a simple homotopy equivalence, take $U = \emptyset$: α is a 0×0 matrix, and so stably $\alpha = \Sigma$.

Now suppose, conversely, that our condition is satisfied: for $\xi, \eta \in RU(\Lambda)$, $\alpha = \xi\Sigma\eta$. Choose r so large that ξ and η belong to $RU_r(\Lambda)$. We have $\alpha = \xi\Sigma\eta = \Sigma(\Sigma^{-1}\xi\Sigma)\eta$. But Σ operates on Λ^{2r} as a finite product of conjugates $\big($by permutations of the summands, which belong to $TU_r(\Lambda)\big)$ of copies of σ, hence is in $RU_r(\Lambda)$: thus so is $\Sigma^{-1}\xi\Sigma$. Thus $\alpha = \Sigma\beta$ for $\beta \in RU_r(\Lambda)$. We can thus suppose (increasing r if necessary) that β is a product of elements of the form: σ, $\nu \in UU_r(\Lambda)$, and $H(\varepsilon)$ for $\varepsilon \in E_r(\Lambda)$ an elementary matrix. It was shown above that multiplying α on the right by σ corresponds to a surgery; by one of the other elements, merely to altering some arbitrary choices: for ε is a product of matrices of the types involved in (T2), (T3) and (T4). Thus (by induction on the length of the expression for β) we can suppose $\alpha = \Sigma$. But this implies that the map $K_{k+1}(M, U) = K_{k+1}(M_0, \partial U) \to K_k(U)$ is a simple isomorphism. It follows, by (2.4), Corollary, that $\phi : M \to X$ is a simple homotopy equivalence, as desired.

Our argument shows, more generally, that under the equivalence relation generated by surgery only the class of α under operation on the left by $TU(\Lambda)$ and on the right by $RU(\Lambda)$ is invariant.

Now return to the groups of finite rank. We stabilised above by adding σ. But (using a permutation of the coordinates) we now have

$$\alpha \oplus \sigma \;\sim\; \sigma \oplus \alpha \;\sim\; (\sigma \oplus \alpha)\sigma = 1 \oplus \alpha \;\sim\; \alpha \oplus 1\,,$$

so that the image of α in $SU(\Lambda)$ $\big($rather than $S'U(\Lambda)\big)$ has a well-defined equivalence class, and the $\big(TU(\Lambda), RU(\Lambda)\big)$-double coset of α is our surgery obstruction. But to obtain a neat result, we need more: that $RU(\Lambda)$ is a normal subgroup of $SU(\Lambda)$; in fact, it contains the commutator subgroup. We will next give an algebraic proof of this fact.

We first develop some purely algebraic properties of the group $RU(\Lambda)$, valid for any ring Λ with involution. First we observe that we obtained an inverse in §5 by changing orientation, and hence the signs of λ and μ. If we do this to a hyperbolic form, the given base is no longer of standard type: we restore it by changing the signs of the f_i. Now if A is a simple automorphism of the standard hyperbolic form H, denote by (H', A') the hyperbolic form and automorphism obtained by changing signs. Then $A \oplus A'$ is a simple automorphism of $H \oplus H'$, and leaves invariant the diagonal submodule, which is a lagrangian. Let M be an automorphism of $H \oplus H'$ which takes the lagrangian $\{e_i, e'_i\}$ to the diagonal lagrangian $\{e_i + e'_i, f_i + f'_i\}$. Then the conjugate of $\alpha \oplus \alpha'$ by M is in $TU(\Lambda)$. To describe M, we can take $i = 1$ (and, in general, use the orthogonal direct sum). We may, e.g., take

$$M = \begin{pmatrix} 1 & 0 & 0 & 0 \\ 1 & 0 & 0 & (-1)^{k+1} \\ 0 & 1 & 1 & 0 \\ 0 & -1 & 0 & 0 \end{pmatrix}.$$

It is easy to show that $M \in RU(\Lambda)$; or one can observe that $M \in SU(\mathbf{Z}) = RU(\mathbf{Z})$. (This follows from [B10] or from (6.1), (6.5), and the fact that surgery can always be performed [K5] in the simply connected, odd-dimensional case).

LEMMA 6.2. *Let $A \in SU_r(\Lambda)$. Then there are matrices T, T' in $TU_{2r}(\Lambda)$ such that $T(\Sigma \oplus A^{-1}) = (A' \oplus \Sigma)T'$. Hence $A \oplus A' \in RU(\Lambda)$.*

Proof We will write down an identity of the desired form. Note that A is already a $2r \times 2r$ matrix: we partition it into $r \times r$ blocks

$$A = \begin{pmatrix} \alpha & \beta \\ \gamma & \delta \end{pmatrix}.$$

Since A is a simple unitary matrix, we have

$$A^{-1} = \begin{pmatrix} \delta^* & (-1)^k\beta^* \\ (-1)^k\gamma^* & \alpha^* \end{pmatrix}.$$

We point out, as a consequence, various identities involving α, β, γ and δ. In particular $\alpha\beta^*$, $\gamma\delta^*$, $\beta^*\delta$ and $\alpha^*\gamma$ are $(-1)^{k+1}$-symmetric: using the fact that A preserves μ as well as λ we can show in fact that each of these matrices has the form $\varepsilon - (-1)^k\varepsilon^*$.

Our identity involves $4r \times 4r$ matrices, or 4×4 blocks of size $r \times r$. We order these by making the standard lagrangian correspond to the first two blocks (order e_i, e_i', f_i, f_i' in the above notation), so in $\Sigma \oplus A^{-1}$, Σ will correspond to rows and columns 1 and 3; A^{-1} to 2 and 4. Then we have

$$\begin{pmatrix} -\beta & \alpha & (-1)^k\alpha & 0 \\ -\delta & \gamma & 0 & \delta \\ 0 & 0 & (-1)^{k+1}\gamma & \delta \\ 0 & 0 & -\alpha & (-1)^k\beta \end{pmatrix} \begin{pmatrix} 0 & 0 & I & 0 \\ 0 & \delta^* & 0 & (-1)^k\beta^* \\ (-1)^k I & 0 & 0 & 0 \\ 0 & (-1)^k\gamma^* & 0 & \alpha^* \end{pmatrix}$$

$$= \begin{pmatrix} \alpha & \alpha\delta^* & -\beta & (-1)^k\alpha\beta^* \\ 0 & 0 & -\delta & I \\ -\gamma & (-1)^k\delta\gamma^* & 0 & \delta\alpha^* \\ (-1)^{k+1}\alpha & \beta\gamma^* & 0 & (-1)^k\beta\alpha^* \end{pmatrix}$$

$$= \begin{pmatrix} \alpha & 0 & -\beta & 0 \\ 0 & 0 & 0 & I \\ -\gamma & 0 & \delta & 0 \\ 0 & (-1)^k I & 0 & 0 \end{pmatrix} \begin{pmatrix} I & \delta^* & -\delta^*\beta & 0 \\ -\alpha & (-1)^k\beta\gamma^* & 0 & \beta\alpha^* \\ 0 & 0 & (-1)^k\gamma^*\beta & \alpha^* \\ 0 & 0 & -\delta & I \end{pmatrix}.$$

It is easily checked that the first and last matrices are in $TU_{2r}(\Lambda)$, by writing each as a product $H(x)(y)$, where $x \in SL_{2r}(\Lambda)$ and y (partitioned into blocks $2r$ square) has the form

$$\begin{pmatrix} I & w + (-1)^{k+1}w^* \\ 0 & I \end{pmatrix}.$$

As to the last assertion, we have already shown that $A \oplus A'$ is in $RU(\Lambda)$, hence so (in turn) are $A \oplus A' \oplus \Sigma$,

$$(A \oplus A' \oplus \Sigma)(1 \oplus T') = (1 \oplus T)(A \oplus \Sigma \oplus A^{-1}),$$

$A \oplus \Sigma \oplus A^{-1}$, and (permuting coordinates and destabilising) $A \oplus A^{-1}$.

Perhaps a word about the discovery of the above identity is in order, since it took the author nearly two years to find it. The above seems in any case simpler than to show directly that $A \oplus A^{-1}$ is in $RU(\Lambda)$ – whereas for the analogous problem with the general linear group, this is comparatively easy [B5]. I then noticed that in the construction below (6.5) of an (M, ϕ, F) determining A, if we change the orientation it is natural to replace A by A', whereas if we interchange the two ends of M, A is replaced by A^{-1}. This can be achieved by suitable application of (T1), (T3) and (T4): we introduce r new trivial copies of $S^k \times D^{k+1}$; then (to move them into the right homotopy class) add "δ^* times" the others, then subtract from the first "α times" the new ones and then remove the first r copies of $S^k \times D^{k+1}$, which by now are trivial. This suggested looking for an identity with

$$T' = H\left(\begin{pmatrix} I & 0 \\ -\alpha & I \end{pmatrix}\begin{pmatrix} I & \delta^* \\ 0 & I \end{pmatrix}\right) \cdot y;$$

and from this it is not too hard to arrive at the above.

THEOREM 6.3. *$RU(\Lambda)$ contains the commutator subgroup $[SU(\Lambda), SU(\Lambda)]$; the quotient $RU(\Lambda)/[SU(\Lambda), SU(\Lambda)]$ is generated by σ, so has order at most 2.*

Proof If A, $B \in SU_r(\Lambda)$, we have

$$A^{-1}B^{-1}AB \oplus I = ((BA)^{-1} \oplus BA)(A \oplus A^{-1})(B \oplus B^{-1}),$$

and each term on the right is in $RU(\Lambda)$ by (6.2). Now since $\sigma^2 \in TU(\Lambda)$, and $RU(\Lambda)$ is (by definition) generated by $TU(\Lambda)$ and σ, the second assertion follows from the inclusion

$$TU(\Lambda) \subset [SU(\Lambda), SU(\Lambda)].$$

Any element of $TU(\Lambda)$ can be written as a product $H(X) \cdot Y$ with $X \in SL(\Lambda)$ and $Y \in UU(\Lambda)$: we consider X and Y separately. Now it is well known that $SL(\Lambda)$ equals its commutator subgroup (see e.g. [B5]), so X is a product of commutators. Since H is a homomorphism, $H(X)$ is a product of commutators when X is.

We will now complete the proof by showing that $UU(\Lambda)$ is contained in the commutator subgroup of $TU(\Lambda)$. We have

$$\begin{pmatrix} I & B \\ 0 & I \end{pmatrix}\begin{pmatrix} C & 0 \\ 0 & C^{*-1} \end{pmatrix}\begin{pmatrix} I & -B \\ 0 & I \end{pmatrix}\begin{pmatrix} C^{-1} & 0 \\ 0 & C^* \end{pmatrix} = \begin{pmatrix} I & B - CBC^* \\ 0 & I \end{pmatrix};$$

this is also more neatly noted since B is the matrix of a simple $(-1)^{k+1}$-hermitian form, and conjugating the first matrix above by the second merely expresses this form with respect to a different basis. Since products in $UU(\Lambda)$ correspond to addition of matrices B, it is enough to show that for any matrix P with a single non-zero entry, $P - (-1)^k P^*$ can be written as a sum of matrices $B - CBC^*$, with B of the form $D - (-1)^k D^*$. Now taking

$$C = \begin{pmatrix} 1 & 0 & -1 \\ 0 & 1 & 0 \\ 0 & 0 & 1 \end{pmatrix}, \quad B = \begin{pmatrix} 0 & 0 & 0 \\ 0 & 0 & P \\ 0 & (-1)^k P^* & 0 \end{pmatrix}$$

we find

$$B - CBC^* = \begin{pmatrix} 0 & (-1)^k P^* & 0 \\ P & 0 & 0 \\ 0 & 0 & 0 \end{pmatrix}$$

and see that by terms of this type we can deal with off-diagonal elements: the diagonal ones are reduced to these by taking

$$C = \begin{pmatrix} 1 & 1 \\ 0 & 1 \end{pmatrix} \quad B = \begin{pmatrix} 0 & 0 \\ 0 & Q \end{pmatrix} \text{ with } CBC^* - B = \begin{pmatrix} Q & Q \\ Q & 0 \end{pmatrix}.$$

This completes the proof of the Theorem. Our result seems to pave the way for an analysis of the unitary and simple unitary groups similar to that performed for the general linear group in [B5].*

We now define the group $L_m(\pi)$ when $m = 2k+1$ as the quotient group $SU(\Lambda)/RU(\Lambda)$: we have shown that this is an abelian group. Given a homomorphism $r : \pi \to \pi'$ which is compatible with w, r induces a homomorphism of group rings $\Lambda \to \Lambda'$ compatible with the involution, and hence (evaluating term by term) homomorphisms of matrix rings, of the group $SU(\Lambda)$, and so finally, $L_m(r) : L_m(\pi) \to L_m(\pi')$. Clearly, this makes L_m a functor.

THEOREM 6.4. *Let (X, W) be a simple Poincaré pair with formal dimension $m = 2k+1 \geqslant 5$, ν a bundle over X, M a compact manifold with boundary, $\phi : (M, \partial M) \to (X, W)$ a map of degree 1 inducing a simple homotopy equivalence $\partial M \to W$, F a stable trivialisation of $\tau_M \oplus \phi^*\nu$. We define a class $\theta \in L_m(\pi_1(X))$ by performing surgery rel ∂M till ϕ is k-connected, and then taking the class of a matrix α as constructed above. Then θ depends only on the bordism class of (M, ϕ, F) relative to ∂M, and the class has an element with ϕ a simple homotopy equivalence if and only if $\theta = 0$. The obstruction θ is natural for inclusion maps.*

Proof We showed in (6.1) that the class of α modulo $RU(\Lambda)$ was a surgery invariant; since by (6.1) $RU(\Lambda)$ is a normal subgroup, this class is just θ. We also proved in (6.1) that surgery was possible if and only if $\theta = 0$. It remains to show naturality. Let $L = \partial M$, $\partial M' = L \cup L'$, and suppose $\phi|\partial M$ extended to a simple homotopy equivalence $(M'; L, L') \to (X'; W, W')$ so that glueing gives a map $(M \cup M', L') \to (X \cup X', W')$ with the same properties as ϕ. Then to compute the surgery obstruction for ϕ' we can use the same U, for the groups K_i are changed only by replacing $\pi_1(X)$ by $\pi_1(X \cup X')$. We obtain also the same matrix α with this coefficient change. But by definition, this represents the image of θ.

We conclude this chapter with a construction technique giving a result analogous to (5.8).

THEOREM 6.5. *Let X^{2k} be a compact smooth or PL manifold, with normal bundle ν, $k \geqslant 3$. Let $A \in SU_r(\Lambda)$. Then we can find a manifold triad $(M, \partial_- M, \partial_+ M)$, a map $\phi : (M, \partial_- M, \partial_+ M) \to (X \times I, X \times 0 \cup \partial X \times I, X \times 1)$ of degree 1, and a stable framing F of $\tau_M \oplus \phi^*\nu$, such that ϕ is the identity on $\partial_- M$, a simple homotopy equivalence on $\partial_+ M$, is k-connected, and has A as invariant. Moreover, θ maps the bordism set (rel ∂M) of all such M bijectively to $L_m(\pi_1(X))$.*

Remark, If $k = 2$, the proof breaks down: it seems that using the methods of [W10] one might realise any $A \in RU(\Lambda)$, but of course these do not lead to surgery obstructions.

Proof First perform surgery on the identity map of X to "kill" r trivial $(k-1)$-

Such results were subsequently obtained by Bak [B1], [B3], Vaserstein [V1] and Sharpe [S7].

spheres (if $A \in SU_r(\Lambda)$). This replaces $X \times I$ by its boundary-connected sum with r copies of $S^k \times S^k$: thus the resulting K_k is a hyperbolic form, the sum of r standard planes. We regard A as a (simple) automorphism of K_k. Thus the classes $A(1 \times S^k)$ generate a lagrangian; by (5.2) we can perform surgery on them, and by the proof of (5.6), the result is mapped to X by a simple homotopy equivalence.

We are more interested in the trace M of the surgeries. All assertions of the lemma are immediate, except that M has invariant A. Now it is clear that $K_k(M)$ is generated by the classes of the k-handles, i.e. by the copies of $S^k \times D^{k+1}$ mentioned above. We remove these (or strictly speaking, smaller concentric copies – but this does not alter the homotopy type) to obtain M_0. Now a preferred base of $K_{k+1}(M_0, \partial U)$ is clearly given by the classes of the attached $(k+1)$-cells. These, by definition, have boundaries $A(1 \times S^k)$. Thus the automorphism A does indeed take a preferred base of $K_{k+1}(U, \partial U)$ to one of $K_{k+1}(M_0, \partial U)$, and A is the invariant, as asserted. The proof of the last assertion is identical with that of the corresponding part of (5.8).

We can obtain some further information about the structure of the group $RU(\Lambda)$ by considering more closely its relation to surgery. For suppose that surgery can be done: then by (6.1) we need only do surgery on k-spheres, and a general position argument shows that these may be chosen disjoint and all the surgeries done simultaneously. But we can use these k-spheres to calculate our invariant α. Surgery will replace it by $\alpha\Sigma$ (recall, Σ denotes the direct sum of copies of σ), and since this corresponds to a complementary pair of lagrangians, it must lie in $TU(\Lambda)\Sigma TU(\Lambda)$. Hence α lies in $TU(\Lambda)\Sigma TU(\Lambda)\Sigma^{-1}$. Since our invariant was only determined up to 2-sided multiplication by $TU(\Lambda)$, we have shown

THEOREM 6.6. *For any $\alpha \in RU_m(\Lambda)$ we can find n such that $\alpha \oplus$ (n copies of σ) has the form $t_1\Sigma t_2\Sigma^{-1}t_3$ with each $t_i \in TU_{m+n}(\Lambda)$.*

Although we have presented the argument in geometrical terms, it can be paraphrased algebraically, so that the above holds without the restrictions imposed on Λ in the geometrical case.

Note also that in the geometrical case this shows that with a suitable choice of framed k-spheres we can take $\alpha = \Sigma t_2 \Sigma^{-1}$ with $t_2 \in TU(\Lambda)$. This remark will be needed later on.

Forms and formations. A '$(-1)^k$-hermitian formation' $(G, \lambda, \mu; H_1, H_2)$ over a ring with involution Λ is a $(-1)^k$-hermitian form (G, λ, μ) over Λ together with an ordered pair of lagrangians H_1, H_2 (which in general need not be free). Formations were introduced by Wall [W9], as 'pairs of subkernels', in the purely algebraic context of the classification of quadratic forms on finite abelian groups. Novikov [N8] initiated the reformulation of the automorphism odd-dimensional surgery obstruction theory of the first edition of this book in terms of formations, which was completed in Ranicki [R1].

By Corollary 5.3.1, for any formation $(G, \lambda, \mu; H_1, H_2)$ *with based f. g. free lagrangians* H_1, H_2 *there exists an automorphism* $\alpha : (G, \lambda, \mu) \to (G, \lambda, \mu)$ *sending* H_1 *to* H_2*. Conversely, every automorphism* α *of a hyperbolic form* (G, λ, μ) *determines a formation* $(G, \lambda, \mu; H, \alpha(H))$*, for any lagrangian* H*. Thus formations with free lagrangians are essentially equivalent to automorphisms of hyperbolic forms. By definition, a formation is simple if the lagrangians are based f. g. free and the automorphism* α *is simple.*

An isomorphism $f : \beta \cong \beta'$ *of formations*

$$\beta = (G, \lambda, \mu; H_1, H_2) \ , \ \beta' = (G', \lambda', \mu'; H_1', H_2')$$

is an isomorphism of forms $f : (G, \lambda, \mu) \cong (G', \lambda', \mu')$ *such that*

$$f(H_1) = H_1' \ , \ f(H_2) = H_2' \, .$$

By definition, a formation $(G, \lambda, \mu; H_1, H_2)$ *is trivial if the lagrangians* H_1, H_2 *are complementary, with* $G = H_1 \oplus H_2$*. A stable isomorphism* $[f] : \beta \cong_s \beta'$ *of formations is an isomorphism* $f : \beta \oplus \gamma \cong \beta' \oplus \gamma'$ *with* γ, γ' *trivial.*

Given a k*-connected* $(2k+1)$*-dimensional normal map* $(\phi, F) : M \to X$ *and a choice of* $\mathbf{Z}[\pi_1(X)]$*-module generators* $\{x_1, x_2, \ldots, x_r\} \subset K_k(M) = \pi_{k+1}(\phi)$ *let* $U = \cup_r (S^k \times D^{k+1}) \subset M$ *etc. The choice determines a simple* $(-1)^k$*-hermitian formation* $\beta = (G, \lambda, \mu; H_1, H_2)$ *with*

$$G = K_k(\partial U) \ , \ H_1 = K_{k+1}(U, \partial U) \ , \ H_2 = K_{k+1}(M_0, \partial U) \, .$$

The $\mathbf{Z}[\pi_1(X)]$*-module* G *is f. g. free of rank* $2r$*,* H_1 *is f. g. free of rank* r*,* H_2 *is stably f. g. free of rank* r*, and*

$$H_1 \cap H_2 = K_{k+1}(M) \ , \ G/(H_1 + H_2) = K_k(M) \, .$$

The stable simple isomorphism class of β *is an invariant of* (ϕ, F) *which is independent of the choice of generators.*

The boundary of a (possibly degenerate) $(-1)^{k+1}$*-hermitian form* $\gamma = (K, \lambda, \mu)$ *over* Λ *is the* $(-1)^k$*-hermitian formation*

$$\partial\gamma = (H, \lambda_H, \mu_H; K, \Gamma_\lambda)$$

with (λ_H, μ_H) *the hyperbolic* $(-1)^k$*-hermitian form on* $H = K \oplus K^*$

$$K^* = \mathrm{Hom}_\Lambda(K, \Lambda) \ , \ \lambda_H\big((x, f), (y, g)\big) = f(y) + (-1)^k \overline{g(x)} \in \Lambda \ ,$$
$$\mu_H(x, f) = f(x) \in Q_k \ (x, y \in K, f, g \in K^*)$$

and $\Gamma_\lambda = \{(x, \lambda(x)) : x \in K\} \subset H$ *the graph lagrangian.*

From the formation point of view $L_{2k+1}(\Lambda)$ *is the Grothendieck group of equivalence classes of simple* $(-1)^k$*-hermitian formations* β *over* Λ*, subject to the equivalence relation*

$\beta \sim \beta'$ *if there exists a stable simple isomorphism*

$$\beta \oplus \partial\gamma \cong_s \beta' \oplus \partial\gamma'$$

for boundary formations $\partial\gamma, \partial\gamma'$ *.*

The surgery obstruction of a k-connected $(2k+1)$-dimensional normal map $(\phi, F) : M \to X$ is the equivalence class of the associated simple $(-1)^k$-hermitian formations β over $\Lambda = \mathbf{Z}[\pi_1(X)]$

$$\theta(M, \phi, F) \; = \; [\beta] \in L_{2k+1}\big(\mathbf{Z}[\pi_1(X)]\big) \; .$$

The next two paragraphs describe an alternative (but equivalent) method of obtaining a formation to represent the surgery obstruction of an odd-dimensional normal map, by replacing the choice of generators for $K_k(M)$ with the trace of the surgeries killing the generators.

Suppose given a $(k+1)$-connected $(2k+2)$-dimensional normal bordism

$$(\psi, G) \; : \; (N^{2k+2}; M, M') \to X \times (I; \{0\}, \{1\})$$

between k-connected $(2k+1)$-dimensional normal maps

$$(\phi, F) \; = \; (\psi, G)| \; : \; M^{2k+1} \to X \; , \; (\phi', F') \; = \; (\psi, G)| \; : \; {M'}^{2k+1} \to X \; .$$

The $\mathbf{Z}[\pi_1(X)]$-module morphisms induced by inclusions

$$\begin{aligned} i \; : \; K_{k+1}(N) \; &= \; L \; \to \; K_{k+1}(N, M') \; = \; K \; , \\ j \; : \; K_{k+1}(N) \; &= \; L \; \to \; K_{k+1}(N, M) \; = \; K^* \end{aligned}$$

are the components of the inclusion of a lagrangian

$$\begin{pmatrix} i \\ j \end{pmatrix} \; : \; L \to H \; = \; K \oplus K^*$$

in the hyperbolic $(-1)^k$-hermitian form (λ, μ) on H, such that $\beta = (H, \lambda, \mu; K, L)$ is a $(-1)^k$-hermitian formation in the stable isomorphism class associated to (ϕ, F), with

$$K \cap L \; = \; \mathrm{Ker}(j) \; = \; K_{k+1}(M) \; , \; H/(K + L) \; = \; \mathrm{Coker}(j) \; = \; K_k(M) \; .$$

Reversing the roles of i and j gives a $(-1)^k$-hermitian formation $\beta' = (H, \lambda, \mu; K^*, L)$ in the stable isomorphism class associated to (ϕ', F'), with

$$K^* \cap L \; = \; \mathrm{Ker}(i) \; = \; K_{k+1}(M') \; , \; H/(K^* + L) \; = \; \mathrm{Coker}(i) \; = \; K_k(M') \; .$$

The $(-1)^{k+1}$-hermitian form

$$i^* j \; : \; L \; = \; K_{k+1}(N) \to L^* \; = \; K^{k+1}(N)$$

(the hessian of the lagrangian L, in the hamiltonian terminology of Novikov [N8]) is the intersection form on the middle-dimensional kernel of $(\psi, G) : N \to X \times I$, with a quadratic refinement determined by the bundle map G as in Theorem 5.2. The boundary $(-1)^k$-hermitian formation $\partial(L, i^*j)$ is stably isomorphic to $\beta \oplus -\beta'$, with an exact sequence

$$0 \to K_{k+1}(M) \oplus K_{k+1}(M') \to L \xrightarrow{i^*j} L^* \to K_k(M) \oplus K_k(M') \to 0 \; .$$

Given a k-connected $(2k+1)$-dimensional normal map $(\phi, F) : M \to X$ and a set of $\mathbf{Z}[\pi_1(X)]$-module generators $\{x_1, x_2, \ldots, x_r\} \subset K_k(M)$ the trace of surgeries on representatives $x_\ell : S^k \times D^{k+1} \subset M$ $(1 \leqslant \ell \leqslant r)$ is a $(k+1)$-connected normal bordism extending (ϕ, F)

$$(\psi, G) \ : \ (N; M, M') \to X \times (I; \{0\}, \{1\})$$

with $(\phi', F') = (\psi, G)| : M' \to X$ k-connected. The formation obtained by working inside M

$$(K_k(\partial U), \lambda, \mu; K_{k+1}(U, \partial U), K_{k+1}(M_0, \partial U))$$

is isomorphic to the formation obtained by working outside M

$$(K_{k+1}(N, M') \oplus K_{k+1}(N, M), \lambda, \mu; K_{k+1}(N, M'), K_{k+1}(N)) \ .$$

Moreover, every $(k+1)$-connected normal bordism (ψ, G) is the trace of surgeries on a set of generators of $K_k(M)$. The surgery obstruction $\theta(M, \phi, F) \in L_{2k+1}(\mathbf{Z}[\pi_1(X)])$ is thus represented by a formation which can be obtained either by choosing a set of generators for $K_k(M)$, or (equivalently) a $(k+1)$-connected $(2k+2)$-dimensional normal bordism (ψ, G) with $(\psi, G)| = (\phi, F) : M \to X$.

The matrix identity of Lemma 6.2 has a counterpart in the formation identity of [R1, 3.3]

$$(G, \lambda, \mu; H_1, H_2) \oplus (G, \lambda, \mu; H_2, H_3) \ = \ (G, \lambda, \mu; H_1, H_3) \in L_{2k+1}(\Lambda) \ .$$

The formation identity was used in [R1, 5.6] to prove that the odd-dimensional L-group $L_{2k+1}(\Lambda) = SU(\Lambda)/RU(\Lambda)$ defined using matrices is isomorphic to the (a priori abelian) L-group $L_{2k+1}(\Lambda)$ of formations. The identities are cognate to the formula of Wall [W24] for the non-additivity of the signature. Given a k-connected 2k-dimensional normal map $(\phi, F) : M^{2k} \to X$ with self-intersection form (λ, μ) on the kernel module $G = K_k(M)$ over $\Lambda = \mathbf{Z}[\pi_1(X)]$ let

$$\psi_i \ : \ (N_i; M, M_i) \to X \times (I; \{0\}, \{1\}) \quad (i = 1, 2, 3)$$

be the three k-connected $(2k+1)$-dimensional normal bordisms to homotopy equivalences

$$\phi_i \ = \ \psi_i| \ : \ M_i \to X \quad (i = 1, 2, 3)$$

with

$$K_{k+1}(N_i, M) \ = \ H_i \subset G \ .$$

For $(i, j) = (1, 2), (2, 3), (3, 1)$ define the unions

$$\psi_{ij} \ = \ \psi_i \cup \psi_j \ : \ (N_{ij}; M_i, M_j) \ = \ (N_i \cup_M N_j; M_i, M_j) \to X \times (I; \{0\}, \{1\}) \ ,$$

which are three k-connected $(2k+1)$-dimensional normal bordisms with homotopy equivalences on the boundary components, with kernel $(-1)^k$-hermitian formations $(G, \lambda, \mu; H_i, H_j)$. Constructing the triple union

$$W^{2k+2} \ = \ N_{12} \times I \cup N_{23} \times I \cup N_{31} \times I$$

as in the diagram

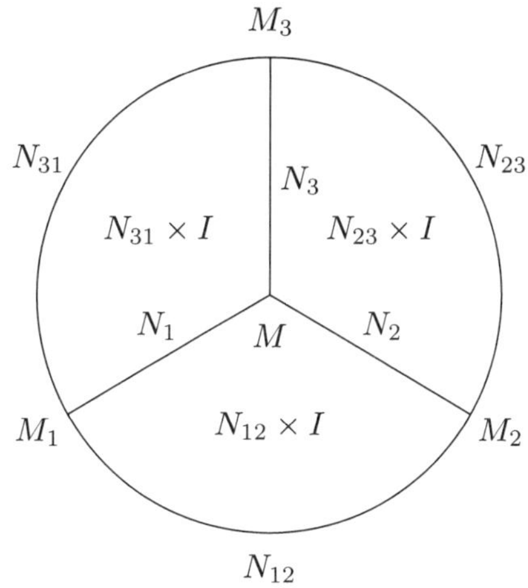

there is obtained a $(k+1)$-connected $(2k+2)$-dimensional normal map

$$(W, \partial W) \to X \times (D^2, S^1)$$

such that the Λ-coefficient surgery obstruction on the boundary is

$$(G, \lambda, \mu; H_1, H_2) \oplus (G, \lambda, \mu; H_2, H_3) \oplus (G, \lambda, \mu; H_3, H_1) \;=\; 0 \in L_{2k+1}(\Lambda)\,.$$

The main result of [R1] is that a $(-1)^k$-hermitian formation β over Λ represents 0 in $L_{2k+1}(\Lambda)$ if and only if β is stably isomorphic to a boundary $\partial\gamma$; this is the formation version of Theorem 6.6. For the formation

$$\beta \;=\; (G, \lambda, \mu; H_1, H_2) \oplus (G, \lambda, \mu; H_2, H_3) \oplus (G, \lambda, \mu; H_3, H_1)$$

the (degenerate) $(-1)^{k+1}$-hermitian form γ can be taken to be (K, λ_K, μ_K) with

$$\begin{aligned} K \;&=\; H_1 \oplus H_2 \oplus H_3\ , \\ \lambda_K&\big((x_1, x_2, x_3), (y_1, y_2, y_3)\big) \\ &=\; \lambda(x_1, y_2) - \lambda(x_2, y_1) + \lambda(x_2, y_3) - \lambda(x_3, y_2) + \lambda(x_3, y_1) - \lambda(x_1, y_3)\ , \\ \mu_K(x_1, x_2, x_3) \;&=\; \lambda(x_1, x_2) + \lambda(x_2, x_3) + \lambda(x_3, x_1)\,. \end{aligned}$$

For $\Lambda = \mathbf{Z}$, $k \equiv 1 (\mathrm{mod}\ 2)$ the signature of this symmetric form is the signature nonadditivity invariant of Wall [W24], which is also known as the Maslov index.

Formations are better suited than automorphisms for describing the kernel structure of a bounded $(2k+2)$-dimensional normal map $(N, M) \to (Y, X)$ such that $M \to X$ is k-connected and $N \to Y$ is $(k+1)$-connected – see §8 below for the relative surgery obstruction theory in this case.

7. The Bounded Odd-dimensional Case

First-time readers may omit this chapter, proceeding directly to §10.

We will now prove (3.1) and (3.2) for pairs (or triads with a sub-pair fixed) in the case $\dim|M| = 2k+1$. This involves breaking less ground than is needed in §8 for the even-dimensional case. For our groups L_m bear a close relation to the groups K_0 and K_1 of [B5], and the relative theory of the K-functors in dimension 1 has been axiomatised by Heller [H12] (also by Bass [B6]) in a form adequate for the present application. This helps with the algebra, and many arguments necessary for the geometry were foreshadowed in [W18, §7].

Before we start, let us observe that although in §5 and §6 we concentrated only on connected manifolds, the deduction of corresponding results in general is trivial. For the components of X can be treated independently; thus the surgery obstruction must lie in the direct sum of groups, one for each component. This tells us how to define $L_m(\pi)$ for a groupoid π of finite type. We can similarly define maps: if π has components with vertex groups π_i, and π' components with vertex groups π'_j, then a morphism $\pi \xrightarrow{r} \pi'$ determines a map $i \longrightarrow j(i)$ of indexing sets and homomorphisms $\pi_i \xrightarrow{r_i} \pi'_{j(i)}$ determined up to conjugacy, and hence homomorphisms $L_m(\pi_i) \xrightarrow{L_m(r_i)} L_m(\pi'_{j(i)})$ (since conjugating by a group element induces the identity on $L_m(\pi)$: in fact we have explicitly factored out the effect of conjugations), and the map

$$\Sigma_i L_m(\pi_i) \;=\; L_m(\pi) \xrightarrow{L_m(r)} L_m(\pi') \;=\; \Sigma_j L_m(\pi'_j)$$

is determined by its components:

$$\begin{aligned} p_j \circ L_m(r) \circ i_i &= 0 && j \neq j(i) \\ &= L_m(r_i) && j = j(i)\,. \end{aligned}$$

If π has vertex groups π_i with integer group rings Λ_i, it is tempting to define it to have integral group ring

$$\Lambda \;=\; \Sigma_i \Lambda_i\,,$$

with coordinate-wise addition and multiplication. We do not adopt this convention, however, as a morphism of groupoids does not induce one of the corresponding rings. Extensions of results to the groupoid case are always straightforward as above; we will not always mention them explicitly, to avoid overmuch notational complication.

Our setup for the next two chapters will be as follows.

HYPOTHESIS 7.1. *We have a map $\phi : (N; M_-, M) \to (Y; X_-, X)$ of degree 1 of a compact manifold triad to a connected simple Poincaré triad, including a simple homotopy equivalence of pairs $(M_-, M_- \cap M) \to (X_-, X_- \cap X)$; a bundle ν over Y, and a stable trivialisation F of $\tau_N \oplus \phi^*\nu$. We wish to do surgery relative to M to make ϕ a simple homotopy equivalence of triads.*

In this chapter we suppose $\dim N = 2k + 1 \geqslant 5$; for our main results we will need $k \geqslant 3$. We assume Y connected and will use $\Lambda = \mathbf{Z}[\pi_1(Y)]$ for coefficients; the extension to the disconnected case goes as above. We cannot, however, assume X connected without losing generality. Let X have components X_i with fundamental group rings Λ_i; let $M_i = \phi^{-1}(X_i)$.

By (1.4) we may suppose, after preliminary surgeries if necessary, that ϕ induces k-connected maps $N \to Y$ and $M \to X$ (hence also $M_i \to X_i$), and that $K_k(N, M) = 0$. As in several places above, we can perform more surgeries to ensure that $K_{k+1}(N, M)$ and $K_k(N)$ are free based Λ-modules, also that $K_k(M_i; \Lambda_i)$ is a free based Λ_i-module. By (5.2.1), we have a simple hermitian form on $K_k(M)$, and by (5.7) $K_{k+1}(N, M)$ is a lagrangian, so that by (5.3), $K_k(M)$ is a hyperbolic form.

Now it follows from the proof of (2.3) that the chain complex (with coefficients Λ_i) of the map $M_i \to X_i$ is chain homotopy equivalent to the single module $K_k(M_i; \Lambda_i)$: indeed, is simply equivalent to it with a preferred basis. It follows that the natural map $K_k(M_i; \Lambda_i) \otimes_{\Lambda_i} \Lambda \to K_k(M_i)$ is an isomorphism respecting bases. Also, we have $K_k(M) = \Sigma_i K_k(M_i)$. Finally, the above isomorphism respects λ and μ too, since the geometrical definitions are changed only by the maps of fundamental groups. So the simple hermitian form on $K_k(M)$ is obtained from those on the $K_k(M_i; \Lambda_i)$ by changing coefficients to Λ and then taking the direct sum. Let $\theta_i \in L_{2k}\big(\pi_1(X_i)\big)$ be the surgery obstruction for M_i; then $\Sigma\theta_i \in \Sigma L_{2k}\big(\pi_1(X_i)\big) = L_{2k}\big(\pi(X)\big)$ is (by definition) the surgery obstruction for M. If $r : \pi(X) \to \pi(Y)$ is induced by inclusion, then as $L_{2k}(r)(\Sigma\theta_i)$ is also induced by tensoring with Λ and taking the direct sum, it is represented by the above form on $K_k(M)$. As this is a hyperbolic form, $L_{2k}(r)(\Sigma\theta_i) = 0$.

Next suppose (N, ϕ, F) cobordant to a normal map (N', ϕ', F') with ϕ' a simple homotopy equivalence of triads. Let W be the cobordism, V the induced cobordism of M, so that

$$\partial V \;=\; M \cup (\partial M \times I) \cup M', \qquad \partial W \;=\; N \cup V \cup (M_- \times I) \cup N'.$$

Write $\psi : (W; M_- \times I, V) \to (Y; X_-, X)$ for the corresponding map, and $V_i = \psi^{-1}(X_i)$. Then $(V_i, \psi \,|\, V_i)$ is a cobordism rel ∂M_i of $(M_i, \phi \,|\, M_i)$ to a simple homotopy equivalence $(M_i', \phi' \,|\, M_i')$. Suppose surgery done to make $\psi \,|\, V_i$ k-connected and $K_k(V_i, M_i; \Lambda_i) = 0$, as we may by (1.4); and further, that $K_{k+1}(V_i, M_i; \Lambda_i)$ is free and based. Then by (5.7), $K_{k+1}(V_i, M_i; \Lambda_i)$ is a lagrangian in $K_k(M_i; \Lambda_i)$. Tensoring with Λ and taking the direct sum, we obtain the lagrangian $K_{k+1}(V, M)$ in $K_k(M)$. But we already have one lagrangian, namely $K_{k+1}(N, M)$. Thus given V as above, we can construct a pair of lagrangians, determining an element of $L_{2k+1}\big(\pi_1(Y)\big)$. It is reasonable to expect that this element represents the obstruction to constructing a cobordism W of $V \cup N$ (whose whole boundary is mapped by a simple homotopy equivalence) to an N' as above. We next justify this: the main theorem (in the case studied in this paragraph) will then follow without difficulty.

LEMMA 7.2. *Let* $\phi : (N; M_-, M) \to (Y; X_-, X)$ *and* $\phi' : (N'; M_+, M) \to (Y'; X_+, X)$ *both satisfy* (7.1), *with* $\dim M = 2k$, *where* ν *and* ν' *agree on* X, *and* ϕ *and* ϕ', F *and* F' *agree on* M. *Suppose preliminary surgeries as above already performed, so that* ϕ *and* ϕ' *induce* k*-connected maps, etc. Glue along* M *and* X *to obtain*

$$\phi'' : (N \cup N'; M_-, M_+) \to (Y \cup Y'; X_-, X_+)$$

and ν'', F''. *Then the obstruction to doing surgery* $\big(\text{rel } \partial(N \cup N')\big)$ *to make* ϕ'' *a simple homotopy equivalence is represented by the pair of lagrangians* $K_{k+1}(N, M)$ *and* $K_{k+1}(N', M)$ *in* $K_k(M)$, *with coefficients* $\Lambda = \mathbf{Z}[\pi_1(Y \cup Y')]$.

Proof Choose a set of disjoint embeddings $f_i : S^k \times D^{k+1} \to \text{Int } N$, and lifts $\widetilde{f}_i$ whose classes generate $K_k\big(N; \mathbf{Z}[\pi_1(Y)]\big)$ and hence also $K_k(N)$. The classes generate $K_k(N \cup N')$, too, since we have the exact sequence

$$K_k(N) \to K_k(N \cup N') \to K_k(N \cup N', N);$$

by excision, the last is $K_k(N', M)$, and this vanishes as a consequence of the preliminary surgeries (1.4). Set $N_1 = \bigcup \text{Im } f_i$, $N_0 = N - \text{Int } N_1$, $N_2 = N'$ and $M_1 = \partial N_1$, $M_2 = M = N \cap N'$; recall that using (2.9) as in §6 we have appropriate maps of degree 1 of all these manifolds, pairs and triads. The surgery obstruction for $N \cup N'$ can be defined by the pair of lagrangians $K_{k+1}(N_1, M_1)$ and $K_{k+1}(N_0 \cup N_2, M_1)$ in $K_k(M_1)$. We must show that the same class is defined by the pair of lagrangians $K_{k+1}(N_0 \cup N_1, M_2)$ and $K_{k+1}(N_2, M_2)$ in $K_k(M_2)$. We will show that each determines the same class as the pair of lagrangians

$$K_{k+1}(N_1 \cup N_2, M_1 \cup M_2) = K_{k+1}(N_1, M_1) \oplus K_{k+1}(N_2, M_2)$$

and $K_{k+1}(N_0, M_1 \cup M_2)$ in $K_k(M_1 \cup M_2) = K_k(M_1) \oplus K_k(M_2)$. By symmetry it is enough to prove one of these.

Consider the following diagram:

$$
\begin{array}{ccccccccc}
 & & 0 & & 0 & & 0 & & \\
 & & \downarrow & & \downarrow & & \uparrow & & \\
0 & \longrightarrow & K_{k+1}(N_0 \cup N_1, M_2) & \longrightarrow & K_k(M_2) & \longrightarrow & K_k(N_0 \cup N_1) & \longrightarrow & 0 \\
 & & \downarrow & & \downarrow & & \uparrow & & \\
0 & \longrightarrow & K_{k+1}(N_0, M_1 \cup M_2) & \longrightarrow & K_k(M_1 \cup M_2) & \longrightarrow & K_k(N_0) & \longrightarrow & 0 \\
 & & \downarrow & & \downarrow & & & & \\
0 & \longleftarrow & K_k(N_1) & \longleftarrow & K_k(M_1) & & & & \\
 & & \downarrow & & \downarrow & & & & \\
 & & 0 & & 0 & & & &
\end{array}
$$

The horizontal sequences are the usual exact sequences; the first vertical sequence is the exact sequence of the triple $M_2 \subset N_1 \cup M_2 \subset N_0 \cup N_1$, using the excision isomorphism $K_{k+1}(N_0, M_1 \cup M_2) = K_{k+1}(N_0 \cup N_1, N_1 \cup M_2)$; the second vertical is the exact sequence of the obvious direct sum decomposition of $K_k(M_1 \cup M_2)$, and the other two maps are inclusion maps. It is easy to verify that the diagram commutes. It follows that we can find a (unique) homomorphism $s : K_k(N_1) \to K_k(M_1)$ which (with the maps in the diagram) gives a morphism between the vertical exact sequences. This must be right inverse to the inclusion map $K_k(M_1) \to K_k(N_1)$. On the other hand, by (5.3) we can find a right inverse $s' : K_k(N_1) \to K_k(M_1)$ to the inclusion such that the image of s' is a lagrangian, complementary to $K_{k+1}(N_1, M_1)$.

Now begin with the pair of lagrangians $K_{k+1}(N_0 \cup N_1, M_2)$ and $K_{k+1}(N_2, M_2)$ in $K_k(M_2)$. We stabilise by adding the complementary pair of lagrangians $s'K_k(N_1)$ and $K_{k+1}(N_1, M_1)$ in $K_k(M_1)$; thus obtaining a pair of lagrangians in $K_{k+1}(N_1, M_1) \oplus K_{k+1}(N_2, M_2)$ as required. The first member is not yet equal to $K_{k+1}(N_0, M_1 \cup M_2)$, but it will suffice to show that it is equivalent to this by an elementary transformation (i.e. one which preserves some lagrangian, so – for any standard basis – is in some conjugate of $TU(\Lambda)$, hence in $RU(\Lambda)$).

Choose a standard basis $\{e_i, f_i\}$ for $K_k(M_2)$ with $\{e_i\}$ a basis for the lagrangians $K_{k+1}(N_0 \cup N_1, M_2)$. Extend $\{e_i\}$ to a basis $\{e_i, e'_j\}$ for the lagrangian $K_{k+1}(N_0, M_1 \cup M_2)$, with e'_j λ-orthogonal to f_i, and extend all these to a standard basis $\{e_i, e'_j, f_i, f'_j\}$ for $K_k(M_1 \cup M_2)$. All these bases, of course, are to be preferred bases. This induces a basis $\{e'_j, f'_j\}$ for $K_k(M_1)$ regarded as a quotient module. Now $\{e'_j\}$ is a basis for $sK_k(N_1)$; a basis for $s'K_k(N_1)$ must have the form $\{e'_j + \Sigma f'_k a_{kj}\}$ (as submodule of $K_k(M_1)$); when we lift to $K_k(M_1 \cup M_2)$, this takes the form

$$\{e'_j + \Sigma f'_k a_{kj} + \Sigma e_i b_{ij} + \Sigma f_i c_{ij}\}\,.$$

As we are only interested in $K_{k+1}(N_0 \cup N_1, M_2) \oplus s'K_k(N_1)$, we make an elementary basis change by subtracting multiples of e_i from the above to get rid of the terms $e_i b_{ij}$. Also, as we do have a lagrangian,

$$0 = \lambda(e_i, e'_j + \Sigma f'_k a_{kj} + \Sigma e_i b_{ij} + \Sigma f_l e_{lj}) = e_{ij}.$$

So $K_{k+1}(N_0 \cup N_1, M_2) \oplus s'K_k(N_1)$ has basis $\{e_i, e'_j + \Sigma f'_k a_{kj}\}$, which can indeed be reduced to the basis $\{e_i, e'_j\}$ of $K_k(N_0, M_1 \cup M_2)$ by an elementary change, viz.

$$e_i \mapsto e_i, \quad f_i \mapsto f_i, \quad e'_j \mapsto e'_j - \Sigma f'_k a_{kj}, \quad f'_j \mapsto f'_j.$$

(That this preserves λ and μ follows since we have a lagrangian). Thus the lemma is proved.

We have been somewhat careless about the order in which the two lagrangians are to be taken, but changing the order will change the obstruction only by a sign, and this does not matter much for our present purposes.

Let $r : \Lambda \to \Lambda'$ be a morphism of rings with involution; let $\eta = \pm 1$. Consider the set of quadruples (G, λ, μ, K), where (G, λ, μ) is a simple η-hermitian form over Λ, and K is a lagrangian in $G \otimes_\Lambda \Lambda'$. Write $(G, \lambda, \mu, K) \sim (G', \lambda', \mu', K')$ if there is a hyperbolic form H, with lagrangian S, over Λ, such that

(1) $(G, \lambda, \mu) \oplus H \oplus (G', -\lambda', -\mu') = H_1$ is a hyperbolic form, with lagrangian S_1.

(2) An automorphism of $H_1 \otimes_\Lambda \Lambda'$ taking $S_1 \otimes_\Lambda \Lambda'$ to $K \oplus (S \otimes_\Lambda \Lambda') \oplus K'$ is (stably) in $RU(\Lambda)$.

LEMMA 7.3. *The relation $\sim$ is an equivalence relation. Direct sum of quadruples is compatible with $\sim$, and induces an abelian group structure on the set of equivalence classes.*

Proof To show $\sim$ reflexive, we take $H = 0$ and S_1 the diagonal in $H \oplus H$ (by (5.4)). The automorphism for (2) is given by the matrix M of §6 (just before (6.2)), which was in $RU(\Lambda)$.

Symmetry is clear: all we need do is to change a few signs where appropriate. As to transitivity, if $G \sim G' \sim G''$, we take the direct sum of the H_1 and H'_1 giving the equivalences, and absorb the term $(G', \lambda', \mu') \oplus (G', -\lambda', -\mu')$ into the "H" term (by proof of reflexivity): the same argument goes for the lagrangians over Λ'.

The second sentence is immediate: the least obvious point concerns inverses, and these are provided as usual by changing signs to obtain $(G, -\lambda, -\mu, K)$. That this is an inverse follows, again, from the proof that $\sim$ is reflexive.

We christen the group so defined as $L_\eta(r)$ (here, η is reckoned mod 4). The case of most direct concern to us is where Λ and Λ' are integral group rings, and then we replace r in the notation by the underlying morphism of groups.

THEOREM 7.4. *Let $r : \Lambda \to \Lambda'$ be a morphism of rings with involution. Then we have a functorial exact sequence (for $\eta = \pm 1$)*

$$L_\eta(\Lambda) \xrightarrow{r_*} L_\eta(\Lambda') \xrightarrow{j} L_\eta(r) \xrightarrow{\partial} L_{\eta-1}(\Lambda) \xrightarrow{r_*} L_{\eta-1}(\Lambda') .$$

Similarly if Λ denotes a finite set of rings.

Proof The homomorphism ∂ is induced by taking a representative (G, λ, μ, K) and forgetting K to obtain a simple η-hermitian form over Λ. Equivalent objects give rise to forms which (by (5.3) and (5.4)) induce the same element of $L_{\eta-1}(\Lambda)$, so ∂ is well-defined; clearly, it is a homomorphism.

To define j, let $\alpha' \in SU_r(\Lambda')$. Take the standard hyperbolic form (G, λ, μ) of rank $2r$ over Λ, with standard lagrangian S, and define

$$K = \alpha'(S \otimes_\Lambda \Lambda') .$$

If we replace α' by $\beta'\alpha'$, with $\beta' \in RU_r(\Lambda)$, K is replaced by $\beta' K$. This gives an equivalent object, since if $H = G \otimes_\Lambda \Lambda'$, then in $(H, \lambda, \mu) \oplus (H, -\lambda, -\mu)$ we have $K \oplus \beta' K \sim K \oplus K \sim$ the diagonal $\sim S \oplus S$ (under the action of $RU_{2r}(\Lambda')$). Stabilising also gives an equivalent object, so j is well-defined. That it is a homomorphism follows since, modulo $RU(\Lambda')$, we know that

$$\begin{pmatrix} \alpha' & 0 \\ 0 & \beta' \end{pmatrix} \sim \begin{pmatrix} \alpha'\beta' & 0 \\ 0 & 1 \end{pmatrix}$$

and our construction is clearly compatible with direct sums. The naturality of j and ∂ is evident from their definitions.

Exactness at $L_\eta(\Lambda')$. If $\alpha \in SU_r(\Lambda)$, we represent $r_*\alpha$ by $\alpha \otimes_\Lambda \Lambda'$, and $jr_*\alpha$ by taking a standard hyperbolic form (G, λ, μ) with lagrangian S, and setting $K = r_*\alpha(S \otimes_\Lambda \Lambda') = (\alpha \otimes_\Lambda \Lambda')(S \otimes_\Lambda \Lambda') = \alpha(S) \otimes_\Lambda \Lambda'$. But now $(G, \lambda, \mu, K) \sim 0$, taking in the definition $H = 0$, $S_1 = \alpha(S)$, and the identity automorphism of $G \otimes_\Lambda \Lambda'$.

Suppose conversely that $\alpha' \in SU_r(\Lambda')$, and $j(\alpha') = 0$. Then with (G, λ, μ) over Λ having lagrangian S, the quadruple $(G, \lambda, \mu, \alpha'(S \otimes_\Lambda \Lambda'))$ is null-equivalent, so for some hyperbolic form (G_1, λ_1, μ_1) over Λ, with lagrangian S_1, we have a lagrangian S_2 of $H = (G, \lambda, \mu) \oplus (G_1, \lambda_1, \mu_1)$ such that an automorphism of $H \otimes_\Lambda \Lambda'$ taking $S_2 \otimes_\Lambda \Lambda'$ to $\alpha'(S \otimes_\Lambda \Lambda') \oplus (S_1 \otimes_\Lambda \Lambda')$ is in $RU(\Lambda')$. Thus modulo $RU(\Lambda')$, we can replace α' by any automorphism of H taking $(S \otimes_\Lambda \Lambda') \oplus (S_1 \otimes_\Lambda \Lambda')$ to $(S_2 \otimes_\Lambda \Lambda')$. If α is an automorphism of H taking $S \oplus S_1$ to S_2 (this exists by (5.3.1)), this shows $\alpha' \sim \alpha \otimes_\Lambda \Lambda'$ and so our class is in the image of r_*.

Exactness at $L_\eta(r)$. First, let $\alpha' \in SU(\Lambda')$. Then $j\alpha'$ is represented by $(G, \lambda, \mu, \alpha'(S \otimes_\Lambda \Lambda'))$, with S a lagrangian in G. So $\partial j\alpha'$ is represented by a hyperbolic form (G, λ, μ), hence is zero. Suppose conversely that (G, λ, μ, K) is such that (G, λ, μ) represents $0 \in L_{\eta-1}(\Lambda)$. Stabilising (add a hyperbolic form H_1, with lagrangian S_1, to G and $S_1 \otimes_\Lambda \Lambda'$ to K) we may suppose (G, λ, μ) a hyperbolic form. Choose a lagrangian S. Then if α' is an automorphism taking $S \otimes_\Lambda \Lambda'$ to K (again, this exists by (5.3.1)), (G, λ, μ, K) represents $j\alpha'$.

Exactness at $L_{\eta-1}(\Gamma)$. This is trivial: a simple η-hermitian form (G, λ, μ) over Λ determines the zero class over Λ' if and only if $G \otimes_\Lambda \Lambda'$ is (stably) a hyperbolic form, i.e. if and only if it possesses (stably) a lagrangian K, and so (G, λ, μ) comes from $L_\eta(r)$.

This proves the algebraic result (a special case of (3.1)): we now go on to establish (3.2) in the case under consideration. Recall that Λ and Λ' in the above will be group rings or finite sets of such, as discussed at the beginning of the chapter.

THEOREM 7.5. *Assume* (7.1)*; perform also further surgeries as in* (1.4) *to make all* K_k *zero or stably free, and extra surgeries to make them all free and based (with the* $K_k(M_i; \Lambda_i)$ *all of the same rank). Let* $r : \pi(X) \to \pi(Y)$ *be the induced morphism,* $\dim N = 2k + 1 \geqslant 5$. *Define an obstruction* θ *in* $L_{2k+1}(r)$ *as the equivalence class of* $(\Sigma_i K_k(M_i; \Lambda_i), \lambda, \mu, K_{k+1}(N, M))$. *Then the conclusions of* (3.2) *hold, viz.* θ *depends only on the bordism class of* (N, ϕ, F)*, vanishes for a simple homotopy equivalence and, for* $k \geqslant 3$*, only if the class contains a simple homotopy equivalence;* $\partial\theta$ *is the surgery obstruction for* $\phi \,|\, M$ *and* θ *is natural for inclusion maps.*

Proof It is evident that θ vanishes for a simple homotopy equivalence, and $\partial\theta$ is by definition the surgery obstruction for $\phi \,|\, M$. Naturality follows, as in (5.6) and (6.4), from the explicit character of the definition of θ in terms of homology. The two remaining assertions are the bordism invariance of θ, and that $\theta = 0$ implies the possibility of surgery.

We begin by observing that (7.2), with the discussion preceding it, shows that if surgery is possible, then θ must vanish. For under this hypothesis we found a lagrangian over Λ (equivalently, set of lagrangians $K_{k+1}(V_i, M_i; \Lambda_i)$ over the Λ_i); and observed that the existence of W implied the vanishing of the surgery obstruction for $N \cup V$; whereas by (7.2) this obstruction is defined by the pair of lagrangians $K_{k+1}(V, M)$ and $K_{k+1}(N, M)$ in $K_k(M)$. Since these are equivalent under $RU(\mathbf{Z}[\pi_1(Y)])$, it follows that the quadruple defining θ is null-equivalent.

We next deduce bordism invariance of θ: to economise notation, we suppress mention of M_-. Let (W, V) be a cobordism of (N, M) to (N', M') and $\psi : (W, V) \to (Y, X)$ extend ϕ and ϕ'; G extend F and F'. Let us suppose all necessary preliminary surgeries performed already. We change an orientation, and regard (W, V) as a cobordism of $(N, M) \cup (N', M')$ to the empty set. As usual, we construct a map $h : (W; N, N') \to (I; 0, 1)$, and compute groups K_k for W, V, etc. using

$$\psi \times h : (W; N, V, N') \to (Y \times I; Y \times 0, X \times I, Y \times 1).$$

Using again the suffix i to distinguish components of X, we see as above that $K_{k+1}(V_i, M_i \cup M'_i; \Lambda_i)$ is a lagrangian in $K_k(M_i; \Lambda_i) \oplus K_k(M'_i; \Lambda_i)$, and that the corresponding lagrangian $K_{k+1}(V, M \cup M')$ in $K_k(M) \oplus K_k(M')$ is equivalent

$\big($under $RU(\Lambda')\big)$ to $K_{k+1}(N \cup N', M \cup M')$. This shows at once, taking $H = 0$ and recalling the change of orientation, that

$$\big(\Sigma K_k(M_i), \lambda, \mu, K_{k+1}(N, M)\big) \sim \big(\Sigma K_k(M_i'), \lambda, \mu, K_{k+1}(N', M')\big),$$

and so that N and N' correspond to the same value of θ.

Finally, let $\theta = 0$. Perform (if necessary) trivial surgeries on the M_i to add hyperbolic forms to the $K_k(M_i; \Lambda_i)$. Then we may suppose given lagrangians $S_i \subset K_k(M_i; \Lambda_i)$ such that $\Sigma S_i \otimes \Lambda'$ is equivalent under $RU(\Lambda')$ to $K_{k+1}(N, M)$. By (5.2), if $k \geqslant 3$ we can represent basis elements of the S_i by framed embeddings of spheres, and so add handles to the $M_i \times I$, obtaining manifolds V_i. By the proof of (5.6), the other end of V_i is mapped to X_i by a simple homotopy equivalence. Now by (7.2), the surgery obstruction for $N \cup V$ (where V is the union of the V_i) is represented by the pair of lagrangians $K_{k+1}(N, M)$ and $K_{k+1}(V, M)$ and so, by hypothesis, vanishes. By (6.4) we can do surgery (with trace W, say) to obtain a simple homotopy equivalence.

REMARK 7.6. It is now easy to obtain the result corresponding to (5.8) and (6.5); viz. that given a compact smooth or PL manifold X of odd dimension $m \geqslant 7$, with normal bundle ν, then θ induces a bijection onto $L_m\big(\pi(X), \pi(\partial X)\big)$ of the set of bordism classes of normal maps (M, ϕ, F), where M is a cobordism of bounded manifolds,

$$\phi : (M; \partial_- M, \partial_c M, \partial_+ M) \to (X \times I; X \times 0, \partial X \times I, X \times I)$$

is a map of quadruples inducing the identity map $\partial_- M \to X \times 0$ and a simple homotopy equivalence $(\partial_+ M, \partial\partial_+ M) \to (X \times 1, \partial X \times 1)$, and F is a stable framing of $\tau_M \oplus \phi^* \nu$ extending the natural one on $\partial_- M$. A proof of a more general result is given in (10.4) below.

8. The Bounded Even-dimensional Case

First-time readers may omit this chapter, proceeding directly to §10.

This chapter was intended to be parallel to §7. However, the algebra is yet more formidable, and I have been unable to obtain a complete description*. This suggests that a different method should be tried; however, I hope that the present partial account will not be devoid of interest.

Although we showed in §6 that surgery was possible if and only if our obstructions vanished, we did not give an explicit description of the way the algebraic properties of the surgery implied that algebraic triviality of the obstruction: this we must now do. We will suppose through the chapter (except where otherwise specified) hypothesis 7.1 with $\dim N \geqslant 6$, and Y connected (as usual, this is innocuous).

We can suppose that ϕ induces a $(k-1)$-connected map $M \to X$ and a k-connected map $N \to Y$, so that the only nonvanishing groups K_i are those in the sequence

$$0 \to K_k(M) \to K_k(N) \to K_k(N, M) \to K_{k-1}(M) \to 0\,.$$

By (2.3), the middle two modules are stably free and s-based; as usual we may suppose (after preliminary surgeries) that they are free and based; by (2.6) the bases may be taken as dual bases (with respect to the usual duality map). Thus the middle map in the above sequence is adjoint to a sesquilinear map

$$\lambda : K_k(N) \times K_k(N) \to \Lambda\,;$$

since λ is induced by cap products, or equivalently by intersection numbers, it is the same as the λ of (5.2), with $G = K_k(N)$. In (5.2) we obtained also a quadratic map $\mu : G \to Q_k$ which appears again here. Note also that λ is not here nonsingular: the above sequence shows that its deviation from being so is measured by the module $K_{k-1}(M)$.

Related to this is the observation that μ now satisfies some identities other than those of (5.2): we will not use these, but think it of interest to explain why this is so. Consider the submodule $K_k(M)$ – or rather, $K_k(M_i; \Lambda_i)$. An element of this determines a framed homotopy class of immersions

$$f : S^k \to M_i\,.$$

**Sharpe [S8] described the relative even-dimensional L-groups $L_{2*}(r : \Lambda \to \Lambda')$ using unitary Steinberg relations. A more systematic description was given in Ranicki [R7, Chapter 2], using chain complexes (see the notes at the end of §17G).*

The self-intersections of such an immersion consist of double circles in M_i: one of these may be covered by two circles in S^k, or doubly covered by one circle in S^k. In the former case, we can get rid of the singularity when we allow an extra dimension, $S^k \to M_i \times I$. In the latter we cannot, but find a corresponding point self-intersection P of S^k in $M_i \times I$, whose g_P is the class of the original double circle in M_i. Clearly $g_P^2 = 1$ (twice the circle comes from S^k, which is simply connected), and orientation considerations show that $w(g_P) = (-1)^{k+1}$. So $\mu(f)$ is a sum of each g_P. Now the equations of (5.2) show, since the class of f is in $K_k(M)$ and hence orthogonal (for λ) to all of $K_k(N)$, that $\lambda(f, f) = 0$ and hence $\mu(f)$ is a sum of elements $g \in \pi_1(Y)$ of order 2 with $w(g) = (-1)^{k+1}$. These are the conditions obtained geometrically, *except* that we found that each $g_P \in \pi_1(X_i)$. This extra condition is not easy to formalise in general (we have not obtained it in full, only a special case of it).

Now suppose for simplicity that $\pi(X) = \pi(Y)$ (so X is connected). We showed in §4 that surgery was possible in this case: we gave there a geometrical proof; we will now express it algebraically. As in §4, choose a preferred base $\{e_i\}$ of $K_k(N, M)$: since fundamental groups are equal, this group is isomorphic to $\pi_{k+1}(\phi)$, so we can represent the e_i by

$$f_i : (D^k \times D^k, S^{k-1} \times D^k) \to (N, M);$$

by §4, we may suppose these to be disjoint embeddings. Denote by ∂f_i the restriction to $S^{k-1} \times D^k$. The classes of these generate $K_{k-1}(M)$, so we can use them to compute the invariant of M. In fact, write V for the union of the images of the f_i, and U for that of the ∂f_i (conforming with the notation of §6): then for M we have the pair of lagrangians $K_k(U, \partial U)$ and $K_k(M_0, \partial U)$ in $K_{k-1}(U)$, where $M_0 = M - \text{Int } U$. Write also $N_0 = N - \text{Int } V$ and $\partial^r V = \partial V - \text{Int } U$ (the 'relative boundary' of V).

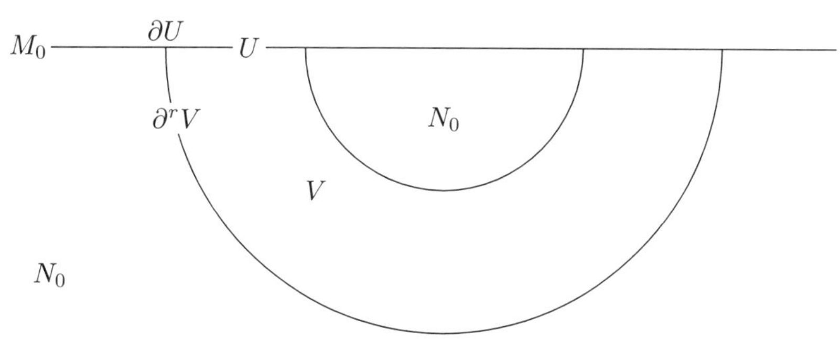

By the construction of V, $\phi \,|\, V$ is homotopic to a map to the base point. Extend this homotopy over N. Then adjust somewhat, writing $Y = Y_0 \cup D^{2k}$, so that $Y_0 \cap D^{2k}$ is a $(2k-1)$-disc on the common boundary and $\phi : (N; N_0, V, M) \to (Y; Y_0, D^{2k}, X)$. We can now extend the K_k notation to N_0, V, etc. (as in §6): note that the homology of D^{2k} etc. vanish in the middle dimensions.

Now inclusions define a simple isomorphism

$$K_{k-1}(\partial U) \cong K_{k-1}(U) \oplus K_{k-1}(\partial^r V)$$

taking the lagrangians $K_k(\partial^r V, \partial U)$, $K_k(U, \partial U)$ to the two summands. We wish to identify the lagrangian $K_k(M_0, \partial U)$, and must first reinterpret the above. Since our construction is that of §4, ϕ induces a simple homotopy equivalence $(N_0, M_0 \cup \partial^r V) \to (Y_0, X_0 \cup D^{2k-1})$, so that the $K_i(N_0)$ and $K_i(N_0, M_0 \cup \partial^r V)$ vanish, with trivial preferred bases. Hence inclusion induces a simple isomorphism $K_k(V, U) \cong K_k(N, M)$, and as V is contractible, this is isomorphic to $K_{k-1}(U)$. Similarly (using excision etc.) we have simple isomorphisms

$$K_k(U, \partial U) \cong K_k(V, \partial^r V) \cong K_k(N, N_0) \cong K_k(N)\,.$$

Moreover, our combined isomorphisms take the intersection pairing of $K_k(N)$ with $K_k(N, M)$ to that of $K_k(V, \partial_r V)$ with $K_k(V, U)$ (since they are induced by inclusion maps) and hence (by an elementary computation in $D^{k+1} \times D^{k+1}$) to that of the lagrangians $K_{k-1}(\partial_r V)$ and $K_{k-1}(U)$ in $K_{k-1}(\partial U)$ (up to sign. In any case, the sign will change if we alter the order of the modules being paired).

Now we have simple isomorphisms

$$K_k(M_0, \partial U) \cong K_k(M_0 \cup \partial^r V, \partial^r V) \cong K_{k-1}(\partial^r V)$$

of our lagrangian onto the second summand above. The projection of the first summand induces a map

$$K_k(N) \cong K_{k-1}(\partial^r V) \cong K_k(M_0, \partial U) \to K_{k-1}(\partial U) \to K_{k-1}(U) \cong K_k(N, M)\,;$$

we claim that the composite is just the map induced by inclusion. For consider the diagram

$$\begin{array}{ccccccccc}
K_k(N, N_0) & \xleftarrow{\cong} & K_k(V, \partial^r V) & \xleftarrow{\cong} & K_k(U, \partial U) & & K_k(V, U) & \xrightarrow{\cong} & K_k(N, M) \\
\uparrow{\scriptstyle\cong} & & \downarrow{\scriptstyle\cong} & & \downarrow & & \downarrow{\scriptstyle\cong} & & \uparrow \\
K_k(N) & & K_{k-1}(\partial^r V) & \longleftarrow & K_{k-1}(\partial U) & \longrightarrow & K_{k-1}(U) & & K_k(N) \\
\downarrow{\scriptstyle\cong} & & \uparrow{\scriptstyle\cong} & & \uparrow & & \uparrow & & \downarrow{\scriptstyle\cong} \\
K_k(N, V) & \xleftarrow{\cong} & K_k(N_0, \partial^r V) & \xleftarrow{\cong} & K_k(M_0, \partial U) & \xrightarrow{\cong} & K_k(M, U) & \longrightarrow & K_k(N, V)\,.
\end{array}$$

In this diagram, all maps are induced by inclusion or are boundary maps; all are simple isomorphisms except the maps involving $K_{k-1}(\partial U)$ and the maps $K_k(M, U) \to K_{k-1}(U)$, $K_k(N) \to K_k(N, M)$. Thus the three squares commute; the two rectangles anticommute, since the paths round them represent the two alternative definitions of the Mayer-Vietoris boundary maps of the triads $(N; N_0, V)$ and $(N; M, V)$: although $M \cup V \neq N$, the relative K_i all vanish. Hence the composite

$$\begin{array}{ccccc}
o & \longrightarrow & o & & o & \longrightarrow & o \\
\uparrow & & \downarrow & & \uparrow & & \\
o & & o & & o & & \\
 & & \downarrow & & \uparrow & & \\
 & & o & \longrightarrow o \longrightarrow & o & &
\end{array}$$

equals the composite

$$\begin{array}{ccccccccc}
 & & & & & & & & o \\
 & & & & & & & & \uparrow \\
o & & & & & & & & o \\
\downarrow & & & & & & & & \uparrow \\
o & \longrightarrow & o & \longrightarrow & o & \longrightarrow & o & \longrightarrow & o
\end{array}$$

which, since the maps along the lower edge are all inclusions, so induce the identity on $K_k(N, Y)$, proving the result.

This explains the relation of the form λ to the invariant of M, in the case $\pi(X) = \pi(Y)$. For we have shown that if we define a form on $K_k(N) \oplus K_k(N, M)$ by requiring the summands to be complementary lagrangians, and the pairing between them the natural one, then the invariant of M is the equivalence class of the pair of lagrangians: $K_k(N)$ itself, and the graph of the natural map $K_k(N) \to K_k(N, M)$ which (as remarked above) is adjoint to λ. We will denote it by $A\lambda$ and its matrix (with respect to a preferred basis of $K_k(N)$ and the dual basis of $K_k(N, M)$) by A. Then the matrix of a transformation taking the first lagrangian above to the second is

$$\begin{pmatrix} I & 0 \\ A & I \end{pmatrix}.$$

This lies in $\Sigma\, TU(\Lambda)\Sigma^{-1}$, in conformity with the remark at the end of §6. Conversely, any element of $\Sigma\, UU(\Lambda)\Sigma^{-1}$ has the above form, and A must be $(-1)^k$-symmetric for the matrix to leave invariant the λ for a $(-1)^{k-1}$-symmetric hyperbolic form, and of the form $P + (-1)^k P^*$ for it to leave invariant μ also. Note that this is the point where the interchange between symmetry and skew-symmetry is effected.

We must now establish corresponding results in the general case: the argument will be similar to the above.

LEMMA 8.1 *Assume* (7.1), *with* $\dim N = 2k \geqslant 6$. *Form the invariants* (*each an equivalence class of pairs of lagrangians*) *of the induced maps* $M_i \to X_i$; *tensor with* Λ *and take the direct sum. Then the result is equivalent* (*in a natural way*) *to the pair above, viz.* $K_k(N)$ *and the graph of* $A\lambda : K_k(N) \to K_k(N, M)$ *is a form on* $K_k(N) \oplus K_k(N, M)$ *in which the summands are lagrangians and the intersection pairing is the natural one.*

Proof There is no need to change the algebra in the above argument, but the geometry breaks down at two points. The first is that the map $\pi_{k+1}(\phi) \to K_k(N, M)$ is no longer an isomorphism; the second, that even if we can choose homotopy classes lifting the e_i, they need not be representable by disjoint embeddings. For each i, we choose a (finite) set $\{x_{ij}\}$ of generators of $K_{k-1}(M_i; \Lambda_i)$.

Let the union of these sets have c elements. Now $x_{ij} \in K_{k-1}(M_i; \Lambda_i) \cong \pi_k(\phi_i)$, where $\phi_i : M_i \to X_i$ is the restriction of ϕ. But the map $N \to Y$ is k-connected, so we can write $x_{ij} = \partial y_{ij}$ with $y_{ij} \in \pi_{k+1}(\phi)$: y_{ij} has image e_{ij} (say) in $K_k(N, M)$. Perform c trivial surgeries (half as many would suffice, really) on N to replace it by its connected sum with c copies of $S^k \times S^k$. Denote the classes of the copies of $S^k \times 1$ and $1 \times S^k$ by a_{ij} and b_{ij}, so that if originally $K_k(N, M)$ had a preferred base $\{c_h\}$, it now has preferred base $\{a_{ij}, b_{ij}, c_h\}$. Replace a_{ij} by $a'_{ij} = a_{ij} + e_{ij}$: this is an elementary basis change since e_{ij} is a linear combination of the c_h. Then $\partial a_{ij} = \partial e'_{ij}$ is the image of x_{ij}. But the x_{ij} are generators. Thus we can subtract suitable linear combinations of the a_{ij} from the c_h to convert them to elements c'_h with $\partial c'_h = 0$.

Now by construction, ∂a_{ij} is the image of x_{ij} in $K_{k-1}(M_i)$. Let $\phi : (N, M) \to (Y, X)$ induce $\psi_0 : M \to X$ and $\psi_1 : N \to Y$. Consider the map of exact sequences, with base-point in M_i,

$$\begin{array}{ccccccccc} \pi_{k+1}(\psi_0) & \longrightarrow & \pi_{k+1}(\psi_1) & \longrightarrow & \pi_{k+1}(\phi) & \longrightarrow & \pi_k(\psi_0) & \longrightarrow & \pi_k(\psi_1) = 0 \\ \downarrow & & \downarrow \cong & & \downarrow & & \downarrow & & \\ 0 \longrightarrow K_k(M) & \longrightarrow & K_k(N) & \longrightarrow & K_k(N, M) & \longrightarrow & K_{k-1}(M) & \longrightarrow & K_{k-1}(N) = 0\,. \end{array}$$

By diagram-chasing, we deduce an exact sequence

$$\pi_{k+1}(\psi_0) \to K_k(M) \to \pi_{k+1}(\phi) \to K_k(N, M) \oplus \pi_k(\psi_0) \to K_{k-1}(M) \to 0\,.$$

Thus $a'_{ij} \in K_k(N, M)$ and $x_{ij} \in \pi_k(\psi_0)$ have a common precursor in $\pi_{k+1}(\phi)$, unique modulo the image of $K_k(M)$. And b'_{ij}, c_h are in the image of $K_k(N) \cong \pi_{k+1}(\psi_1)$. So all our classes come from homotopy groups. Observe for the first ones that our argument implies that any map $S^{k-1} \to M$ representing x_{ij} can be extended to $(D^k, S^{k-1}) \to (N, M)$ representing a_{ij}. As to the other classes, we can now introduce a (trivial) boundary in any desired M_i, so as to have maps of discs instead of spheres. This overcomes the first difficulty mentioned above.

As to the second, we can now apply (1.3) to represent our classes by framed immersions (of discs – in some cases, if preferred – spheres). If these are in general position, the intersections and self-intersections will form a finite point set. Ignoring them in the above argument will only affect K_1 and K_{2k-1}; as far as the middle dimensions go, we reach the same conclusion, provided we use coefficients Λ throughout.

The above result is somewhat analogous to (5.7). We now seek an analogue (8.4) to (7.2): to show how, given two manifolds N with the same boundary M, to combine their properties to find the simple hermitian form on the union. We will fix notation as follows. We have simple Poincaré triads $(Y_1; W_1, X)$ and $(Y_2; W_2, X)$ meeting in X, so that glueing gives a simple Poincaré triad $(Y_1 \cup Y_2; W_1, W_2)$: we write $Y = Y_1 \cup Y_2$. We have a bundle ν over Y; maps of degree 1

$$\phi_a : (N_a; L_a, M) \to (Y_a; W_a, X) \quad (a = 1, 2)$$

where N_a is a manifold, $N_1 \cap N_2 = M$ and $N_1 \cup N_2 = N$, say. The induced maps $L_a \to W_a$ are to be simple homotopy equivalences, and the maps ϕ_a agree on M, so combine to give a map $\phi : N \to Y$. Finally, we have a stable framing F of $\tau_N \oplus \phi^*\nu$. Our problem is, to calculate the surgery obstruction for ϕ in terms of invariants associated to ϕ_1 and ϕ_2. In the course of the argument, we will find another proof (8.4.1) that forms on appropriate $K_k(N_1)$ and $K_k(N_2)$ induce (stably) isomorphic pairs of lagrangians.

We will also need some notation for all the components and fundamental groups. Let X have components X_i, with fundamental group rings R_i. Let Y_a $(a = 1, 2)$ have components Y_{aj}, with fundamental group rings Λ_{aj}. We suppose Y connected, and denote its fundamental group ring by Λ. Choose base points $x_i \in X_i$, $y_{aj} \in Y_{aj}$. Each X_i is on the boundary of one Y_{1j} and one Y_{2j}: join x_i to the corresponding points y_{aj} by paths in Y. By 'general position', we may suppose these paths embedded disjointly except at their ends; they form a graph Γ with vertices the y_{aj} and edges corresponding to the x_i. The paths induce homomorphisms

$$\pi_1(X_i) \;=\; \pi_1(X_i; x_i) \to \pi_1(Y_{aj}; x_i) \cong \pi_1(Y_{aj}; y_{aj}) \;=\; \pi_1(Y_{aj})$$

of fundamental groups, and hence of their group rings. We can use van Kampen's theorem to compute $\pi_1(Y)$ in terms of the others. Note also that Γ is a retract of Y, and hence connected. Approximately, we can map a collar neighbourhood $X_i \times I$ of X_i, via projection on I, onto the edge through X_i, and the rest of Y_{aj} onto y_{aj} (rigorously, this involves replacing Y by a homotopy equivalent complex, and then applying homotopy extension several times).

We will also write $N_{aj} = \phi_a^{-1}(Y_{aj})$; we may suppose N_{aj} connected, and construct a graph Γ as above in N.

LEMMA 8.2. *After performing preliminary surgeries on M and N_1, we may suppose that ϕ induces a $(k-1)$-connected map $M \to X$ and a k-connected map $N_1 \to Y_1$, and that $K_k(N_1, M; \Lambda_1)$ has a free (preferred) basis represented by framed immersions $(D^k, S^{k-1}) \to (N_1, M)$ whose boundaries are disjoint embeddings $S^{k-1} \to M$ whose classes generate the groups $K_{k-1}(M_i; R_i)$.*

This follows immediately from (1.4) and the proof of (8.1). Note that we can start with any (finite) set of disjoint framed embeddings $S^{k-1} \to M$ whose classes give generators, and use only these and some further trivial embeddings. Observe also that connectedness of N_1 is not important: we can, as usual, take the components separately.

We can now argue in the same way with N_2 (further surgeries can be supposed to keep M fixed), and extend the same framed embeddings $S^{k-1} \to M$ on the 'other side'.

LEMMA 8.3. *After preliminary surgeries, we may construct a (finite) set of framed immersions*

$$(S^k; D^k_-, D^k_+; S^{k-1}) \to (N; N_1, N_2; M)$$

such that the induced framed embeddings $S^{k-1} \to M$ *generate the groups* K_{k-1} *of the various components of* M*; and for each component* N_{aj} *of* N_1 *or* N_2*, the immersions which meet* N_{aj} *provide a preferred free base for* $K_k(N_{aj}, N_{aj} \cap M; \Lambda_{aj})$.

Proof By (8.2), if we start with framed embeddings $S^{k-1} \to M$ which generate the appropriate K_{k-1} groups, we can extend to framed immersions $(D^k, S^{k-1}) \to (N_{aj}, M)$ giving a preferred free base of $K_k(N_{aj}, N_{aj} \cap M; \Lambda_{aj})$ at the expense of performing some surgeries to replace N_{aj} by its connected sum with some copies of $S^k \times S^k$, and adjoining trivial framed embeddings of S^{k-1} in some components (we may choose where) of $N_{aj} \cap M$.

To complete the matching up, we must just be careful how many trivial embeddings of S^{k-1} we want in each component of M. Choose a maximal tree T in the graph Γ, and order the Y_{aj} so that T is formed by starting with the first, and attaching at each stage an edge joining the existing tree to the first vertex not in it. At present, we have plenty of framed immersions of (D^k, S^{k-1}), but some trivial spheres in M span a disc only in N_1 or N_2, but not both. We start with the last N_{aj}, add to it a suitable number of copies of $S^k \times S^k$, each increasing the rank of $K_k(N_{aj}, N_{aj} \cap M; \Lambda_{aj})$ by 2, and thus requiring two new discs. One of these can be attached to each trivial sphere in $N_{aj} \cap M$ which does not already span a disc in N_{aj}; as we were forced to have an even number, there may be one left over, which we assign to have boundary in that M_i such that x_i is the last edge of T.

Proceed thus with the N_{aj}, till we have done the operation with the second. As each edge x_i has two ends, each M_i has now been considered, and all spheres span a disc on each side, with the possible exception of one sphere in the component M_i corresponding to the first edge of the tree T. We show that this exceptional case cannot occur.

For if it does, the total number of discs is odd. But this is the sum of the ranks of the $K_k(N_{aj}, N_{aj} \cap M; \Lambda_{aj})$, or equivalently, of the $K_k(N_{aj}, N_{aj} \cap M)$, and hence of $K_k(N_1, M)$ and $K_k(N_2, M)$. But this latter is dual to $K_k(N_2)$, and the short exact sequence

$$0 \to K_k(N_2) \to K_k(N) \to K_k(N_1, M) \to 0$$

now shows that $K_k(N)$ has odd rank, contradicting [W18, (4.7)]. This concludes the proof of the lemma.

We will from now on use the immersions provided for us by (8.3). We may suppose given paths joining the images of S^{k-1} to the base points in the M_i; and paths joining base points giving an embedding of Γ in N as we had in Y. As each $\Gamma \cap N_{aj}$ is contractible, we can regard it as a somewhat enlarged base point for N_{aj}, and hence obtain lifts of our discs. But Γ is not in general contractible: we will restrict ourselves for simplicity to the case when it is, as this suffices for our application. Then our immersions and paths do define elements of $K_k(N)$, $K_k(N_1, M)$ and $K_k(N_2, M)$: we will use the bases of $K_k(N_1, M)$ and $K_k(N_2, M)$ provided by these elements; by construction these are preferred bases. We use the dual bases of $K_k(N_1)$ and $K_k(N_2)$: since the duality for N_a is defined over the rings Λ_{aj}, these bases come from bases of $K_k(N_{aj};\Lambda_{aj})$.

Now we have a hermitian form (λ,μ) on $K_k(N_a)$: we will denote by A_a the matrix of λ_a, which is also the matrix of the map $K_k(N_a)\to K_k(N_a,M)$ with respect to the chosen bases.

LEMMA 8.4. *The immersions define elements of $K_k(N)$, which split the exact sequence*

$$0\to K_k(N_1)\to K_k(N)\to K_k(N_2,M)\to 0$$

and thus define a preferred basis of $K_k(N)$. With respect to this basis, intersections λ on $K_k(N)$ have a matrix

$$A=\begin{pmatrix} A_1 & I \\ (-1)^k I & (-1)^k B\end{pmatrix},$$

where B has the form $C+(-1)^k C^*$. Moreover, if $D=I-BA_1$, then D is the matrix of a simple isomorphism and $A_1=-A_2D$.

Proof The leading term is the matrix of intersection numbers on $K_k(N_1)$, which is by definition A_1. The upper right hand entry is the unit matrix, since we chose the basis of $K_k(N_1)$ dual to the basis provided by our immersions (for $K_k(N_1,M)$). The lower left hand entry appears by symmetry. If we now denote the remaining term by $(-1)^kB$, it follows by (5.2) that B has the form stated.

It also follows from §5 that A is the matrix of a simple isomorphism. Since

$$\begin{pmatrix} I & 0 \\ (-1)^{k-1}B & I\end{pmatrix}\begin{pmatrix} A_1 & I \\ (-1)^k I & (-1)^k B\end{pmatrix}=\begin{pmatrix} A_1 & I \\ (-1)^k(I-BA_1) & 0\end{pmatrix},$$

it follows that $D=I-BA_1$ also defines a simple isomorphism. It also follows that

$$A^{-1}=\begin{pmatrix} 0 & (-1)^kD^{-1} \\ I & (-1)^{k-1}A_1D^{-1}\end{pmatrix}\begin{pmatrix} I & 0 \\ (-1)^{k-1}B & I\end{pmatrix}$$

$$=\begin{pmatrix} -D^{-1}B & (-1)^kD^{-1} \\ I+A_1D^{-1}B & (-1)^{k-1}A_1D^{-1}\end{pmatrix}.$$

Now the dual of the above exact sequence is

$$0\leftarrow K_k(N_1,M)\leftarrow K_k(N)\rightleftarrows K_k(N_2)\leftarrow 0.$$

Note, however, that this splitting does not coincide with that induced by our geometrically constructed map $K_k(N_1,M)\to K_k(N)$, and the matrix A_{-1} itself represents the change of basis. Thus intersection numbers with respect to the new basis have matrix $(A^{-1})^*A(A^{-1})=(-1)^kA^{-1}$. In particular, the matrix A_2 of intersection numbers on $K_k(N_2)$ is $-A_1D^{-1}$.

COROLLARY 8.4.1. *Take the pairs of lagrangians provided by* (8.1) *from* N_1 *and* N_2*; change the orientation of* N_2*. Then the matrix* $\begin{pmatrix} I & -B \\ 0 & I \end{pmatrix}$ *gives a (simple) isomorphism of hyperbolic forms taking the first pair to the second.*

For since B has the form $C + (-1)^k C^*$, our matrix does give a (simple) isomorphism of hyperbolic forms, which clearly takes the (standard) lagrangian $K_k(N_1)$ to $K_k(N_2)$ (even preserving the chosen bases). That it takes the graph of $A\lambda_1$ to that of $-A\lambda_2$ follows from the identity

$$\begin{pmatrix} I & -B \\ 0 & I \end{pmatrix}\begin{pmatrix} I \\ A_1 \end{pmatrix} = \begin{pmatrix} I - BA_1 \\ A_1 \end{pmatrix} = \begin{pmatrix} I \\ -A_2 \end{pmatrix} D\,.$$

We have not discussed above the corresponding forms μ. As usual, it is sufficient to describe what happens on basis elements. On those coming from $K_k(N_1)$, of course, we have the form μ_1. The rest determine the diagonal terms of a suitable matrix C above; they must be calculated geometrically.

In applying this result, we need to know something about the matrix B of intersections of the immersions constructed in (8.3); also about their self-intersections. Since the equatorial spheres S^{k-1} are disjointly embedded, these each split as a sum of two terms, one coming from intersections in N_1, the other from intersections in N_2. With the definitions above, the first term comes from the relevant Λ_{1j}; the second from some Λ_{2j}: we do not get arbitrary elements of Λ. Also if (for example) $\pi_1(X_1) \cong \pi_1(Y_{2i})$ for some i, then we can (by §4) manoeuvre each framed immersion $(D^k, S^{k-1}) \to (Y_{2i}, X_i)$ to be an embedding, and all these are disjoint, at the expense of introducing extra intersections in the 'N_1-part' of the corresponding spheres.

I do not see at present how to avoid these geometrical difficulties. One possible procedure is as follows. Given the (connected) manifold N_1, one should not regard λ_1 and μ_1 as an adequate system of invariants. Instead, choose framed immersions $(D^k, S^{k-1}) \to (N_1, M)$ as in (8.2), with boundaries disjointly embedded, and defining a preferred basis of $K_k(N_1, M; \Lambda_1)$. One can then take the intersections and self-intersections *of these discs* as extra invariants. These are not well-defined, even by the homotopy classes of the discs: some intersections $\big($coming from $\pi_1(M_i)\big)$ can be 'pushed off the edge' M_i of N. Also, they are not independent of each other and of λ_1 and μ_1: thus further investigation is needed. I will not pursue such investigations: as mentioned earlier, I now believe that there is probably a better way around these difficulties. See §17G.

9. Completion of Proof in the General Case

First-time readers may omit this chapter, proceeding directly to §10.

We now introduce an entirely different approach to our main theorem. We begin by defining a rather complicated bordism group, and then proceed to interpret it. Our main theorem will then follow by applying the special case in §4. We start with the simplest case.

Let us define an 'object' to consist of the following:

> a simple Poincaré pair (Y, X) and bundle ν over Y, compact manifold N with boundary M, with $\dim N = n$,
>
> a map $\phi : (N, M) \to (Y, X)$ of pairs of degree 1, including a simple homotopy equivalence $M \to X$,
>
> a stable framing F of $\tau_N \oplus \phi^*\nu$, and finally
>
> a map $\omega : Y \to K$, such that w_Y factorises as $\pi_1(Y) \xrightarrow{\omega_*} \pi_1(K) \to \{\pm 1\}$.

Here K is a *CW* complex which will usually be taken to be an Eilenberg-MacLane space of type $(\pi, 1)$; but is, anyway, fixed. The fundamental classes $[N]$, $[Y]$ (with $\phi_*[N] = [Y]$) constitute part of the structure of the object θ: if we change their signs, we obtain a new object which I will denote by $-\theta$. Given two objects θ_1, θ_2 (all the above, except K, acquire subscripts 1, 2 in our notation), we may suppose after inessential changes that the various sets involved are disjoint, and then define $\theta_1 + \theta_2$ by taking the unions for Y, X, N and M, and the obvious bundles and maps for ν, ϕ, F and w. The sum operation is commutative and associative, and admits a zero element: the object with Y (hence, also, N, M and X) empty.

We will now define a relation on our sets of objects. We write $\theta \sim 0$ to denote that we can construct the following:

> a simple Poincaré triad $(Z; Y, Y_+)$ with $Y \cap Y_+ = X$, and a bundle μ over Z extending ν,
>
> a compact manifold triad $(P; N, N_+)$ with $N \cap N_+ = M$,
>
> a map $\psi : (P; N, N_+) \to (Z; Y, Y_+)$ of degree 1 extending ϕ, and inducing a simple homotopy equivalence $N_+ \to Y_+$,
>
> a stable framing G of $\tau_P \oplus \psi^*\mu$, stably extending F, and an extension of ω to a map $\Omega : Z \to K$, such that w_Z factorises as $\pi_1(Z) \xrightarrow{\Omega_*} \pi_1(K) \to \{\pm 1\}$.

We further write $\theta_1 \sim \theta_2$ if $\theta_1 + (-\theta_2) \sim 0$. Observe that this is consistent with our former definition in the case $\theta_2 = 0$, for $\theta_1 + (-0) \sim 0$ means just $\theta_1 \sim 0$.

LEMMA 9.1. *$\sim$ is an equivalence relation.*

Proof By multiplying N, M, Y and X by I, we see that $\sim$ is reflexive. Symmetry follows by changing the orientation of P. Now we merely need the usual glueing argument which shows that any sort of cobordism gives an equivalence relation (cf. [W13, VA, 1.1]). Here, given constructs as above (suffixed by 1 and 2) to show $\theta_1 + (-\theta_2) \sim 0$ and $\theta_2 + (-\theta_3) \sim 0$ respectively, we can set $Z_3 = Z_1 \cup Z_2$ (assuming that $Z_1 \cap Z_2 = Y_2$ with no superfluous intersection), $Y_{+3} = Y_{+1} \cup Y_{+2}$ (where $Y_{+1} \cap Y_{+2} = X_2$). Since μ_1 and μ_2 both extend ν_2 on the intersection, they combine to give a bundle μ_3. So we have a triad $(Z_3; Y_1 \cup Y_3, Y_{+3})$, which is a Poincaré triad by (2.7). Similarly we construct a compact manifold triad $(P_3; N_1 \cup N_3, N_{+3})$ as for ordinary cobordism. The maps ψ_1 and ψ_2 agree on $P_1 \cap P_2 = N_2$, so combine to give a map ψ_3; similarly we get G_3 and ω_3, and it is immediate that $N_{+3} \to Y_{+3}$ is a simple homotopy equivalence. The result is then established.

We denote the set of equivalence classes by $L_n^1(K)^*$. Evidently, the relation $\sim$ is compatible with $+$ (defined by disjoint union), so the set $L_n^1(K)$ inherits a commutative and associative addition operation with a zero. Moreover, the '$-$' operation gives additive inverses, so $L_n^1(K)$ is an additive abelian group. If $f : K \to L$ is a continuous map, then replacing a representative object over K by one over L in which ω is altered to $f \circ \omega$ is compatible with $+$ and $\sim$, so defines a homomorphism $L_n^1(f) : L_n^1(K) \to L_n^1(L)$. Evidently we have a covariant functor from CW complexes to abelian groups. Moreover, only the homotopy class of f is relevant – since if θ_2 differs from θ_1 only by changing ω_1 in its homotopy class, $\theta_1 \sim \theta_2$. To prove this, we multiply (Y, X) and (N, M) by I, extend ν, ϕ and F by product maps, and extend $\omega_1 \cup \omega_2$ by a homotopy: this constructs a cobordism of the desired kind.

Next, suppose K connected and consider the set of objects which satisfy the additional requirements:

$$Y \text{ is connected, and the map } \omega_* : \pi_1(Y) \to \pi_1(K) \text{ is an isomorphism.}$$

We call such objects *restricted objects.* On these we impose stricter relations. Write $\theta \approx 0$ if we have the data for $\theta \sim 0$ but, in addition, Z is connected, and the map $\Omega_* : \pi_1(Z) \to \pi_1(K)$ is an isomorphism. Analogously we write $\theta_1 \approx \theta_2$ if we have the data for $\theta_1 + (-\theta_2) \sim 0$ satisfying these two extra conditions. Here, there is no object 0, and so no overlap between our two uses of the symbol $\approx$.

LEMMA 9.2. *$\approx$ is an equivalence relation. The objects θ with $\theta \approx 0$ constitute an equivalence class.*

Proof The verification of the first assertion proceeds as for Lemma 1: the only difference is that a certain map must be shown to be an isomorphism: the proof is immediate, using van Kampen's theorem. For the second, we must show that $\theta_1 \approx \theta_2 \approx 0$ implies $\theta_1 \approx 0$. Almost the same argument applies here: we glue

** See Farrell and Hsiang [F5, p.102] for a revised definition of $L_n^1(K)$ in the nonorientable case, replacing the orientation map $w : \pi_1(K) \to \{\pm 1\}$ by a principal $\mathbf{Z}_2$-bundle over K. This ensures that the isomorphism of 9.4.1 is functorial.*

together systems representing 'cobordisms' of θ_1 to θ_2 and of θ_2 to zero.

We write $L_n^2(K)$ for the set of equivalence classes of restricted objects under $\approx$. We make no claims yet concerning group structure here, but content ourselves to note that there is a natural map $L_n^2(K) \to L_n^1(K)$, defined by forgetting the extra conditions. The relevance of all this to our problem can now be indicated. If θ is an object, then (N, M, ϕ, F) defines a class $\xi(\theta)$ of degree 1 in $\Omega_*(Y, \nu)$ relative to $\phi \mid M$.

THEOREM 9.3. *Let θ be a restricted object with $n \geqslant 5$. Then $\xi(\theta)$ has a representative with ϕ a simple homotopy equivalence of pairs if and only if $\theta \approx 0$.*

Proof First suppose $\xi(\theta)$ has such a representative. Then the cobordism between it and (N, M, ϕ, F) consists of a manifold triad $(P; N, N_+)$, an extension of ϕ to a map $\psi' : (P, M) \to (Y, X)$ inducing a simple homotopy equivalence $N_+ \to Y$, and a (stable) extension of F to a stable framing G of $\tau_P \oplus \psi^*\nu$. Now define a Poincaré triad by taking $Y \times I$, and identifying $X \times I$ with X (so obtaining the mapping cylinder rel X of the identity map of Y) giving Z, say: take $(Z; Y \times 0, Y \times 1)$. It is now easy (invent a map $P - M \to I$) to cover ψ' by a map $\psi : (P; N, N_+) \to (Z; Y \times 0, Y \times 1)$ of degree 1 of triads. Defining Ω to be the composite of ω and the projection of Z on Y, we have now constructed a cobordism which shows $\theta \approx 0$.

Conversely, let $\theta \approx 0$: use the notation of the definition. Consider the map

$$\psi : (P; N, N_+) \to (Z; Y, Y_+) \ :$$

a map of degree 1 of Poincaré triads, provided with the usual bundle μ over Z and stable framing G of $\tau_P \oplus \psi^*\mu$. We have induced simple homotopy equivalences $M \to X$ and $N_+ \to Y_+$; moreover, inclusion induces an isomorphism of $\pi_1(Y)$ on $\pi_1(Z)$. Hence by Theorem 3.3, as $n \geqslant 6$ we can perform surgery to make ψ a simple homotopy equivalence of triads. In particular, we can do surgery for ϕ too. This proves the theorem.

This shows that $L_n^2(K)$ is of interest for our problem, whereas we saw above that $L_n^1(K)$ has good algebraic and functorial properties. The crucial step in the present development is to show that these two are essentially the same.

THEOREM 9.4. *Let $n \geqslant 4$, and let K have a finite 2-skeleton. Then the natural map $L_n^2(K) \to L_n^1(K)$ is a bijection.*

Proof We first consider surjectivity. Suppose given $(Y, X, \nu, N, M, \phi, F, \omega)$ representing an element of $L_n^1(K)$: we seek another representative for which Y is connected and $\omega_* : \pi_1(Y) \to \pi_1(K)$ is an isomorphism. We will construct this by performing surgery on Y. The crucial result for this is (2.8), or rather its relativisation.

The first step in our simplification of the given object is to replace (Y, X) by (Y', X) where $Y' = Y_0 \cup_{\delta H} H$ is as in the comment following (2.8), and modify appropriately the various maps and bundles: clearly, we obtain an equivalent object (cf. the argument above that ω can be replaced by a homotopic map).

We will next perform surgery on the map $\omega' = \omega \,|\, H : H \to K$. Let η be the line bundle over K determined by the homomorphism $w : \pi_1(K) \to \{\pm 1\}$. Since H has the homotopy type of a graph, it follows easily that $\tau_H \oplus \omega'^* \eta$ is trivial: choose a framing f. We can now apply (1.2), and deduce that by a finite sequence of surgeries (adding 1- and 2-handles to $H \times I$, to give a manifold J) we can change H to a manifold H' such that the map $H' \to K$ induces an isomorphism of fundamental groups.

Let $Z = Y_0 \times I \cup J = Y \times I \cup$ 1-handles and 2-handles; $\partial Z = Y \times 0 \cup X \times I \cup Y_+ \times 1$, where $Y_+ = Y_0 \cup H'$. The map $\omega : Y \to K$ restricted to H was shown above (using (1.2)) to extend over J, thus we obtain a map $\Omega : Z \to K$. We will identify $X \times I$ to X in Z, and still write Z for the result. I claim that $\Omega_* : \pi_1(Y_+) \to \pi_1(K)$ is an isomorphism. For we have maps

$$\pi_1(K) \;\cong\; \pi_1(H') \to \pi_1(Y_+) = \pi_1(Y_0) *_{\pi_1(\partial H)} \pi_1(H') \xrightarrow{\Omega_*} \pi_1(K)\,,$$

whose composite is the identity. The first map is surjective since by (2.8)

$$\pi_1(\partial H) = \pi_1(H) \to \pi_1(Y') = \pi_1(Y_0)$$

is surjective. It follows at once that Ω_* is an isomorphism.

We also claim that $(Z; Y, Y_+)$ is a simple Poincaré triad. For it follows from (2.7) that $(Y_0, \partial H)$ is a simple Poincaré pair. But Z is obtained by glueing $(Y_0, \partial H) \times I$ to J along $\partial H \times I$, and all the required dualities now follow from (2.10), which permits multiplying by I, and (2.7)(i), which permits glueing.

We must extend ν over Z. But there is an s-isomorphism of $\nu \,|\, H$ with $\omega'^* \eta$ (since bundles over a 1-complex are classified by the first Stiefel-Whitney class), and the construction of J provided an extension of this bundle. Glueing, we have the desired extension of ν.

Finally, we must construct a cobordism P of N to some N_+, and appropriate extensions ψ of ϕ and G of F. It will be simpler to describe this construction in the case when J is obtained from $H \times I$ by attaching a single 1- or 2-handle; the argument, however, is the same in the general case. First replace ϕ by a homotopic map (this is clearly permissible) which is transverse regular on the attaching sphere, S^0 or S^1, of the handle. The preimage of this sphere is then a framed submanifold S of dimension 0 or 1. Since ϕ is then automatically t-regular on some neighbourhood of S^0 or S^1, we can assume after, if necessary, shrinking the neighbourhood $S^0 \times D^n$ or $S^1 \times D^{n-1}$ on which surgery is performed, that ϕ is t-regular on all of this set, and that its preimage has components homeomorphic to D^n or to $S^1 \times D^{n-1}$. Moreover, the restriction of ϕ to a component of type D^n is a diffeomorphism onto a component of $S^0 \times D^n$, and the restriction to $S^1 \times D^{n-1}$ is represented by the product of an immersion $S^1 \to S^1$ and the identity map of D^{n-1}.

We first consider a 1-handle. Since ϕ has degree 1, and is t-regular at S^0, the total multiplicity of the preimage of each component is $+1$. Let $S^0 = \{x, y\}$: choose $\xi \in \phi^{-1}\{x\}$, $\eta \in \phi^{-1}\{y\}$ each with multiplicity $+1$, perform surgery on the corresponding discs, and extend ϕ using the identity map on $D^1 \times D^n$, and F as usual. The other points of the preimage can be arranged in complementary pairs, the two members of a pair having the same image, but mapped with opposite degrees. We add a further copy of $D^1 \times D^n$ along each such pair, and extend ϕ by the projection on D^n, composed with $D^n \subset H$. Since the degrees were opposite, we can again extend F.

Now consider a 2-handle: this comes from a framed embedding of S^1 in Y, and we suppose ϕ transverse to S^1, so that the preimage is a framed 1-submanifold of N. Since N is connected, any two components of this can be joined by an arc α; since $\pi(N) \to \pi_1(Y)$ is surjective (because ϕ has degree 1) we can suppose that α projects to a nullhomotopic loop in Y. A standard argument now shows how to modify ϕ (first do a homotopy to make $\phi(\alpha)$ a point) so that $\phi^{-1}(S^1)$ has one less component. Eventually, we can suppose $T = \phi^{-1}(S^1)$ connected. Since ϕ is transverse to S^1, the map $T \to S^1$ has the same degree as $N \to Y$, viz. 1; so a final homotopy makes $\phi \,|\, \phi^{-1}(S^1 \times D^{n-1})$ a homeomorphism. We can now attach a 2-handle to N by the 'same' attaching map, and there is a unique extension of ν over this handle such that F also extends.

This completes the proof that our map is surjective. We will not go into the detailed proof of injectivity, for the same arguments suffice. In fact, if we are given two restricted objects which are equivalent in the unrestricted sense, we must construct a restricted equivalence. This construction is the same as above, only we must do surgery on Z and P instead of Y and N: the same arguments are used to justify the steps. The theorem follows.

We pause here to note that the above proof is not valid if K is a space other than a CW complex with finite 2-skeleton; clearly however, the important condition is that $\pi_1(K)$ be finitely presented.

COROLLARY 9.4.1. *The group $L_n^1(K)$ is isomorphic to $L_n(\pi_1(K), w)$ if $n \geqslant 5$.*

Proof We have a bijection $L_n^2(K) \to L_n^1(K)$, by the theorem. We can also define a homomorphism $L_n^1(K) \to L_n(\pi_1(K), w)$ as follows: given an object θ, we perform surgery on ϕ to make it k-connected (where $n = 2k$ or $2k+1$), and then compute (by §4 or §5) a surgery obstruction in $L_n(\pi_1(Y), w \circ \omega_*)$, and take its image by ω_* in $L_n(\pi_1(K), w)$: this amounts to computing the surgery obstructions with coefficients in $\mathbf{Z}[\pi_1(K)]$. The result is evidently additive for disjoint unions, and changes sign on going from θ to $-\theta$. Thus to obtain a homomorphism of the group $L_n^1(K)$, it is sufficient to show that if $\theta \sim 0$ we must necessarily obtain the zero surgery obstruction. But by (5.7) and (8.1) the surgery obstruction has zero image in $L_n(\pi_1(Z), w \circ \Omega_*)$, so *a fortiori* in $L_n(\pi_1(K), w)$.

Now Theorem 3 shows that the composite map

$$L_n^2(K) \to L_n^1(K) \to L_n(\pi_1(K), w)$$

is injective, hence (by Theorem 4) so is the second factor. As to surjectivity, let us begin with a constant map $S^{n-1} \to K$, and perform surgery (using the bundle η on K) to obtain a map $a : A^{n-1} \to K$ for which $a_* : \pi_1(A) \cong \pi_1(K)$ (and w satisfies the usual condition). Now by (5.8) and (6.5), any element of $L_n\big(\pi_1(A), w \circ a_*\big)$ is represented by a map of the type

$$\phi : (B, A \cup A') \to (A \times I, A \times 0 \cup A \times 1)$$

(with appropriate ν and F). Set $\omega = a \circ p_1 : A \times I \to K$: then we obtain an object (of restricted type) with the given image in $L_n\big(\pi_1(K), w\big)$.

At this stage, we have merely achieved a reformulation – indeed a weakening – of our original results. However, we now extend the above in three ways, all possible because we have arranged the totality of problems in a convenient way. First, we will relativise. This will enable us to complete the proofs of (3.1) and (3.2), except for periodicity. Next, we will consider the natural transformations of functors induced by multiplying K (and the other complexes) by some fixed closed compact manifold. Finally, we will consider some particular natural transformations induced by taking maps transverse regular on a submanifold of K, in the case when K itself is a manifold: very special cases of these ideas are used in §12 to obtain some computational results.

Let K be a CW n-ad, $w^1 \in H^1(|K|; \mathbf{Z}_2)$ $\big($if $|K|$ is connected, w^1 is equivalent to a homomorphism $w : \pi_1(|K|) \to \{\pm 1\}\big)$. An object (of type n) over (K, w^1) shall consist of:

> a simple Poincaré $(n+1)$-ad X, a bundle ν over $|X|$, and a compact manifold $(n+1)$-ad M,
>
> a map $\phi : M \to X$ of $(n+1)$-ads of degree 1 inducing a simple homotopy equivalence of n-ads $\partial_n M \to \partial_n X$,
>
> a bundle μ over $|X|$ and a stable framing F of $\tau(|M|) \oplus \phi^*\mu$, and
>
> a map $\omega : X \to s_n K$ of $(n+1)$-ads with $w^1(|X|) = \omega^* w^1$.

Changing the signs of the fundamental classes in an object θ gives a new object $-\theta$. Given two objects θ_1, θ_2 over (K, w^1) we define the object $\theta_1 + \theta_2$ by taking the disjoint union of the corresponding X, M, ϕ, ν and F, and using the two maps ω from θ_1 and θ_2. We set $\theta \sim 0$ if we can find:

> a simple Poincaré $(n+2)$-ad Y, and a compact manifold $(n+2)$-ad N with $X = \partial_{n+1} Y$, $M = \partial_{n+1} N$,
>
> an extension of ϕ to a map $\psi : N \to Y$ of degree 1 inducing a simple homotopy equivalence of $(n+1)$-ads $\partial_n N \to \partial_n Y$,
>
> a bundle ν over $|Y|$ extending μ, and an extension of F to a stable framing G of $\tau(|N|) \oplus \psi^*\nu$, and
>
> a map $\Omega : Y \to s_n^2 K$ of $(n+2)$-ads, extending ω, with $w^1(|Y|) = \Omega^* w^1$.

Again writing $\theta_1 \sim \theta_2$ for $\theta_1 + (-\theta_2) \sim 0$ gives an equivalence relation, and the set $L_m^1(K, w^1)$ of equivalence classes has a natural abelian group structure: the proof is the same as before. Clearly, L_m^1 is a functor on the category of CW

n-ads (with map w^1) and homotopy classes of maps.

An object will be called *restricted* if for each $\alpha \subset \{1, 2, \ldots, n-1\}$ the maps $\omega_{\alpha n} : X_{\alpha n} \to K'_{\alpha n} = K_\alpha$ induce equivalences $\pi(\omega_{\alpha n})$ of corresponding fundamental groupoids. Again, if θ is a restricted object, we write $\theta \approx 0$ if we have the data above for $\theta \sim 0$ satisfying the additional requirement that for each $\alpha \subset \{1, 2, \ldots, n-1\}$ the map

$$\Omega_{\alpha,n,n+1} : T_{\alpha,n,n+1} \to s_n^2 K_{\alpha,n,n+1} = K_\alpha$$

induces equivalences of corresponding fundamental groupoids. We define $\theta_1 \approx \theta_2$ analogously. As before, one shows that $\approx$ is an equivalence relation and that the objects ≈ 0 form an equivalence class. We obtain a natural map of the set $L_m^2(K, w^1)$ of equivalence classes to $L_m^1(K, w^1)$.

Theorems 9.3 and 9.4 generalise to

THEOREM 9.5. *If $m - n \geqslant 3$, and $|K|$ has finite 2-skeleton, the natural map $L_m^2(K, w^1) \to L_m^1(K, w^1)$ is a bijection. If $m - n \geqslant 4$, then given a restricted object θ we can perform surgery to make ϕ a simple homotopy equivalence of $(n+1)$-ads if and only if $\theta \approx 0$ ($\Longleftrightarrow \theta \sim 0$, by the first part).*

Proof These results follow by induction from Theorems 3 and 4. In each case we wish to perform surgery on an $(n+1)$-ad: to perform surgery on $X_{(\alpha)}$ (or $M_{(\alpha)}$), we may suppose by induction that for all $\beta \subseteq \alpha$ we have already performed surgery on $X_{(\beta)}$ (or $M_{(\beta)}$). Then considering $X_{(\alpha)}$ (or $M_{(\alpha)}$) modulo $\partial X_{(\alpha)}$ or $\partial M_{(\alpha)}$, we have the hypotheses necessary for the proof in Theorem 9.3 or 9.4. Our assertions now follow.

We now obtain the first payoff for our cobordism approach with a painless derivation of exact sequences.

THEOREM 9.6. *For any CW $(n+1)$-ad K, and $1 \leqslant i \leqslant n$, we have a natural exact sequence*

$$\cdots \to L_m^1(\partial_n K, w^1) \xrightarrow{i_*} L_m^1(\delta_n K, w^1) \xrightarrow{j_*} L_m^1(K, w^1) \xrightarrow{d_*} L_{m-1}^1(\partial_n K, w^1) \to \ldots$$

More precisely, i_* is induced by inclusion, j_* is the composite of a natural equivalence $L_m^1(\delta_n K, w^1) \xrightarrow{\alpha} L_m^1(\sigma_n \delta_n K, w^1)$ and a morphism induced by the inclusion $\sigma_n \delta_n K \subset K$. The sequence is functorial in (K, w^1).

Proof The natural transformation α is obtained by regarding an object of type $(n-1)$ as an object of type n by adjoining (as new distinguished subset) the empty set. It is immediate that this is a natural equivalence. Similarly we define d_* by taking a map $\phi : M \to X$ of $(n+1)$-ads, and considering the restriction $\partial_n \phi : \partial_n M \to \partial_n X$: the same construction goes for cobordism, as we have a map of equivalence classes, clearly a homomorphism. It thus remains only to prove exactness of the sequence.

This is now a standard argument [W13, Part VA, (2.3)] obscured only by the complexity of our definitions. An object of type $(n-1)$ over $\partial_n K$ is trivial, when regarded as object over $\delta_n K$, if and only if there is a cobordism to the empty set. But such a cobordism can be regarded just as an object of type n over K, with the given object as boundary. This proves exactness at $\partial_n K$.

Exactness at K follows since $\partial_* j_* = 0$ (trivially), and if an object of type n over K has trivial boundary, we can take a cobordism to zero of the boundary and glue it to the object (as in §1, glueing of Poincaré complexes (and $(n+1)$-ads) is justified by (2.7)) to obtain a new object, cobordant to the given one, which has $\partial_n M$ empty and so is the image by α of an object of type $(n-1)$ over $\delta_n K$.

Finally $j_* i_* = 0$, for if $\phi : M \to X$ is (part of) an object of type $(n-1)$ over $\partial_n K$, and we consider its image, an object of type n over K, then a cobordism to zero is provided by $\phi \times 1 : M \times I \to X \times I$, where we set

$$\begin{aligned} \partial_j(M \times I) &= \partial_j M \times I \qquad 1 \leqslant j \leqslant n-1 \\ \partial_n(M \times I) &= M \times 1 \\ \partial_{n+1}(M \times I) &= M \times 0, \text{ and similarly for } X. \end{aligned}$$

Conversely, an object of type $(n-1)$ over $\delta_n K$ is in the kernel of j_* if, regarded as an object of type n over K, it admits a cobordism $\psi : N \to Y$ to zero. Since $\partial_n \partial_{n+1} \psi$ is empty, ψ defines a cobordism (over $\delta_n K$) of the given object (essentially $\partial_{n+1}\psi$) to an object represented by $\partial_n \psi$. But the latter is an object over $\partial_n K$.

This proves the theorem: functorial character is obvious. As in §3, we could use faces other than $\partial_i K$ to obtain exact sequences: the proof is the same, apart from indexing complications. Also note that boundary operators corresponding to different faces commute: they do so already at the level of representative objects. The above proof could also be carried out for restricted objects and the functor L^2, but by (9.5) this gives nothing new.

Let π be a groupoid (of finite type) of type $\mathbf{2}^n$. We define an Eilenberg-MacLane $(n+1)$-ad K of type $(\pi, 1)$ to be a CW $(n+1)$-ad such that each $K\{\alpha\}$ is a (finite) disjoint union of Eilenberg-MacLane spaces of type $(G, 1)$ for appropriate groups G, such that $\pi(K) = \pi$. Existence of these follows from the existence of Eilenberg-MacLane spaces, and from the mapping cube construction of §0. It also follows easily that for any CW n-ad K, there is a map of n-ads, unique up to homotopy, from K to any n-ad of type $(\pi(K), 1)$; in particular, these last are unique up to homotopy.

If π is a groupoid of type $\mathbf{2}^{n-1}$ (with each $\pi\{\alpha\}$ of finite type), and w a function as in §3, we define $L_m(\pi, w)$ by choosing a CW n-ad $K(\pi, 1)$ of type $(\pi, 1)$ and setting $L_m(\pi, w) = L_m^1(K(\pi, 1), w)$.

LEMMA 9.7. *Let $m - n \geqslant 4$, and K be a CW $(n+1)$-ad. Then the homomorphism $L_m^1(K, w^1) \to L_m(\pi(K), w)$ is an isomorphism.*

Proof We use induction on n. If $n = 1$, $m \geqslant 5$. Here, K is just a CW complex; as usual, the components act independently and we may suppose K connected. In this case, the result was proved as (9.4.1) above (where we also showed that this definition of L_m agrees with the old one when $n = 0$).

The general case now follows by induction on n on applying the Five Lemma to the commutative exact diagram

$$\begin{array}{ccccccccc} L^1_m(\partial_i K) & \longrightarrow & L^1_m(\delta_i K) & \longrightarrow & L^1_m(K) & \longrightarrow & L^1_{m-1}(\partial_i K) & \longrightarrow & L^1_{m-1}(\delta_i K) \\ \downarrow & & \downarrow & & \downarrow & & \downarrow & & \downarrow \\ L_m(\pi(\partial_i K)) & \to & L_m(\pi(\delta_i K)) & \to & L_m(\pi(K)) & \to & L_{m-1}(\pi(\partial_i K)) & \to & L_{m-1}(\pi(\delta_i K)). \end{array}$$

We may now conclude the proof of (3.1) and (3.2), except for the statement concerning periodicity. Indeed, (3.1) (with the above definition of $L_m(\pi)$) is an immediate consequence of (9.6).

Proof of (3.2). Given (X, ν, M, ϕ, F) we define a restricted object by taking $K = K(\pi(X), 1)$ and ω the canonical map. If X is an n-ad, then to conform with our definition we must make M, X and K into $(n+1)$-ads by applying σ_n (i.e. defining $\partial_n = \phi$ in each case). By (9.5), surgery is possible if and only if this object defines the zero element of $L^2_m(K)$; but again (again by (9.5)), it suffices to consider the corresponding element of $L^1_m(K) = L_m(\pi(X))$. If, however, X is an $(n+1)$-ad, and $\partial_n\phi$ a simple homotopy equivalence, we simply take $K = K(\pi(X), 1)$ and ω canonical, and have at once a restricted object. The result again follows by (9.5).

The surgery obstruction for $\partial_n\phi$ follows at once from this argument. It remains to show naturality for inclusion maps. Suppose that $M \subset M'$, $X \subset X'$, and $\phi' : M' \to X'$ induces $\phi : M \to X$ (as in the statement in §3). Then ϕ defines an object θ over $K(\pi(X), 1)$, and ϕ' an object θ' over $K(\pi(X'), 1)$. Also, ϕ defines an object θ'' over $K(\pi(X'), 1)$. The usual glueing argument now shows that $\theta' \sim \theta''$, and the class in $L_m(\pi(X'))$ defined by ϕ' is that of θ', hence of θ'', which is the image under the natural map of the class of θ in $L_m(\pi(X))$, as required.

We now introduce products for the $L_m(\pi)$. Our principal objective is to obtain an adequate geometrical formulation of the periodicity phenomenon.

First observe that if θ is an object (of any type) over K, and L is a closed (smooth or PL) manifold, we obtain an object $\theta \times L$ of the same type over $K \times L$ by multiplying X and M by L, ϕ and ω by the identity map of L, and modifying μ, F and w appropriately. In fact more generally, if θ has type n and L is a manifold $(m+1)$-ad, multiplication produces an object of type $(m+n)$. One might perhaps expect to obtain an object of type $(m+n)$ by multiplying objects of types m and n, but this does not happen with our definitions: we do not then have a simple homotopy equivalence on a face of codimension one, only on one of codimension two. It may perhaps be possible to refine our results to obtain a product structure by 'blowing up' this face to a prismatic neighbourhood (as in an example in §3), but we will not attempt this here.

We return to the simplest case, when L is a closed oriented manifold. By projecting $K \times L$ on K, we can then regard $\theta \times L$ as an object over K. If $\theta \sim 0$, multiplying by L shows that $(\theta \times L) \sim 0$. Also if L bounds a compact oriented (smooth or PL) manifold R, we can multiply the $(n+1)$-ads M and X by the 2-ad (R, L) to obtain $(n+2)$-ads N and Y; using again the obvious ψ, ν, G and Ω, this again implies $(\theta \times L) \sim 0$. As our product clearly distributes over disjoint unions of θ's or L's, it follows that

PROPOSITION 9.8. *For any (K, w^1) there is a natural bilinear map*

$$\Omega_r \times L_s^{\,1}(K, w^1) \to L_{r+s}^{1}(K, w^1)\,.$$

We will not determine the pairing in this paper*. We think that if 'objects' were generalised to permit M also to be a simple Poincaré $(n+1)$-ad, the groups L would be changed only by isomorphisms. If this is the case, Ω_r can be replaced by the cobordism group of oriented Poincaré complexes, which is probably much smaller. It may well be that only the signature of L is relevant;† certainly this is the case modulo torsion (as can be seen by modifying the proofs below). Here, we confine ourselves to studying periodicity.

THEOREM 9.9. *Suppose $m \geqslant 5$ and π a group. Then the map*

$$L_m(\pi) \to L_{m+4}(\pi)$$

induced by multiplying by $P_2(\mathbf{C})$ coincides with the periodicity isomorphism of §5 or §6.

Proof First take $m = 2k$. Any element of $L_m(\pi)$ can be represented by a map $\phi : M \to X$ of degree 1 (with ν and F as usual) inducing a simple homotopy equivalence on the boundary. We may suppose ϕ k-connected and $K_k(M)$ free, so the obstruction is represented by a simple hermitian form on $K_k(M)$. We choose a preferred basis of $K_k(M)$, and represent it by immersions $f_i : S^k \times D^k \to M$ in general position.

Now multiply by $P = P_2(\mathbf{C})$. By the Künneth theorem (over $\mathbf{Z}$), $K_i(M \times P)$ vanishes except for $i = k$, $k+2$, $k+4$, and

$$K_k(M \times P) \;\cong\; K_{k+2}(M \times P) \;\cong\; K_{k+4}(M \times P) \;\cong\; K_k(M)\,.$$

We will first perform surgery to kill the basis elements of $K_k(M \times P)$ and show that this leads to a manifold N whose only nonvanishing group K_i is $K_{k+2}(N) = K_{k+2}(M \times P) \cong K_k(M)$. Next we will show that this isomorphism preserves simple hermitian forms; the result will clearly follow from this.

The f_i induce immersions

$$f_i \times 1 : S^k \times D^k \times P \to M \times P\,.$$

Let $S^2 \subset P$ be a projective line, with tubular neighbourhood N (the normal bundle, of course, is nontrivial). We have induced immersions of $S^k \times S^2$ which are not, however, in general position. Since two projective lines in P intersect

* *See the notes at the end of §17G for the surgery product formula.*

† It appears that the Euler characteristic of L is relevant also: see §13, especially (13A.5). For Poincaré cobordism, see §17C.

just once with intersection number $+1$, we can move these to immersions $g_i : S^k \times S^2 \to M \times P$ in general position, with intersections and self-intersections corresponding exactly to those of the f_i in M. We may also suppose the $g_i(S^k \times 1)$ disjointly embedded: they inherit framings from the framings of the f_i and of a point in P.

Next perform framed surgery on the spheres $g_i(S^k \times 1)$: suppose W formed by attaching corresponding $(k+1)$-handles to $M \times P$: W is a cobordism of this to N, say. By construction,

$$K_{k+1}(W, M \times P) \to K_k(M \times P)$$

is an isomorphism (and a simple one); since other $K_i(W, M \times P)$ vanish, the only nonvanishing $K_i(W)$ are with $i = k+2$, $k+4$. By duality,

$$K_{k+4}(M \times P) \;\cong\; K_{k+4}(W) \to K_{k+4}(W, N)$$

is also a (simple) isomorphism. We deduce that the only nonvanishing $K_i(N)$ is $K_{k+2}(N) \cong K_{k+2}(W) \cong K_{k+2}(M \times P) \cong K_k(M)$.

The proof will now be concluded by showing that the surgery 'converts' our immersions g_i into framed immersions of S^{k+2}, with the same intersection and self-intersection invariants as the g_i. As these represent the corresponding homology classes, it will follow that our isomorphism of $K_k(M)$ on $K_{k+2}(N)$ preserves simple hermitian forms.

Recall that the definition of the framing of the framed immersions f_i $\big(\text{in}(1.1)\big)$ was by compatibility of framings on an added disc D^{k+1}, mapped to X by a component of the relative homotopy class defining f_i. Since S^2 has nontrivial normal bundle in P, if we contemplate attaching to $M \times P \times I$ a thickening of $D^{k+1} \times S^2$, this also will have to have nontrivial normal bundle, but it is induced by projection on S^2.

In fact we attach $D^{k+1} \times S^2$ in two stages, each $D^{k+1} \times D^2$. We performed framed surgery on $g_i(S^k \times 1)$, using the framing f_i and the normal framing of 1 in P. Decompose the latter as the sum of a normal framing of 1 in S^2 (with neighbourhood D^2_-, say) and a normal framing of D^2_- in P. The resulting $(k+1)$-handle is a thickening of $D^{k+1} \times 1$, but can also be thought of as a thickening of $D^{k+1} \times D^2_-$. Thus g_i in $M \times P$ is converted to the sphere $g_i(S^k \times D^2_+) \cup (D^{k+1} \times \partial D^2_+)$ in N (in the smooth case, we will have to round off a number of corners for this argument). This is framed as follows. To construct 'ν' for $X \times P$, we must take the direct sum of the original ν, and a normal bundle ρ for P. Correspondingly to F we add a trivialisation of $\tau_P \oplus \rho$. Over each of D^2_-, D^2_+ the latter becomes (stably) the natural framing of the tangent bundle; hence also over each of $D^{k+1} \times D^2_-$, $D^{k+1} \times D^2_+$, which are thus framed to satisfy the conditions of (1.1). Hence we do have framed immersions in N, as stated.

Now suppose m is odd; write $m = 1$, $k \geqslant 3$. We represent an element of $L_m(\pi)$ by a $(k-1)$-connected map $\phi : M \to X$, and $K_{k_1}(M)$ by disjoint embeddings

$f_i : S^{k-1} \times D^k \to M$, with images forming U and $M_0 = M - \text{Int } U$. Then, algebraically, the element is represented by the pair of lagrangians $K_k(U, \partial U)$ and $K_k(M_0, \partial U)$ in $K_{k-1}(\partial U)$.

On multiplying by P we find that $K_i(M \times P)$ is nonzero for $k - 1 \leqslant i \leqslant k + 4$. As above, we perform surgery on spheres $f_i(S^{k-1}) \times$ point, thus killing $K_{k-1}(M \times P)$ and again this converts the $f_i(S^{k-1}) \times S^2$ into framed spheres $g_i(S^{k+1} \times D^{k+2})$, whose classes generate K_{k+1} of the resulting manifold N', which is naturally isomorphic to $K_{k+1}(M \times P) \cong K_{k-1}(M)$.

We cannot yet compute in N', as $K_k(N') \neq 0$: we must next kill this. We claim that $K_k(N')$ is stably free and based; in fact, is canonically isomorphic to $K_k(M_0, \partial U) = K_k(M, U)$. For if W' is the cobordism of $M \times P$ to N', then W' is formed from $M \times P \times I$ by adding k-handles via the f_i, i.e. to the k-cells of U. Thus $K_k(W') = K_k(M \times P, U) \cong K_k(M, U)$. As W' is formed from N' by adding $(k + 4)$-handles, $K_k(W') = K_k(N')$.

We may suppose, as usual, $K_k(M_0, \partial U)$ free and based; then so is $K_k(N')$. Perform surgery on the elements of a free basis, choosing our spheres disjoint from the g_i, thus obtaining a cobordism W of N' to N, say. As, by construction, $K_{k+1}(W, N') \to K_k(N')$ is an isomorphism, and by duality $K_{k+3}(N') \to K_{k+3}(W)$ also is, we find that $K_i(N)$ vanishes except for $i = k + 1$, $k + 2$; and in these dimensions,

$$K_i(N) \cong K_i(W) \cong K_i(N') \cong K_i(W') \cong K_i(M \times P) \cong K_{i-2}(M).$$

Since the surgery from N' to N left the g_i alone, these still define embeddings in N, which generate $K_{k+1}(N) = K_{k+1}(N')$. Let the union of their images be V. We define an isomorphism of $K_{k-1}(\partial U)$ on $K_{k+1}(\partial V)$ by taking the class of $f_i(S^{k-1} \times 1)$ to that of $g_i(S^{k+1} \times 1)$, and the class of $f_i(1 \times S^{k-1})$ to that of $g_i(1 \times S^{k+1})$. Clearly, this carries $K_k(U, \partial U)$ to $K_{k+2}(V, \partial V)$ by a simple isomorphism. If we can show that it does the same for $K_k(M_0, \partial U)$ and $K_{k+2}(N_0, \partial V)$, the result will be established.

Well, consider $(M_0, \partial U) \times P$. Certainly the natural isomorphism $K_{k-1}(\partial U) \cong K_{k+1}(\partial U \times P)$ induces a (simple) isomorphism of $K_k(M_0, \partial U)$ on $K_{k+2}(M_0 \times P, \partial U \times P) = K_{k+2}(M \times P, U \times P)$. Here we can replace $U \times P$ by the product of U with a (tubular or regular) neighbourhood T of S^2 in P. Now surgery is performed inside $U \times T$, and has the effect of replacing each copy of $S^{k-1} \times D^k \times T$ by a manifold V' of the homotopy type of $S^2 \vee S^{k+1} \vee S^{k+3}$: here, the 2-sphere comes from T, the $(k + 3)$-sphere is the transverse sphere of the surgery, and the $(k + 1)$-sphere is the one constructed earlier, and used to define V. If we shrink the tubes V a little to have V in the interior of V', then $V' - \text{Int } V$ gives a cobordism of ∂V to $\partial V'$. It is easily verified that the inclusion of either end in this cobordism induces an isomorphism of $(k + 1)$st homology (which is free of rank 2). More precisely, the generators of $H_{k+1}(\partial V') = H_{k+1}\big(\partial(U \times T)\big)$ represented by $S^{k-1} \times 1 \times S^2$ and $1 \times S^{k-1} \times S^2$ correspond by this isomorphism to the generators of $H_{k+1}(\partial V)$ represented by $S^{k+1} \times 1$ and $1 \times S^{k+1}$: we have exactly the isomorphism above. Moreover, it follows that $K_{k+2}(M \times P, U \times T)$

is the isomorphic image of

$$\begin{aligned} K_{k+2}\big(M \times P - \text{Int }(U \times T), \partial(U \times T)\big) &= K_{k+2}(N' - \text{Int } V', \partial V') \\ &\cong K_{k+2}(N' - \text{Int } V, \partial V)\,. \end{aligned}$$

Now write $W_0 = W - (\text{Int } V) \times I$. Since W_0 is obtained from N_0' by adding $(k+1)$-handles, and hence from N_0 by adding $(k+3)$-handles, it follows that any element of the above lagrangian is also in the image of $K_{k+2}(N_0, \partial V)$, and hence that the two lagrangians coincide.

Finally, the induced isomorphism $K_k(M_0, \partial U) \to K_{k+2}(N_0, \partial V)$ is simple. For firstly, the $K_*(M)$ are the homology groups of the chain complex

$$0 \to K_k(M_0, \partial U) = K_k(M, U) \xrightarrow{d} K_{k-1}(U) \to 0$$

and (by definition of the preferred base for the first module), if we regard this as a based complex, it has the right simple homotopy type. Hence for $K_*(M \times P)$, we may take the sum of three copies of this complex (with dimensions shifted by 0, 2, 4). Surgery to obtain W' abolished $K_{k-1}(U)$ in dimension $k-1$; by duality, if (instead of the above) we use the dual chain complex

$$0 \to K_k(U, \partial U) = K_k(M, M_0) \xrightarrow{d^*} K_{k-1}(M_0) \to 0$$

in dimensions $k+4$, $k+3$, to get N' we abolish the former. Thus to compute $K_*(N')$ we use the based chain complex

$$0 \to K_{k-1}(M_0) \xrightarrow{0} K_k(M_0, \partial U) \xrightarrow{d} K_{k-1}(U) \xrightarrow{0} K_k(M_0, \partial U) \to 0$$

in dimensions $k+3$, $k+2$, $k+1$ and k. Now we performed surgery to get from N' to N precisely by killing the elements of a preferred base of $K_k(M_0, \partial U)$: thus we can abolish this from the chain groups, to find ones suitable for $K_*(W)$. Dually, to go from here to $K_*(N)$ we abolish $K_{k-1}(M_0)$. Thus $K_*(N)$ is to be computed from the base chain complex

$$0 \to K_k(M_0, \partial U) \xrightarrow{d} K_{k-1}(U) \to 0\,.$$

As we used the given base of $K_{k-1}(U) \cong K_{k+1}(V)$ for our construction, the preferred class of bases of $K_{k+2}(N_0, \partial V) \cong K_k(M_0, \partial U)$ must also coincide with that induced by the isomorphism. This completes the proof.

THEOREM 9.10. *Let π be an object of type $\mathbf{2}^n$ of $\mathcal{G}pd$. Then for $m \geqslant n+5$, multiplication by $P_2(\mathbf{C})$ induces an isomorphism $L_m(\pi) \to L_{m+4}(\pi)$.*

Proof For $n = 0$, the result follows from (9.9) by taking a direct sum, one term from each component of π. For higher values of n, the result follows by induction and the Five Lemma, applied to the diagram

$$\begin{array}{ccccccccc} L_m(\partial_n \pi) & \longrightarrow & L_m(\delta_n \pi) & \longrightarrow & L_m(\pi) & \longrightarrow & L_{m-1}(\partial_n \pi) & \longrightarrow & L_{m-1}(\delta_n \pi) \\ \downarrow & & \downarrow & & \downarrow & & \downarrow & & \downarrow \\ L_{m+4}(\partial_n \pi) & \longrightarrow & L_{m+4}(\delta_n \pi) & \longrightarrow & L_{m+4}(\pi) & \longrightarrow & L_{m+3}(\partial_n \pi) & \longrightarrow & L_{m+3}(\delta_n \pi) \end{array}$$

in which the sequences are exact by (9.6), the vertical morphisms are all induced by multiplication with $P_2(\mathbf{C})$, so the diagram commutes; and the outer four vertical maps are isomorphisms by the induction hypothesis.

Part 2

Patterns of Application

10. Manifold Structures on Poincaré Complexes

In this chapter we describe the main application of the results of §3: to the classification of smooth or PL manifolds with a given simple homotopy type.The general theory is given here; some more detailed calculations will be found in later chapters. This application is originally due to Browder [B15] (for the existence statements) and to Novikov [N5] (for the uniqueness results): both these authors, however, considered only closed, simply connected manifolds. The results were extended by the author [W17] to manifolds with boundary, and reformulated by Sullivan (Princeton thesis, 1965, see also [S20], [S21], [S22]; again for the uniqueness results only), in an improved version which we will use (cf. the difference between (10.1) and (10.2) below). Compare also [W2].

First, some notation. Let X be a simple Poincaré complex (or pair, or n-ad), with formal dimension m. Then a simple homotopy equivalence $\phi : M \to X$, where M is a compact smooth (or PL) manifold (or manifold n-ad) will be said to define a smooth (or PL) structure on X. (Sullivan calls PL structures h-triangulations). The map $\phi' : M' \to X$ defines the same structure if there is a diffeomorphism (PL homeomorphism) $h : M \to M'$ such that $\phi \simeq \phi' \circ h$. Thus a structure is an equivalence class of pairs (M, ϕ). We denote the set of structures on X by $\mathscr{S}^{\text{Diff}}(X)$ (resp. by $\mathscr{S}^{PL}(X)$), or, if we do not wish to distinguish, simply by $\mathscr{S}(X)$*. Sometimes when X is a pair, ∂X (or if X is an $(n+1)$-ad, $\partial_n X$) may already be a manifold (manifold n-ad), and we wish to consider only those ϕ which induce diffeomorphisms (or PL homeomorphisms) of ∂M on ∂X (or $\partial_n M$ on $\partial_n X$). We say these define smooth (or PL) structures of X relative to $\partial_n X$. In this case, for $\phi : M \to X$ and $\phi' : M' \to X$ to define the same structure we require a diffeomorphism (PL homeomorphism) $h : M \to M'$ such that $\phi \simeq \phi' \circ h$ rel ∂M (or rel $|\partial_n M|$). Let $\mathscr{S}(X, \partial X)$ (or $\mathscr{S}(X, \partial_n X)$) denote the set of equivalence classes.

Our primary objective is to compute the sets $\mathscr{S}(X)$ and $\mathscr{S}(X, \partial X)$ in all cases. The reader should be warned that in spite of the simplifications which occur when all the spaces are simply connected, there seems to be no natural procedure in the general case for endowing these sets with group structures. Our computation, such as it is, depends on the results of §3; thus to construct an element of $\mathscr{S}(X)$, we must first construct an object (M, ϕ, ν, F). The discussion at the beginning of §3 collected these into bordism classes forming groups $\Omega_m(X, \nu)$, and showed that we need only consider bordism classes of degree 1.

* *The generalised Poincaré conjecture (Smale [S10]) in the PL and TOP categories gives $\mathscr{S}^{PL}(S^m) = \mathscr{S}^{TOP}(S^m) = \{*\}$ for $m \geqslant 5$. The classification of exotic spheres by Kervaire and Milnor [K5] gives $\mathscr{S}^{\text{Diff}}(S^m) = \theta_m = \pi_m(PL/O)$ for $m \geqslant 5$.*

LEMMA 10.1_A. *For any finite simplicial complex X, and vector bundle (PL bundle) ν^k over X, there is a natural isomorphism between the bordism group $\Omega_m(X,\nu)$ and the stable homotopy group of the Thom space $\{S^{m+k} : X^\nu\}$. Moreover, if X is a Poincaré complex and $w_1(\nu) = w_1(X)$, the isomorphism preserves degree.*

Proof We use standard transversality arguments, first applied to this problem by Novikov and Browder. Let $\chi \in \Omega_m(X,\nu)$ have a representative (M,ϕ,F). Then F is a stable trivialisation of $\tau_M \oplus \phi^*\nu$: add to ν a trivial bundle of large enough dimension to make this a trivialisation, and also to make $k > m+1$. Let E be the total space of the disc bundle over M associated to $\phi^*\nu$ (in the PL case we have a disc bundle by [H16] since we have added a trivial bundle): the tangent bundle of E can be identified with $\tau_M \oplus \phi^*\nu$, so is trivialised by F. By the immersion classification theorem [H15], [H3], there is a corresponding immersion of E in $\mathbf{R}^{m+k}$. As $k > m+1$, if we deform this to be in general position, the zero cross-section M will be embedded; hence also some neighbourhood of it. We now shrink E down into this neighbourhood, so can suppose E itself embedded. Now $E \subset \mathbf{R}^{m+k} \subset S^{m+k}$, so we have a map which collapses ∂E and everything outside E to a point, thus $S^{m+k} \to E/\partial E = M^{\phi^*\nu}$. Also, ϕ induces a map $M^{\phi^*\nu} \to X^\nu$; composing, we have $S^{m+k} \to X^\nu$. We can add a further trivial bundle ε^1 to ν; this will replace E by $E \times I$, F by the product framing, the embedding by the product embedding $E \times I \subset \mathbf{R}^{m+k} \times I \subset S^{m+k+1}$, and hence also replaces $S^{m+k} \to M^{\phi^*\nu}$ by the suspended map $S^{m+k+1} \to \Sigma M^{\phi^*\nu} = M^{\phi^*\nu+\varepsilon}$.

A similar construction shows that the (stable) homotopy class of the result depends on the bordism class of X but not on any intervening choices. For given two sets of choices (the second denoted by primes) we have a bordism $\partial W = M \cup M'$, $\psi : W \to X$; a stable framing of $\tau_W \oplus \psi^*\nu$ extending the given framings; again we add trivial bundles to ν (we have already seen that we need only suspend the corresponding final maps on account of this); obtain a disc bundle D over W which has framed tangent bundle, and then an immersion, and hence an embedding, of D in $\mathbf{R}^{m+k} \times I$ extending the given embeddings $E \subset \mathbf{R}^{m+k} \times 0$, $E' \subset \mathbf{R}^{m+k} \times 1$. The Thom construction now provides a homotopy $S^{m+k} \times I \to M^{\phi^*\nu}$ between the two maps previously constructed.

We have defined a map $\Omega_m(X,\nu) \to \{S^{m+k}, X^\nu\}$; next we prove it a homomorphism. If we start with a disjoint union of manifolds M_i mapped to X, we obtain a disjoint union of bundles E_i, which we can embed in different hemispheres of S^{m+k}. The map from E_1 sends (say) the upper hemisphere to the base point in $M^{\phi^*\nu}$; that from E_2 collapses the lower hemisphere. The map obtained from the union is given by combining the two above; but this is a standard definition of addition in homotopy theory.

We can replace X by a homotopy equivalent open manifold – for example, an open regular neighbourhood for some simplicial embedding of X in Euclidean space – which has smooth and PL structures. Then X^ν is also a manifold, except at the point at infinity. Thus given a map $S^{m+k} \to X^\nu$, we can make it transversal to the zero cross-section X [T2], [W45]. The preimage M is then a manifold mapping to X, with normal bundle induced from ν. If we suppose ν a disc bundle (as we may, on adding a trivial bundle), we can take E to be a tubular neighbourhood of X and the collapsing map $c : S^{m+k} \to E/\partial E$ to agree with the (transversalised) f in a neighbourhood of M. Since the complement of M is mapped to a contractible set, $c \simeq f$. Hence our map is onto. As in the first part of the proof, we can relativise the argument to show our map an isomorphism: if (M, ϕ, F) gives rise to a nullhomotopic map f, by making the nullhomotopy transverse to X we obtain an (N, ψ, G) which has it as boundary.

The fact that our isomorphism preserves degree follows, since a collapsing map $S^{m+k} \to E/\partial E = M^{\phi^*\nu}$ has degree 1, from the commutative diagram (with **Z** coefficients)

$$\begin{array}{ccc} H_{m+k}(M^{\phi^*\nu}) & \longrightarrow & H_{m+k}(X^{\nu}) \\ \uparrow\cong & & \uparrow\cong \\ H^t_m(M) & \xrightarrow{\phi_*} & H^t_m(X) \end{array}$$

in which the vertical maps are Thom isomorphisms.

The relative forms of this result can be proved by the same method (indeed, once we have connected maps, the Five Lemma can be used to prove them isomorphisms). A remark in §0 shows that bordism of n-ads is really bordism of suitable pairs.

LEMMA 10.1_B. *For any finite simplicial $(n+1)$-ad X and vector bundle (PL bundle) ν^k over $|X|$, there is a natural isomorphism between the bordism group $\Omega_m(X, \nu)$ and the stable homotopy group $\pi^s_{m+k}(|X|^\nu, |\partial X|^\nu)$. This preserves degree when the latter is defined.*

LEMMA 10.1_C. *Given a manifold n-ad M^{m-1}, map $\phi : M \to \partial_n X$, and stable trivialisation F of $\tau_{|M|} \oplus \phi^*\nu$, we have a corresponding natural bijection of $\Omega_m(X, \partial_n X, \nu)$ on the set of s-homotopy classes of s-extensions*

$$(D^{m+k+1}; D^{m+k}_+, D^{m+k}_-) \to \left(|X|^\nu; \bigcup{}_{1\leqslant i\leqslant n}|\partial_i X|^\nu, |\partial_n X|^\nu\right)$$

of the map of D^{m+k}_- obtained by applying (10.1_B) to (M, ϕ, F).

We omit the proofs of these results, and turn to their application to our problem. First consider the simplest case of (10.1_A). By a result of Spivak [S14], if X is a Poincaré complex, there exists a spherical fibration ${\nu_X}^k$ over X, unique up to suspension and fibre homotopy equivalence, such that $\pi^s_{m+k}(X^\nu)$ contains maps of degree 1. Thus by (10.1_A), for a smooth or PL bundle ν, $\Omega_m(X, \nu)$ contains maps of degree 1 if and only if ν is stably fibre homotopy equivalent to ν_X. More precisely, by [W21, 3.6], a class of degree 1 in $\pi^s_{m+k}(X^\nu)$ induces a stable fibre homotopy equivalence of ν on ν_X, unique up to fibre homotopy, and the set of such classes maps bijectively to the set of stable fibre homotopy equivalences of ν on ν_X.

This leads to a reformulation of the theory, in which ν is no longer fixed, and use is made of ν_X. For instead of considering pairs (ν, x): ν a vector or PL bundle over X, $x \in \pi^s_{m+k}(X^\nu)$, we are now led to consider pairs (ν, h): ν as before, h a (stable) fibre homotopy equivalence of ν on ν_X. We wish to consider h as a reduction of the structural group of ν_X. Some notation is needed first.

Write G_k for the space of homotopy equivalences of S^{k-1} on itself, BG_k for its classifying space [S19]. Let PL_k and BPL_k be as in Milnor's paper [M11], and O_k be the orthogonal group, BO_k its classifying space. With suspension, these define three sequences; we denote the limits ("mapping telescopes") as $k \to \infty$ by omitting the k. Then we have a commutative diagram of maps of classifying spaces

$$\begin{array}{ccccc} BO_k & \xrightarrow{a_k} & BPL_k & \xrightarrow{b_k} & BG_k \\ \downarrow & & \downarrow & & \downarrow \\ BO & \xrightarrow{a} & BPL & \xrightarrow{b} & BG \end{array}$$

where a_k corresponds to the functor from vector bundles to PL bundles defined by triangulating the bundle [L3], and b_k to the functor from PL bundles (identifiable with microbundles by [K16]) to spherical fibrations defined by deleting the zero cross-section; a and b are limits (non-unique) as $k \to \infty$.

In this chapter we are concerned only with the stable classifying spaces on the lower row. It will make for convenience in notation and expression to write as though $O \to PL \to G$ were inclusions of topological groups. Such arguments can always be justified by referring back to the maps of classifying spaces above. In particular, we think of ν_X as a bundle with group G_k – or rather, stably, as having group G. To give a vector bundle or PL bundle ν, and a stable fibre homotopy equivalence $h : \nu \to \nu_X$, is now the same as reducing the structural group of ν_X from G to O or PL. Such reductions correspond bijectively to homotopy classes of sections of the associated bundle with fibre G/O or G/PL.

The reductions contain the information about tangent spaces necessary for considering the structure problem. Thus we call them *tangential structures* on X, and denote the set of them all by $\mathscr{T}^{\mathrm{Diff}}(X)$ and $\mathscr{T}^{PL}(X)$ (or, ambiguously, $\mathscr{T}(X)$). Similarly if X is a Poincaré $(n+1)$-ad, we can regard $(|X|, |\partial X|)$ as a Poincaré pair and obtain a Spivak fibration ν_X over $|X|$; we write $\mathscr{T}^{\mathrm{Diff}}(X)$ and $\mathscr{T}^{PL}(X)$ for the sets of equivalence classes of reductions of the structural group of ν_X from G to O resp. PL. Slightly more complicated is the relative case, in which we suppose given (or deduce from other data) a reduction of the structural group over $\partial_n X$. We then define $\mathscr{T}(X, \partial_n X)$ as the set of homotopy classes rel $\partial_n X$ of sections of the appropriate bundle over X (with fibre G/O or G/PL) which extend the given section over $\partial_n X$.

To fit our notation with a good notation for smoothing theory, it is perhaps desirable to regard $\mathscr{S}^{\mathrm{Diff}}(X)$, $\mathscr{S}^{PL}(X)$, $\mathscr{T}^{\mathrm{Diff}}(X)$ and $\mathscr{T}^{PL}(X)$ as abbreviations for $\mathscr{S}^{G/O}(X)$, $\mathscr{S}^{G/PL}(X)$, $\mathscr{T}^{G/O}(X)$ and $\mathscr{T}^{G/PL}(X)$. One then has the obvious notations $\mathscr{S}^{PL/O}$ for concordance classes of smoothings of PL structures and $\mathscr{T}^{PL/O}$ for corresponding reductions of the tangent bundle. The main result of smoothing theory (see e.g. [H6]) then asserts that the natural map $\mathscr{S}^{PL/O}(X) \to \mathscr{T}^{PL/O}(X)$ is bijective. This chapter can be regarded as stating the corresponding results for G/PL and G/O. This notation also fits well with the topological case: observe that one of the main results of Kirby and Siebenmann [K9] is that $\mathscr{S}^{TOP/PL}(X) \to \mathscr{T}^{TOP/PL}(X)$ is bijective, for $\dim X \geqslant 5$ (6 if X has boundary)*.

We can now reformulate (10.1). If X is a Poincaré complex, we define the degree 1 bordism set of X, say $\Omega d(X)$, to be the set of equivalence classes of quadruples (M, ϕ, ν, F): M a smooth (PL) manifold, $\phi : M \to X$ of degree 1, ν a vector (PL) bundle over X, and F a stable trivialisation of $\tau_M \oplus \phi^*\nu$. Here we regard two quadruples (suffixed by 1 and 2) as equivalent if we can find a bordism $\partial N = M_1 - M_2$, $\psi : N \to X$ extending ϕ_1 and ϕ_2; and, in addition, a stable isomorphism of ν_1 on ν_2 and a stable trivialisation of $\tau_N \oplus \psi^*\nu$, extending F_2 and also (with respect to the above stable isomorphism) F_1. Similarly we define the degree 1 bordism sets of pairs and $(n+1)$-ads: the definition differs from that of $\Omega_m(X, \nu)$ only in staying with degree 1 maps, and allowing ν to vary.

PROPOSITION 10.2. *If X is a Poincaré complex or $(n+1)$-ad, there is a natural bijection of $\Omega d(X)$ on $\mathscr{T}(X)$. If also we are given a representative (M, ϕ, ν, F) of an element of $\Omega d(\partial_n X)$, defining (by the above) an element of $\mathscr{T}(\partial_n X)$, we obtain a bijection of the corresponding sets $\Omega d(X, \partial_n X)$ and $\mathscr{T}(X, \partial_n X)$.*

This follows from (10.1) on reformulating the result in the manner described above. We will now abandon the notation Ωd, and regard an element of $\mathscr{T}(X)$ both as a class of sections of a bundle and as a bordism class of degree 1 maps. We now apply the results of §3.

Let X be a simple Poincaré $(n+1)$-ad, with formal dimension m.

THEOREM 10.3. *There is an exact sequence provided $m \geqslant n+5$,*

$$\mathscr{S}(X) \xrightarrow{\eta} \mathscr{T}(X) \xrightarrow{\theta} L_m\big(\pi(X)\big).$$

Moreover, if $\partial_n X$ has already a manifold structure the sequence

$$\mathscr{S}(X, \partial_n X) \xrightarrow{\eta} \mathscr{T}(X, \partial_n X) \xrightarrow{\theta} L_m\big(\pi(\delta_n X)\big)$$

is exact, if $m \geqslant n+4$.

Proof If we identify $\mathscr{T}(X)$ with the degree 1 bordism set, there is an obvious (forgetful) map from $\mathscr{S}(X)$ to $\mathscr{T}(X)$. Theorem 3.2 states that a given element of $\mathscr{T}(X)$ is in the image if and only if a certain obstruction in $L_m\big(\pi(X)\big)$, depending only on the class in $\mathscr{T}(X)$, is zero. But this defines a map θ and proves exactness. Exactly the same argument is valid in the relative case. Note that the relative case implies the other (taking $\partial_n X$ empty).

Moreover, Kirby and Siebenmann [K11] showed that $TOP/PL \simeq K(\mathbf{Z}_2, 3)$, so that $\mathscr{S}^{TOP/PL}(X) = \mathscr{T}^{TOP/PL}(X) = H^3(X; \mathbf{Z}_2)$.

Since the sets $\mathscr{S}$ and $\mathscr{T}$ are not in general groups*, it is worth emphasising that the meaning of exactness is that the image of η is exactly $\theta^{-1}(0)$.

The above result does not constitute a computation of the set $\mathscr{S}(X)$, even theoretically, since we have as yet no estimate of when two elements of $\mathscr{S}(X)$ have the same image under η. However, given two structures which induce the same tangential structure, we can apply the relative form of (10.2) to $X \times I$ (an $(n+1)$-ad if X is an n-ad) to obtain a degree 1 map $N \to X \times I$ (with corresponding ν, F) such that $\phi \mid \partial N$ defines the given structures at $X \times 0$ and $X \times 1$. The surgery obstruction for ϕ (relative to these parts) lies in $L_{m+1}(\pi(X))$; if this vanishes, we can perform surgery (keeping the ends fixed) to obtain a simple homotopy equivalence. This provides an s-cobordism between the ends: by the s-cobordism theorem (see [M5], [K3], [S17] or [W13, IV]) we have a diffeomorphism, so the two structures are the same. Both these arguments depend, as usual, on the relevant dimensions exceeding four. To deduce an exact sequence formally, we modify this argument slightly.

THEOREM 10.4. *Let X be a manifold $(n+1)$-ad of dimension $m \geqslant n+4$, with stable normal bundle ν, and consider normal maps (N, ϕ, F): N a manifold $(n+3)$-ad, $\phi : N \to X \times (I; 0, 1)$ of degree 1, inducing diffeomorphisms (PL homeomorphisms) $\partial_{n+1}N \to X \times 0$ and $\partial_n N \to \partial_n X \times I$ and a simple homotopy equivalence $\partial_{n+2}N \to X \times 1$, and F a stable trivialisation of $\tau_N \oplus \phi^*\nu$ extending the natural trivialisation over $\partial_{n+1}N$. Then with the obvious definition of bordism for such normal maps (N, ϕ, F) the set of bordism classes is mapped bijectively by the surgery obstruction (rel $|\partial_n N| \cup |\partial_{n+1}N| \cup |\partial_{n+2}N|$) onto the group $L_{m+1}(\pi(\delta_n X))$. If $m = n+3$, the map is bijective.*

In the 'obvious' definition of bordism we follow our usual convention (of [W13, VA, Chapter 1]) of imposing restrictions on the bordisms precisely analogous to those imposed on the manifolds.

Proof We proceed by induction on n. For $n = 1$, the result was established in (5.8) (m odd) and (6.5) (m even). Thus we may assume $n > 1$, and the result already established for $(n-1)$. Let $\theta \in L_{m+1}(\pi(\delta_n X))$.

By the induction hypothesis, applied to $\partial_{n-1}X$, we can construct an $(n+2)$-ad, which we label $\partial_{n-1}N$, $\partial_{n-1}N \to \partial_{n-1}X \times I$ of degree 1 and F, with surgery obstruction $\partial_{n-1}\theta$. Attach X to $\partial_{n-1}N$ along the intersection of $\partial_{n-1}X$: the result defines an object mapping to $X \cup (\partial_{n-1}X \times I)$ (which is simply homotopy equivalent to X), inducing simple homotopy equivalences on ∂_{n-1} and on ∂_n. By naturality, the surgery obstruction for this object is $i_*\partial_{n-1}\theta = 0$, where $i : \partial_{n-1}\delta_n X \to \delta_{n-1}\delta_n X$. Thus we can perform surgery to obtain a manifold $(n+1)$-ad X' simply homotopy equivalent to X. Denote the trace of the surgery by N'.

Define a map $N' \to I$ with $X \to 0$, $X' \to 1$, and $\partial_{n-1}N \xrightarrow{\phi'_{n-1}} \partial_{n-1}X \times I \to I$. Then N, and the compound $N' \xrightarrow{\phi'} X \times I$ of the maps into X and I already

*See the note at the end of this chapter for the group structure in the topological category.

constructed, are the first two members of a normal map of the type under consideration. Moreover, if θ^i is the surgery obstruction for N', then $\partial_{n-1}\theta' = \partial_{n-1}\theta$, so by (3.1) we have $\theta - \theta' = \alpha \circ j_*(\theta'')$ for some $\theta'' \in L_{m+1}\big(\pi(\delta_{n-1}\delta_n X)\big)$.

We now apply the induction hypothesis again, this time to X', regarded as an n-ad by amalgamating the former ∂_{n-1} and ∂_n (both of which are now to be kept fixed). We obtain a normal map (N'', ϕ', F''), mapped to $X' \times 1$, with surgery obstruction θ''. Now form N by glueing N' along X' to N''. We have natural maps to X and to I (mapping N' to $[0, \frac{1}{2}]$ and N'' to $[\frac{1}{2}, 1]$), and a corresponding F, also obtained by glueing; by naturality, the surgery obstruction for N'', regarded as $(n+3)$-ad, is $\alpha \circ j_*(\theta'')$; that for N is the sum of this with the surgery obstruction, θ' for N, i.e. θ. This proves surjectivity. Injectivity follows from a relativised form of the same argument, (constructing a cobordism) which we leave to the reader. The dimensional condition can be weakened by one for this.

As a corollary to this theorem we find that the exact sequence of (10.3) can be extended a term to the left, as follows.

THEOREM 10.5. *Let X be a simple Poincaré $(n+1)$-ad of dimension $m \geqslant n+4$, with a prescribed structure on $\partial_n X$. Then the group $L_{m+1}\big(\pi(\delta_n X)\big)$ operates on $\mathscr{S}(X, \partial_n X)$ so that two elements are in the same orbit under the group if and only if they have the same image (by η) in $\mathscr{T}(X, \partial_n X)$.*

Proof Given a structure x on $(X, \partial_n X)$, we represent it by a manifold M: given also $\theta \in L_{m+1}\big(\pi(\delta_n X)\big)$, construct N as in (10.4), starting with M and having surgery obstruction θ. Then $\partial_{n+2}N$ depends, up to s-cobordism and hence up to diffeomorphism, only on the bordism class θ and the original structure x, and hence determines a new structure, $(x + \theta)$ say, on $(X, \partial_n X)$. Since surgery obstructions add for unions, $(x + \theta) + \theta' = x + (\theta + \theta')$, and clearly $x + 0 = x$, so we have a genuine group operation. The exactness statement was established in the remarks preceding (10.4). Note that if $m = n + 3$, one has a partial operation. A result is still obtained, provided the definition of equivalence for manifold structures is weakened to s-cobordism, no longer known to imply diffeomorphism.

In the proof of (10.4) Poincaré complexes have vanished from the scene. In fact, it is convenient here to distinguish two problems. First, we wish to know whether X is simply homotopy equivalent to a smooth or PL manifold at all – i.e. whether $\mathscr{S}(X)$ is empty or not. Second, if it is not empty, we wish to compute it. Now (10.3) gives the full story about the first question – we have to compute the set $\mathscr{T}(X)$ and the map θ and determine whether the image contains zero or not (or correspondingly in the relative case). The second question leads further. Note that in discussing it we may choose a manifold structure $\phi : M \to X$, and then replace X by M in the subsequent discussion. Thus from here on we may suppose X a manifold $(n+1)$-ad. (This can be regarded as choosing a base point in $\mathscr{S}(X)$).

The new viewpoint leads to some simplification. First we consider tangential structures: these are homotopy classes of sections of a certain bundle. If this were a principal bundle, we could argue that existence of a section implied triviality of the bundle and more precisely, that choosing a section gave rise to a natural bijection between classes of sections and classes of maps from base to fibre.

Rather than embark on technicalities necessary to discuss these bundles, we outline a direct argument. First observe that taking direct sums of bundles, or joins of spherical fibrations, leads to (weakly) homotopy associative and homotopy commutative (weak) H-space structures on BO, BPL and BG. Clearly, the natural maps $BO \to BPL \to BG$ are (weak) H-maps. Hence the mapping fibres G/O and G/PL have induced (weak) H-space structures, with the same properties, i.e., for any finite complex X, the sets of homotopy classes of maps $[X, G/O]$ and $[X, G/PL]$ have the structure of abelian groups.

Now a reduction of the (stable) structural 'group' of a spherical fibration ξ from G to O or PL is defined by giving a vector bundle or PL bundle ν, and a fibre homotopy equivalence $h : \nu \to \xi \oplus \varepsilon^r$ (for some r). Write, temporarily, $R(\xi)$ for the set of equivalence classes of such reductions. Then $R(\varepsilon^r) = [X, G/O]$ or $[X, G/PL]$, where X is the base space of ε^r, and contains the natural reduction O as the zero element of the group. More generally, taking direct sums gives a natural map $R(\xi) \times R(\eta) \to R(\xi \oplus \eta)$ and this 'addition' is commutative in a natural sense, and associative. Now for any $x \in R(\xi)$, defined say by $j : \nu \to \xi \oplus \varepsilon^r$, we can choose an inverse bundle τ to ν (assuming, say, X is a finite complex) and a bundle isomorphism $t : \tau \oplus \nu \to \varepsilon^m$ (for some m), which is then also a fibre homotopy equivalence. Consider τ as a spherical fibration; the identity map defines an element y of $R(\tau)$, and $x + y$ is represented by

$$1 \oplus h : \tau \oplus \nu \to \tau \oplus \xi \oplus \varepsilon^r$$

and hence also by $k = (1 \oplus h) \circ t^{-1} : \varepsilon^m \to \tau \oplus \xi \oplus \varepsilon^r$. If we use k itself to identify $\tau \oplus \xi \oplus \varepsilon^r$ with the trivial fibration ε^m, we can write $x + y = 0$. Since all our reductions are defined only up to stable equivalence, it is now easy to verify that adding x and y respectively defines inverse bijections between $R(\varepsilon^m)$ and $R(\xi)$. A corresponding argument applies also to the relative case.

The special case in which X is a (perhaps bounded) manifold, and we take x the reduction of structural groups defined by the normal bundle, gives

LEMMA 10.6. *For X a smooth manifold $(n+1)$-ad, we have natural bijections of $\mathscr{T}^{\text{Diff}}(X)$ on $[X, G/O]$ and of $\mathscr{T}^{\text{Diff}}$ on the set of homotopy classes rel $\partial_n X$ of maps $X \to G/O$ which take $\partial_n X$ to the base point. The corresponding result holds also in the PL case.*

Using this bijection, one can give an attractive definition of the map $\eta : \mathscr{S}(X) \to \mathscr{T}(X) \cong [X, G/O]$ or $[X, G/PL]$, following [S22]. Let $\phi : M \to X$ be a simple homotopy equivalence of manifold $(n+1)$-ads* (for this argument

*It is enough to have a homotopy equivalence, and use engulfing instead of the s-cobordism theorem. See [S22].

again, the cases $n \geqslant 2$ are effectively the same as $n = 1$).

Let ϕ^{-1} be a homotopy inverse to ϕ, ν_M the normal bundle of M, $\nu = (\phi^{-1})^*\nu_M$, so that we can identify ν_M with $\phi^*\nu$, and F the natural framing of $\tau_M \oplus \nu_M$. Then $\eta(M, \phi, \nu, F) \in \mathscr{T}(X) \cong [X, G/O]$ or $[X, G/PL]$, as defined above, is a stable trivialisation of $\tau_X \oplus \nu$. To obtain this more geometrically, we approximate $(\phi, 0) : M \to X \times D^k$ (k large enough) by an embedding ϕ' with normal disc bundle ν', with total space E: since $\tau_M \oplus \nu' \cong (\phi')^*(\tau_X \oplus \varepsilon^k)$, ν' is indeed (stably) equivalent to $\phi^*(\tau_X \oplus \nu)$. The inclusion of E in $X \times D^k$ is a simple homotopy equivalence, hence the complement of Int E is an s-cobordism, and so a collar on ∂E (for this argument, see [W13, IV, Chapter 6]). Thus we can expand the tube, and take $E = X \times D^k$. Now the projection $\partial E = X \times S^{k-1} \to S^{k-1}$ defines a fibre homotopy trivialisation of ν'; (ν', h) defines a class of maps of M to G/O or G/PL, and hence, composing with ϕ^{-1}, a map of X. It is easy to show that this is the same class as we constructed above.

We enter here a caveat to the reader. In the situation of (10.6), we have the map

$$\theta : [X, G/PL] \to L_m\big(\pi(X)\big)$$

of abelian groups which satisfies, by definition, $\theta(0) = 0$. However, θ is NOT in general a homomorphism†. The result fails even in the closed, simply connected case with $4|m$, as one readily sees by computing with Pontrjagin classes (the simplest example is the quaternion projective plane).

However, there is one important case in which θ is a homomorphism. Let X be a manifold $(n+1)$-ad; consider structures on $X \times I$ inducing the identity structure at each end, and correspondingly for tangential structures. We have, by (10.6), a bijection

$$(X \times I, X \times \partial I) \to [X \times I / X \times \partial I, G/PL] \;=\; [\Sigma X, G/PL],$$

and so can also regard θ as a map from this.

PROPOSITION 10.7. *Let X^m be a manifold $(n+1)$-ad; then $\theta : [\Sigma X, G/PL] \to L_{m+1}\big(\pi(X)\big)$ is a homomorphism. More generally, if we consider tangential structures on $X \times I$ inducing the standard structure on $X \times \partial I \cup \partial_n X \times 1$, the corresponding $\theta : [\Sigma(X/\partial_n X), G/PL] \to L_{m+1}\big(\pi(\delta X)\big)$ is a homomorphism.*

Proof It is well known that the same group structure is obtained in $[\Sigma X, G/PL]$ whether we use the H-space structure of G/PL or the co-H space structure of ΣX; for this proof we will use the latter. Given, then, two maps of ΣX, or rather of $X \times I$, to G/PL, with $X \times \partial I \to *$, we combine them by mapping $X \times [0, \frac{1}{2}]$ using the first, and $X \times [\frac{1}{2}, 1]$ using the second; thus obtaining a product map. Now we similarly apply transversality (as in(10.1)) in two stages, and obtain a cobordism of X to itself which is obtained by glueing together two other such. But then it is immediate that the surgery obstruction for the sum cobordism is the sum of the obstructions for the two parts. This proves the result.

† *See the note at the end of this chapter for the algebraic properties of the surgery obstruction function θ in the topological category.*

This case is important because it appears in

THEOREM 10.8. *Let X^m be a manifold $(n+1)$-ad, $m \geqslant n+4$ ($\partial_n X$ may be empty). Then there is an exact sequence*

$$\mathscr{S}(X \times I, X \times \partial I \cup \partial_n X \times I) \xrightarrow{\eta} [\Sigma(X/\partial_n X), G/PL]$$
$$\xrightarrow{\theta} L_{m+1}\big(\pi(\delta_n X)\big) \xrightarrow{\partial} \mathscr{S}(X, \partial_n X)\,.$$

Proof The exactness of (η, θ) was proved in (10.3) (even with $m \geqslant n+3$). The map ∂ is defined by operating (as in (10.5)) on the identity structure. Now if we take an element α of $[\Sigma(X/X_n), G/PL] \cong \mathscr{T}(X \times I, X \times \partial I \cup \partial_n X \times I)$, we can represent it by $\phi : M \to X \times I$, inducing a diffeomorphism onto $X \times \partial I \cup \partial_n X \times I$. Then $\theta(\alpha)$ is the surgery obstruction, which can be alternatively interpreted, by (10.4), as a suitable cobordism class. Now $\partial\theta(\alpha)$ is defined as the induced structure on $X \times I$; however, by the above, this is the identity structure.

Suppose conversely that $\partial x = 0$. Then x is represented by $\phi : M \to X \times I$, inducing a diffeomorphism onto $X \times 0 \cup \partial_n X \times I$ and a simple homotopy equivalence onto $X \times I$; however, the hypothesis $\partial x = 0$ implies that we can suppose the latter also a diffeomorphism. But then the cobordism class (rel $X \times \partial I \cup \partial_n X \times I$) defines by (10.2) an element of $\mathscr{T}(X \times I, X \times \partial I \cup \partial_n X \times I)$ which maps to x.

Observe that if $m = n+3$, the result continues to hold if $L_{m+1}\big(\pi(\delta_n X)\big)$ is replaced by the subset (the image in (10.4)) on which ∂ is defined. This subset always contains $\operatorname{Im}\theta$.

The results (10.3), (10.5) and (10.8) constitute our theoretical computation of $\mathscr{S}(X)$. We can collect them, mnemonically, into an 'exact sequence'

$$\mathscr{S}(X \times I, X \times \partial I \cup \partial_n X \times I) \xrightarrow{\eta} [\Sigma(X/X_n), G/PL] \xrightarrow{\theta} L_{m+1}\big(\pi(\delta_n X)\big)$$
$$\xrightarrow{\partial} \mathscr{S}(X, \partial_n X) \xrightarrow{\eta} [(X/X_n), G/PL] \xrightarrow{\theta} L_m\big(\pi(\delta_n X)\big)\,,$$

but the reader should refer back for the precise meaning of 'exactness' at each point in the sequence. It is natural to ask whether the sequence can be continued to the left. It can, and doing so offers no difficulty, but we will not elaborate further here. I am informed by D. Sullivan that this sequence for X closed, simply connected is to be found in his (Princeton, 1965) thesis. See also §17A.

We have emphasised above the exact sequence principally useful for computing $\mathscr{S}(X)$. However, we also have the exact sequences (3.1) of the groups L, the (cohomology-type) exact sequences of maps into G/PL, and (as is easily seen) exact sequences of structure sets also. These fit together naturally to form commutative exact diagrams. For example if $X = (M^m, \partial M)$ is a compact manifold pair of dimension $m \geqslant 6$, and we use the natural structures to give base points to all our sets, so that we can talk of exact sequences of pointed sets, we have a commutative exact diagram in the shape of a braid

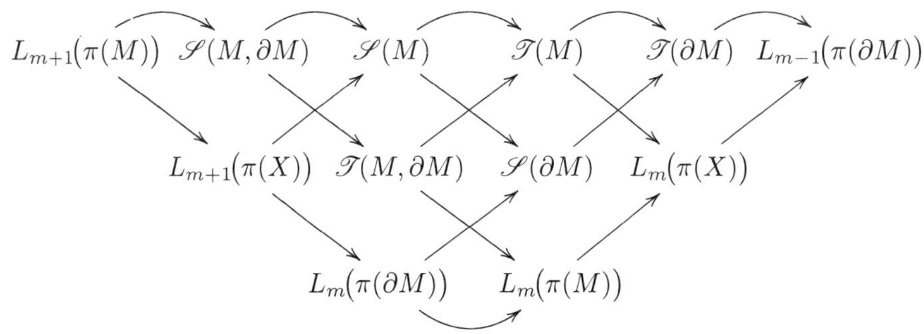

(of course, this can be extended to the left). Three more exact sequences are obtained here if we introduce a mixed type of structure, consisting of a (manifold) structure on ∂M and a tangential structure on M. This would appear in the middle of the diagram. Clearly by such devices a very large number of exact sequences can be obtained.

The algebraic surgery exact sequence. The classifying space G/TOP has two H-space structures, the one given by Whitney sum, and the 'characteristic variety' structure due to Sullivan [S24]. For a closed m-dimensional manifold X the surgery obstruction function in the topological category

$$\theta^{TOP} \ : \ \mathscr{T}^{TOP}(X) \ = \ [X, G/TOP] \to L_m\big(\pi(X)\big)$$

is a homomorphism of abelian groups with respect to the latter structure. Quinn [Q4] interpreted θ^{TOP} as a geometric assembly map A. See Ranicki [R9] for the algebraic surgery exact sequence

$$\cdots \to H_m(X;\mathbf{L}_\bullet) \xrightarrow{A} L_m\big(\pi(X)\big) \to \mathscr{S}_m(X) \to H_{m-1}(X;\mathbf{L}_\bullet) \to \dots$$

which is defined for any space X, with $\mathbf{L}_\bullet$ an Ω-spectrum of simplicial sets of quadratic forms and formations over $\mathbf{Z}$ such that $\mathbf{L}_0 \simeq G/TOP$. A normal map $(\phi, F) : M \to X$ of closed m-dimensional manifolds has a normal invariant $[\phi, F]_{\mathbf{L}} \in H_m(X;\mathbf{L}_\bullet)$ which is the cobordism class of the sheaf of kernel quadratic forms and formations over $\mathbf{Z}$ of the various normal maps $(\phi, F)| : N = \phi^{-1}(Y) \to Y$ ($Y \subset X$ submanifold), with

$$\mathscr{T}^{TOP}(X) \to H_m(X;\mathbf{L}_\bullet) \ ; \ (\phi, F) \to [\phi, F]_{\mathbf{L}}$$

a bijection. A simple m-dimensional Poincaré complex X has a total surgery obstruction $s(X) \in \mathscr{S}_m(X)$ such that $s(X) = 0$ if (and for $m \geqslant 5$ only if) X is simple homotopy equivalent to an m-dimensional topological manifold. The image $[s(X)] \in H_{m-1}(X;\mathbf{L}_\bullet)$ is the obstruction to a TOP reduction of the Spivak normal fibration; if $[s(X)] = 0$ then the surgery obstructions $\theta(\phi, F) \in L_m\big(\pi(X)\big)$ of normal maps $(\phi, F) : M \to X$ from closed topological manifolds define a coset of $\mathrm{Im}(A) \subseteq L_m\big(\pi(X)\big)$ with image $s(X) \in \mathscr{S}_m(X)$. If X is an m-dimensional topological manifold the topological surgery exact sequence of

(10.8) *is isomorphic to the algebraic surgery exact sequence*

$$\begin{array}{ccccccccc}
\cdots \longrightarrow & L_{m+1}(\pi(X)) & \longrightarrow & \mathscr{S}^{TOP}(X) & \longrightarrow & \mathscr{T}^{TOP}(X) & \xrightarrow{\theta^{TOP}} & L_m(\pi(X)) \\
& \| & & \downarrow\cong & & \downarrow\cong & & \| \\
\cdots \longrightarrow & L_{m+1}(\pi(X)) & \longrightarrow & \mathscr{S}_{m+1}(X) & \longrightarrow & H_m(X;\mathbf{L}_\bullet) & \xrightarrow{A} & L_m(\pi(X)) .
\end{array}$$

A simple homotopy equivalence of m-dimensional manifolds $f : M \to X$ has a structure invariant $s(f) \in \mathscr{S}_{m+1}(X) = \mathscr{S}^{TOP}(X)$, such that $s(f) = 0$ if (and for $m \geqslant 5$ only if) f is homotopic to a homeomorphism.

Rigidity. By definition, a topological manifold X is rigid if any homotopy equivalence of manifolds $f : M \to X$ is homotopic to a homeomorphism, or equivalently if $\mathscr{S}^{TOP}(X) = \{1\}$ consists of the unique element $1 : X \to X$. For example, every surface is rigid in this sense. One of the early achievements of surgery (due to Novikov [N5], [N7], originally in the differentiable category) was the systematic construction of homotopy equivalences of manifolds $f : M \to X$ of dimension $\geqslant 5$ which were not homotopic to homeomorphisms; such X are not rigid. The first examples were simply connected, e.g. products of spheres $S^p \times S^q$ $(p, q \geqslant 2)$ – see §13A for the simply connected surgery obstruction groups $L_*(1)$. The surgery exact sequence allows the computation of the structure set $\mathscr{S}^{TOP}(X)$ for many classes of non-simply connected manifolds. There are two types of results:

(a) *non-rigidity results, proving that $\mathscr{S}^{TOP}(X) \neq \{1\}$ by algebra,*

(b) *rigidity results, proving that $\mathscr{S}^{TOP}(X) = \{1\}$ by geometry.*

The surgery obstruction theory developed in this book has been used to obtain both types of results for non-simply connected manifolds – see §§ 13, 14 for examples of non-rigidity with finite fundamental group, and §15 for examples of rigidity with infinite fundamental group. The strongest form of the Borel and Novikov conjectures is that every aspherical Poincaré complex $X = K(\pi, 1)$ is homotopy equivalent to a rigid topological manifold M. The study of these conjectures is one of the most active areas of research in the topology of high-dimensional manifolds – see the notes at the end of §§ 15B, 17H for some references.

11. Applications to Submanifolds

This chapter applies the general theory of Part 1 to the classification of submanifolds using the language of Poincaré embeddings. However, the method is perhaps best understood in terms of the splitting of simple homotopy equivalences.

Consider first the case of a manifold V and a submanifold $M \subset V$. For any manifold W every map $f : W \to V$ is homotopic to a map (also denoted by f) which is transverse at $M \subset V$, with $N = f^{-1}(M) \subset W$ a submanifold. If f is a simple homotopy equivalence, the restriction $f| : N \to M$ need not be a simple homotopy equivalence (or even a homotopy equivalence). A simple homotopy equivalence $f : W \to V$ is said to split at $M \subset V$ if $f| : N \to M$ can be chosen to be a simple homotopy equivalence. For example, if f is homotopic to a homeomorphism then f splits at $M \subset V$. Thus any obstruction to splitting a simple homotopy equivalence of manifolds $f : W \to V$ at a submanifold $M \subset V$ is an obstruction to f being homotopic to a homeomorphism.

More generally, suppose that V is an $(m+q)$-dimensional Poincaré complex, and that $M \subset V$ is an m-dimensional Poincaré subcomplex, so that $V = C \cup M(p)$ for a $(q-1)$-spherical fibration ξ over M with projection $p : E \to M$, and with $C \cap M(p) = E$ (e.g. V could be a manifold and $M \subset V$ could be a submanifold, as above, with C the closure of the complement of a tubular neighbourhood of M in V and ξ the sphere bundle of the normal bundle). By definition, a simple homotopy equivalence $f : W \to V$ from an $(m+q)$-dimensional manifold W splits at $M \subset V$ with respect to a bundle reduction of ξ if f is homotopic to a map which is transverse regular at M with the restrictions $f| : f^{-1}(M) \to M$, $f| : f^{-1}(C) \to C$ simple homotopy equivalences.

The L-groups of Part 1 are generalised in this chapter to the LS-groups, which are the surgery obstruction groups for splitting simple homotopy equivalences for $m \geqslant 5$. In the first instance, the LS-groups are defined geometrically, by analogy with the geometric definition of the L-groups in §9. For the applications to manifold embedding theorems it is necessary to actually compute the LS-groups, using appropriate combinations of algebra and topology.

The π-π-theorem of §4 is used to prove that the LS-groups depend only on the pushout square Φ of fundamental groupoids

$$\begin{array}{ccc} \pi(E) & \longrightarrow & \pi(C) \\ \downarrow & \Phi & \downarrow \\ \pi(M) & \longrightarrow & \pi(V) \end{array}$$

with an exact sequence

$$\cdots \to L_{m+q+1}\big(\pi(C) \to \pi(V)\big) \to LS_m(\Phi) \to L_m\big(\pi(M)\big) \\ \to L_{m+q}\big(\pi(C) \to \pi(V)\big) \to \cdots .$$

A simple homotopy equivalence $f : W \to V$ (as above) has a splitting obstruction $s_M(f) \in LS_m(\Phi)$, such that $s_M(f) = 0$ if (and for $m \geqslant 5$ only if) f splits at $M \subset V$. The splitting obstruction has image the ordinary surgery obstruction of the m-dimensional normal map given by the restriction of f to the transverse inverse image at $M \subset V$

$$[s_M(f)] \;=\; \theta(f\,| : f^{-1}(M) \to M) \in L_m\big(\pi(M)\big).$$

See the note at the end of this chapter for the connection with the structure invariant $s(f) \in \mathscr{S}^{TOP}(V) = \mathscr{S}_{m+q+1}(V)$ in the case when V is an $(m+q)$-dimensional manifold and $M \subset V$ is a codimension q submanifold.

For codimension $q \geqslant 3$ $\pi(C) \cong \pi(V)$ and the splitting obstruction is just the ordinary surgery obstruction

$$s_M(f) \;=\; \theta(f\,| : f^{-1}(M) \to M) \in LS_m(\Phi) \;=\; L_m\big(\pi(M)\big).$$

See the note at the end of this chapter for a discussion of surgery in codimension $q = 2$.

For surgery in codimension $q = 1$ there are three cases to be considered, which are dealt with in §12.

We next consider the application of surgery to the problems of construction and classification of submanifolds of a fixed PL or smooth manifold. The results to be obtained are again best formulated in terms of a 'triangulation or smoothing' of initial homotopy-theoretic data. It is misleading to regard this as a complete solution to the problem of embeddings: the problems raised seem to the author in some respects to be harder than the original geometrical problems. We hope to give more positive results on this point elsewhere.

We need a notion of 'submanifold' appropriate for Poincaré complexes. The definition is found by collecting all the invariants of a submanifold in the PL or smooth case which are expressible in homotopy terms. This was essentially done by W. Browder in [B20]. The formulation below has also been made by N. Levitt (Princeton thesis, 1967. See also [L16] and §17C).

Let M^m, V^{m+q} be (simple) Poincaré complexes. Then an *embedding* of M in V shall consist of:

a $(q-1)$-spherical fibration ξ, with projection $p : E \to M$,

an $(m+q)$-dimensional (simple) Poincaré pair (C, E), and

a (simple) homotopy equivalence $h : C \cup M(p) \to V$, where $M(p)$ is the mapping cylinder of p, and $C \cap M(p) = E$.

When other types of embedding are also under discussion, we shall refer to this as a *simple Poincaré embedding.*

In most cases, the above set of conditions is redundant.

LEMMA 11.1. *Assume either that $q \geqslant 3$ or that $q = 2$ and $\partial_\xi : \pi_2(M) \to \pi_1(S^1)$ is surjective. Then the existence of a (simple) homotopy equivalence $h : C \cup M(p) \to V$ implies that (C, E) is a (simple) Poincaré pair.*

Proof The hypothesis implies that $p_* : \pi_1(E) \to \pi_1(M)$ is an isomorphism, hence so is the map of fundamental groups induced by the inclusion $E \subset M(p)$. By van Kampen's theorem, the inclusion $C \subset C \cup M(p)$ also induces an isomorphism of fundamental groups. But $C \cup M(p)$, equivalent to V, is a (simple) Poincaré complex. By the cutting theorem (2.7 (ii)), to show that (C, E) is an $(m+q)$-dimensional (simple) Poincaré pair, it is enough to show that $\big(M(p), E\big)$ is. Now it is shown by Spivak [S14] that for any Poincaré complex M and spherical fibration ξ, $\big(M(p), E\big)$ is a Poincaré pair. Indeed, one can so arrange that the diagram

$$\begin{array}{ccccc} & & H^t_{m-*}(M) & & \\ & {\scriptstyle [M]\cap}\nearrow & & \searrow{\scriptstyle \Phi} & \\ H^*(M) & & \xrightarrow{\quad [M(p)]\cap \quad} & & H^{t'}_{m+q-*}\big(M(p), E\big) \end{array}$$

commutes, where Φ is the Thom isomorphism. In describing $M(p)$ and E as finite complexes, we suppose made the natural choices, so that $M \subset M(p)$ is a simple homotopy equivalence, and the Thom isomorphism is induced by a chain map $C_*(M) \to C_*\big(M(p), E\big)$ which is a simple equivalence. Then $[M]\cap$ and Φ above are induced by simple equivalences of chains, hence so is $[M(p)]\cap$. Thus we have a simple Poincaré pair.

The following remark may mitigate any feeling that this proof involves 'cheating'. Since ξ is a fibration, E and $M(p)$ are not finite CW complexes, though we can find a finite complex E_0 and homotopy equivalence $E \to E_0$, and the natural map $M(p) \to M$ is a homotopy equivalence. We can suppose (using a mapping cylinder) $E_0 \subset M$. Then (W, E_0) need not be a simple Poincaré pair; let the map $C^*(M) \to C_*(M, E_0)$ have torsion τ. Then if $E_0 \to E_1$ is a homotopy equivalence (of finite complexes) with torsion τ', (M, E_1) is a simple Poincaré pair if and only if $\tau' = \tau$.

We next show how PL and smooth embeddings of manifolds determine Poincaré embeddings of the underlying Poincaré complexes.

THEOREM 11.2. *Suppose V^{m+q} a PL or smooth closed manifold, M^m a PL or smooth submanifold and (in the PL case) that M is locally flat in V (this is true automatically if $q \geqslant 3$ [Z1]). Then the embedding determines a simple Poincaré embedding of M in V.*

Proof Let N be a regular neighbourhood resp. a (closed) tubular neighbourhood of M in V, C the closure of its complement, $E = \partial N = \partial C$. The inclusion of M in N is a simple homotopy equivalence; let r be a homotopy inverse (in the smooth case, we can let r be the projection). We will be finished if we can find a spherical fibration ξ with $(M(p_\xi), E_\xi) \simeq (N, E)$. In the smooth case, N is already a disc bundle over V, and E the corresponding sphere bundle. In the PL case the existence of ξ follows from the main result of Spivak [S14], if $q \geqslant 3$. A better proof for our purposes is to observe that N is a block bundle over M in the sense of Rourke and Sanderson [R17, I], and that any block bundle (with zero section deleted) is homotopy equivalent to a spherical fibration (see again [R17]).

Before we can discuss smoothing theorems, we must say a little more about structure groups. A Poincaré embedding has, as part of the definition, a 'normal' spherical fibration, classified by a map into BG_q. A smooth submanifold of a smooth manifold has a normal vector bundle, classified by a map to BO_q. A locally flat PL submanifold of a PL manifold does not in general have a normal PL bundle [R18], although it does so stably, so to have a good theory we must use the normal block-bundles referred to above. The functor assigning to a finite simplicial complex the set of isomorphism classes of q-block bundles over it is a representable homotopy functor [R17, I] : we denote the classifying space by $B\widetilde{PL}_q$. There are maps, corresponding to natural transformations which we have already discussed,

$$BO_q \to BPL_q \to B\widetilde{PL}_q \to BG_q .$$

One can form direct sums of block bundles; also we have $\lim\limits_{q \to \infty} B\widetilde{PL}_q = BPL$. As in §10, we shall speak of reductions of the structural group of a spherical fibration from G_q to $\widetilde{PL}_q$ and O_q. Here there is one extra feature. Given a Poincaré embedding i of M^m in V^{m+q}, the Spivak normal fibration ν_M is the sum of $i^*\nu_V$ and the normal fibration ξ of M in V. For evidently there is a collapsing map

$$V^{h^{-1}} \simeq C \cup M(p) \to \frac{C \cup M(p)}{C} = \frac{M(p)}{E} = M^\xi$$

of degree 1. Similarly we obtain $V^{\nu_V} \to M^{\xi \oplus i^* \nu_V}$ of degree 1. Since the former is reducible, so is the latter. Now by the argument of (10.6), given stable reductions of the structural groups of ξ and ν_V (hence also of $i^*\nu_V$) from G to O or PL, we can add to obtain a corresponding reduction for ν_M. Observe that a (non-stable) reduction from G_q to O_q or $\widetilde{PL}_q$ induces a stable reduction in an obvious way. The tangential structure on V and a reduction of the group of ξ from G_q to O_q or $\widetilde{PL}_q$ induce a tangential structure on M.

We are now ready for the main theorem on smoothing Poincaré embeddings of codimension $q > 2$. The idea of this theorem is due to W. Browder [B20].

THEOREM 11.3. *Suppose given a closed smooth (PL) manifolds M^m, V^{m+q}, with $m + q \geqslant 5$, $q \geqslant 2$, and a simple Poincaré embedding $(\xi, (C, E), h)$ of M in V. If $q = 2$, assume that $(h \mid C)_* : \pi(C) \cong \pi(V)$. Suppose given also a reduction (ν, j) of the structural group of ξ to O_q ($\widetilde{PL}_q$) such that the tangential structure induced on M coincides with the natural one. Then there is a smooth* *(locally flat) PL embedding $i : M \to V$ inducing the given Poincaré embedding and the reduction (ν, j).*

Proof We first adjust notation to identify ξ with ν by the fibre homotopy equivalence j, and so let ξ be a smooth vector bundle or PL block bundle, E the total space of its sphere bundle, and $M(p)$ of its disc bundle, and M^ξ the Thom space $M(p)/E$.

Consider the map $h^{-1} : V \to C \cup M(p)$. Now $M(p)$ is a smooth or PL manifold, and M a submanifold with normal (block) bundle ξ. We can therefore adjust h^{-1} to be transverse to M (see [T2] in the smooth case, [R17, II] in the PL case – or [W45] if ξ is a PL bundle. The transversality of [A5] is of no use here). Then M has preimage M', say, a submanifold of V. The normal bundle, ξ', of M' in V is induced from ξ. Also, since h^{-1} is a homotopy equivalence and the induced map $f : M' \to M$ was obtained by transversality, f has degree 1. Again, since h^{-1} is a homotopy equivalence, ν_V is induced by it from some bundle over $C \cup M(p)$, and so $\nu_V \mid M'$ is induced by f from the restricted bundle over M. Thus the tangential conditions are satisfied for (M', f) to define an element of the degree 1 bordism set which we identify by (10.2) with $\mathscr{T}(M)$. Clearly, the tangential structure obtained is the induced structure described above.

But we assumed that this coincided with the natural tangential structure on M, so (M', f) is cobordant to the identity map. Thus there exist a cobordism L of M' to M, an extension $\phi : L \to M$ of f and 1_M, and a stable trivialisation of $\tau_L \oplus \phi^*(i^*\nu \oplus \xi)$. Let A be the total space of the disc bundle (or block bundle) $\phi^*\xi$: the restriction to M' is the normal (block-) bundle of M' in V, so we can identify the part, A', of A over M' with a (tubular or regular) neighbourhood of M' in V. We attach A to $V \times I$ by making such an identification in $V \times I$, obtaining W say.

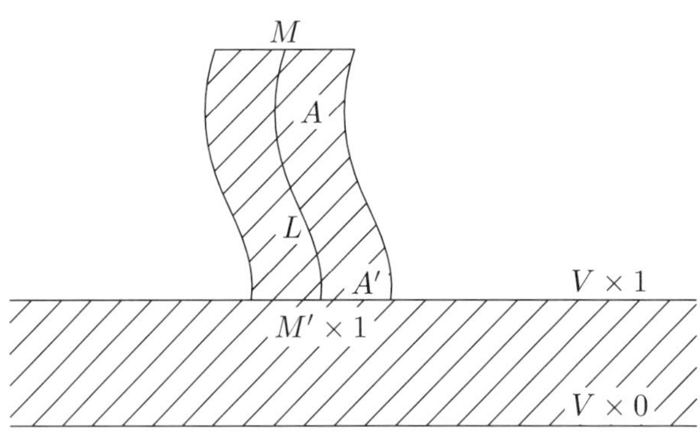

The idea of the proof is now to do surgery on W. We first define a simple Poincaré triad Y by letting $|Y|$ be the mapping cylinder of $h^{-1} : V \to C \cup M(p)$, and $Y\{1\} = V \cup M(p)$, (where V is at the lower end of the cylinder), $Y\{2\} = C$, and $Y\{\ \} = E$. Correspondingly we regard W as a manifold triad with $W\{1\} =$

$V \times 0 \cup M(p)$, $W\{\ \} = E$, and $W\{2\}$ is the rest of the boundary of W, i.e. the union of $(V \times 1) - A'$ and the sphere bundle of $\phi^*\xi$. A map of triads, $g : W \to Y$ is defined by mapping $V \times I$ by the natural map into the mapping cylinder and A to $M(p)$ by a block-bundle map covering ϕ: we may suppose that these agree on A' (indeed, this requirement was imposed on the identification of A' with a subset of V). Evidently g is a map of degree 1, and $\partial_2 g$ is a simple homotopy equivalence (even a smooth or PL homeomorphism). Since $|W|$ is homotopy equivalent to V, we can take the bundle ν_V over it, and seek a trivialisation of $\tau_W \oplus g^*\nu_V$. On $V \times I$ we use the natural trivialisation. On A', which is homotopy equivalent to L, we have $g^*\nu_V = \phi^* i^* \nu_V$, and $\tau_W = \tau_L \oplus \phi^*\xi$ (since A' is the total space of $\phi^*\xi$) ; moreover, we have stably identified τ_L with $\phi^*\tau_M$. It is thus enough to trivialise $i^*\nu_V \oplus \xi \oplus \tau_M$, and this we have already done in discussing tangential structures on M. Again by this discussion the trivialisations agree on A'.

We wish to apply Theorem 3.3. For this we need to assume that $m + q \geqslant 5$, and that $\pi(Y\{2\}) \to \pi(Y\{12\})$ is an isomorphism, i.e. that $\pi(C) \to \pi(V)$ is so. If $q > 2$ it follows by van Kampen's theorem since $\pi(E) \to \pi\big(M(p)\big)$ is an isomorphism, using the exact homotopy sequence of the fibration (once for each component of M). The hypotheses of (3.3) hold, thus by that theorem we can perform surgery to obtain a simple homotopy equivalence of triads.

Let W' be the manifold triad obtained by surgery. Then $|W'|$ is simply homotopy equivalent to Y, hence to $V \times I$, and so is an s-cobordism of $V \times 0$ to $W'\{2\} \cup M(p)$. By the s-cobordism theorem [K3], [W13, IV], it is diffeomorphic to V. But $M \subset M(p)$ is embedded as a locally flat submanifold, hence also in V, with a normal (block) bundle ξ, and the closed complement is simply homotopy equivalent to $W'\{2\}$, hence to $Y\{2\} = C$; the attaching map is also as in Y, hence as prescribed. Our embedding thus induces the given Poincaré embedding.

The hypotheses can be somewhat simplified in the PL case.

COROLLARY 11.3.1. *Suppose given closed PL manifolds M^m, V^{m+q} with $m + q \geqslant 5$, $q \geqslant 3$, and a simple Poincaré embedding of M in V. Then there is a PL embedding $M \to V$ which induces it.*

We reduce to the theorem by showing that there is a unique reduction of the structural group of the normal fibration ξ of the Poincaré embedding from G_q to $\widetilde{PL}_q$ such that the tangential structure induced on M coincides with the natural one. Indeed, our earlier discussion has shown that this last condition determines a stable reduction from G to PL. Our assertion now follows from the known result [R17, III] that for $q \geqslant 3$ the diagram

$$\begin{array}{ccc} B\widetilde{PL}_q & \longrightarrow & BPL \\ \downarrow & & \downarrow \\ BG_q & \longrightarrow & BG \end{array}$$

is, up to homotopy, that of a pullback (induced fibration), so a stable reduction determines an unstable one.

The corresponding result in the smooth case is only valid in the so-called metastable range.

COROLLARY 11.3.2. *The result of* (11.3.1) *holds for smooth manifolds and embeddings, provided* $2q \geqslant m+3$.

For the corresponding diagram

$$\begin{array}{ccc} BO_q & \longrightarrow & BO \\ \downarrow & & \downarrow \\ BG_q & \longrightarrow & BG \end{array}$$

is exactly $(2q-2)$-connected (cf. [H4]).

Remark. The idea of extending (11.3) to cover some of the cases $q=2$ is partly due to A. J. Casson. The result above implies the surgery results obtained in [C10].

The above discussion concerns only the absolute case, but the results are quite easy to relativise. Although other definitions are possible, the only type we shall consider is that of embeddings $f: M \to V$ of n-ads. Thus if M has boundary, we will insist that $f(\partial M) \subset \partial V$. Now if M is a Poincaré $(n+1)$-ad, and ξ a spherical fibration $p: E \to |M|$, with $M(p)$ the mapping cylinder of p, then $M(p)$ supports a Poincaré $(n+2)$-ad N, where if p induces $q: M(p) \to M$ we set

$$\left.\begin{aligned} N\{\alpha\} &= p^{-1}M\{\alpha\} \\ N\{\alpha, n+1\} &= q^{-1}M\{\alpha\} \end{aligned}\right\} \quad \text{if } \alpha \subset \{1, 2, \ldots, n\}.$$

The proof that Poincaré duality holds uses the results of Spivak [S14] and induction. As in (11.1), if M is a simple Poincaré $(n+1)$-ad, there is a natural way to give N the structure of a simple Poincaré $(n+2)$-ad.

Now given (simple) Poincaré $(n+1)$-ads M^m, V^{m+q}, a Poincaré embedding of M in V consists of:

a $(q-1)$-spherical fibration ξ over $|M|$, determining N as above,

a (simple) Poincaré $(n+2)$-ad C, with $\partial_{n+1} = \partial_{n+1}N$, and

a (simple) homotopy equivalence $h: C \cup N \to V$ of $(n+1)$-ads,

where $|C| \cap |N| = \partial_{n+1}N$ and $|\partial_i(C \cup N)| = |\partial_i C| \cup |\partial_i N|$ for $1 \leqslant i \leqslant n$. The above results relativise as follows (the next proof is left to the reader).

LEMMA 11.1 *relative. In the definition above, the fact that the* $(n+2)$*-ad* C *satisfies (simple) Poincaré duality follows from the other conditions, provided* $q \geqslant 3$ *or* $q = 2$ *and*

$$h|C : \pi(\delta_{n+1}C) \cong \pi(V).$$

To abbreviate the next results we define a *locally flat embedding* of (smooth or PL) manifold $(n+1)$-ads to be a (smooth or PL) embedding $f : M^m \to V^{m+q}$ such that at each point $P \in |M|$, lying on say r of the faces $|\partial_i M|$, there is a (smooth or PL) local coordinate system at $f(P)$, mapping V to $\{x \in \mathbf{R}^{m+q} : x_i \geqslant 0 \text{ for } 1 \leqslant i \leqslant r\}$, $f(P)$ to 0, the faces of V containing $f(P)$ to the subsets $x_i = 0\,(1 \leqslant i \leqslant r)$ (so there are r of these), and $f(M)$ to the subset defined by $x_i = 0$ for $m < i \leqslant m+q$.

THEOREM 11.2 *relative. Any locally flat* (PL *or smooth*) *embedding of* (PL *or smooth*) *manifold* $(n+1)$*-ads determines a simple Poincaré embedding.*

The necessary relative form of the tubular neighbourhood theorem was shown by Cerf [C12]; in the PL case one uses the results about regular neighbourhoods [H29] and block bundles [R17, I] in bounded manifolds, and induction.

THEOREM 11.3 *relative. Given* (PL *or smooth*) *manifold* $(n+1)$*-ads* M^m, V^{m+q} *with* $m+q \geqslant n+4$, $q \geqslant 2$; *a simple Poincaré embedding* (ξ, C, h) *of* M *in* V; *and a reduction of the structural group of* ξ *to* $\widetilde{PL}_q$ *or* O_q *such that the induced tangential structure on* M *coincides with the natural one, then provided if* $q=2$ *that* $(h\,|\,C)_* : \pi(\delta_{n+1}C) \cong \pi(V)$, *there is a locally flat* (PL *or smooth*) *embedding of* M *in* V *inducing the given Poincaré embedding and reduction of* ξ.

Further, if such an embedding is already given for $\partial_n M \to \partial_n V$, *it can be extended to one of* M *in* V, *if* $m+q \geqslant n+3$, $q \geqslant 2$, *and* (*if* $q=2$)

$$(h\,|\,C)_* : \pi(\delta_n \delta_{n+1} C) \cong \pi(\delta_n V)\,.$$

As this is a substantial extension of (11.3), we outline the proof. Clearly the first version is a consequence of the second and induction on n. More thoroughly, we can induct on all faces, and see that it is enough to extend to M an embedding already constructed on the union of all its faces. Amalgamating the faces (and, in the smooth case, rounding some corners), this shows that it suffices to consider the case of a Poincaré pair. The proof of this case follows that given above for (11.3) with the following modifications. The map h, hence also h^{-1}, can be taken to be a (smooth or PL) homeomorphism on the boundary. We can make this transverse to M leaving it fixed on the boundary. Then $\partial M' = \partial M$, and M and M' have the same tangential structure relative to the boundary. So we can find a manifold tetrad L with $|\partial_2 L| = M'$, $|\partial_1 L| = \partial M \times I$, and $|\partial_0 L| = M$; again form the block bundle A over L inducing A' over L'. When we glue this to $V \times I$ to obtain a manifold tetrad W, one face is obtained from $\partial V \times I$ by glueing along $B = (A' \cap \partial V) \times I$ a copy of $B \times I$. The result (apart from rounding corners in the smooth case) is homeomorphic again to $\partial V \times I$. This face will be kept fixed in the surgery; the rest we treat as before. At the final stage of the argument we use the relative form of the s-cobordism theorem.

COROLLARY 11.3.1 *relative. If* $q \geqslant 3$, *then the reduction of the structural group of* ξ *is redundant in the* PL *case; also in the smooth case if* $2q \geqslant m+3$.

The proof of this is as before. For the last clause, m can be replaced by the geometrical dimension of $|M|$ (or of $|M|$ relative to $|\partial_n M|$ in the case when we are extending an embedding) – this is the least dimension of a CW complex homotopy equivalent to $|M|$ (or, of a CW complex rel $|\partial_n M|$ homotopy equivalent rel $|\partial_n M|$ to $|M|$).

Corollary 11.3.3. *The embeddings obtained in* (11.3) *are unique up to concordance. They are unique up to isotopy if $q \geqslant 3$ in the PL case, or if $2q \geqslant m+4$ in the smooth case.*

Uniqueness up to concordance follows from the relative case by extending embeddings of $M \times \partial I$ to $M \times I$. We now appeal to [H27] in the PL case and [H1] in the smooth case to obtain isotopies.

Another important special case (discovered during the summer of 1966 by Casson and Sullivan; also by Browder and Haefliger [H5], all in the simply connected case) is

Corollary 11.3.4. *Let M^m and V^{m+q} be compact PL manifolds, $q \geqslant 3$, $\partial M = \emptyset$, and $f : M \to V$ a homotopy equivalence. Then f is homotopic to a PL embedding.*

If $m \leqslant 2$, the result holds by a general position argument. Now if we replace the inclusion $\partial V \subset V$ by an equivalent fibration, the fibre is homotopy equivalent to S^{q-1}, as in [S14]. Let ξ be the induced spherical fibration over M. Then taking $C = \partial V \times I$, $E = \partial V \times 0$ and h to be induced by the projection of ξ, we have defined a (not simple) Poincaré embedding of M in V. We must construct a simple Poincaré embedding; the result will then follow from (11.3.1). Let f have Whitehead torsion τ. Construct homotopy equivalences $h_1 : \partial V \to A$, $h_2 : A \to B$ with torsions $(-1)^{m+q-1}\tau^*$, $-\tau$ respectively; let C be the union of the mapping cylinders of h_1, h_2^{-1}. Setting $\partial_2 C = B$, $\partial_1 C = \partial V$, we find that C is a simple Poincaré triad. There are induced homotopy equivalences of B with the total space E of ξ and of $\big(M(p) \cup C, \partial V\big)$ with $(V, \partial V)$. The latter is simple, since $M \to M(p) \cup C$ also has torsion τ (by construction). The argument of the cutting theorem (2.7) now shows that the homotopy equivalence $B \to E$ is simple. Thus we have a simple Poincaré embedding and the result follows.

Corollary 11.3.4 *relative. Let M^m be a compact PL $(n+1)$-ad, V^{m+q} a compact PL $(n+2)$-ad, $q \geqslant 3$, and $f : M \to \partial_{n+1} V$ a homotopy equivalence whose restriction $\partial_n M \to \partial_n \delta_{n+1} V$ is a locally flat embedding. Then f is homotopic rel $\partial_n M$ to a locally flat embedding.*

In all the above results we have given a Poincaré embedding $M \to V$ and manifold structures on M and V and sought to construct a locally flat embedding of M in V. A different version of the problem is when V has a manifold structure but no manifold structure on M is prescribed in advance, and we seek a manifold structure on M, and a locally flat embedding of it, with that structure, in V.

In the case $q \geqslant 3$ it is simple to describe the relation between these two problems, using the proof of (11.3). We suppose given a reduction (ν, j) of the group of the normal spherical fibration ξ to O_q or PL_q. This induces a tangential structure χ on M. For further progress to be possible, it is necessary and sufficient that χ correspond to a manifold structure, i.e. (if $m \geqslant 5$) that $\theta(\chi) = 0$. If this condition is satisfied, we can apply (11.3) to M with some manifold structure inducing the tangential structure χ, and obtain a locally flat embedding.

We turn now to consideration of the cases $q = 1$, 2. Before commencing our detailed discussion, it is important to remark that – in contrast to the properties of $\widetilde{PL}_q$ used in (11.3.1) when $q \geqslant 3$ – we have:

For $q = 1$ or 2, the maps

$$BO_q \to BPL_q \to B\widetilde{PL}_q \to BG_q$$

are homotopy equivalences.

The case $q = 1$ is trivial; that $BO_2 \to BG_2$ is a homotopy equivalence is well-known. The case of $\widetilde{PL}_2$ follows from [W19] on noting that (see [R17]) $\lim_{m \to \infty} BPL_{m+q,m} = B\widetilde{PL}_q$, up to homotopy. The result on PL_2 is due to Akiba [A1] and G. P. Scott [S1]. Thus there is no need to reduce the structural group of ξ in discussing these cases. In particular, a tangential structure on V induces one on M. This shows that a more satisfactory result is obtained by considering the case where no manifold structure on M is given in advance.

Let us now consider the problem of smoothing embeddings in codimension 1 or 2 using the method of (11.3). We assume given, then, a closed PL manifold V^{m+q} and a simple Poincaré embedding $\big(\xi, (C, E), h\big)$ of M^m in V. As observed above, this induces a tangential structure χ on M; we assume further that $\theta(\chi) = 0 \in L_m\big(\pi(M)\big)$, and that $m \geqslant 5$. Then M has manifold structures inducing χ. (Of course, for $m \leqslant 4$ we may simply suppose given a manifold structure on M for the next part of the argument). We now construct a cobordism L, a block bundle A, and a manifold – or rather manifold triad – W.

We now wish to perform surgery as in (11.3) to obtain an s-cobordism. However, we must now appeal to (3.2) rather than (3.3) to see that there is an obstruction in $L_{m+q+1}\big(\pi(C) \to \pi(V)\big)$ to performing the desired surgery.

We next investigate the effect of the choices made in the construction on the cobordism class of W. Thus suppose the entire construction performed twice, and let us try to construct a cobordism between them. Transversality extends easily to give a cobordism of M_1' to M_2'. Now we have cobordisms L_i of the manifolds M_i' to manifolds M_i. Glueing these three together, we have a cobordism (with ν, F as usual) from M_1 to M_2, and wish to perform surgery to make it an s-cobordism. For $m \geqslant 4$, the obstruction to this lies in $L_{m+1}\big(\pi(M)\big)$. If this vanishes, the support of the surgeries gives a cobordism of suitable character from L_1 to L_2, and we glue a block bundle (or, as $q \leqslant 2$, disc bundle) over this to $V \times I \times I$ to obtain the desired cobordism from W_1 to W_2. Now the cobordism class of W determines an obstruction in $L_{m+q+1}(\pi(C) \to \pi(V))$, and if this vanishes (and $m \geqslant 4$) we can perform the final surgery and obtain our embedding.

Observe that (if $m \geqslant 5$) by (10.4) any element of $L_{m+1}(\pi(M))$ can occur above (we glue a suitable cobordism of M_1 onto L_1 to define L_2). Thus in order to obtain a complete result, we must investigate the effect of the choice of an element of $L_{m+1}(\pi(M))$ on the final surgery obstruction in $L_{m+q+1}(\pi(C) \to \pi(V))$.

The above arguments suggest that there exists an obstruction group G for our problem and an exact sequence

$$\cdots \to L_{m+1}(\pi(M)) \to L_{m+q+1}(\pi(C) \to \pi(V)) \to G \to L_m(\pi(M)) \to \dots .$$

We propose to give a direct proof of this, using the methods of §9: we wish to define an obstruction group G which depends only on the dimension, fundamental groupoids, and maps w involved. We must begin by listing these, and studying the extent of their mutual dependence.

Note that we have two different functions w for $\pi(M)$: that obtained from the Poincaré structure on M, which we denote by w^M, and one induced from V (or equivalently, from $M(p)$), which we denote by w^V. The product $w^M w^V = w^\xi$, say, corresponds in the manifold case to the first Stiefel class of the normal bundle; in general to that of ξ. All other w are induced (as in Part 1) from w^V. The spaces we have are M itself, the total space E of a $(q-1)$-spherical fibration ξ over M (as $q = 1$ or 2, the fibre is then S^0 or S^1), C (which contains E), and $V \simeq C \cup_E M(p)$. Thus $\pi(M)$ is closely related to $\pi(E)$; $\pi(C)$ only by a morphism $\pi(E) \to \pi(C)$, and $\pi(V)$ is then determined by the van Kampen theorem (see e.g. Brown [B38]), which states that the diagram

$$\Phi : \quad \begin{array}{ccc} \pi(E) & \longrightarrow & \pi(C) \\ \downarrow & & \downarrow \\ \pi(M) & \longrightarrow & \pi(V) \end{array}$$

is a 'pushout' diagram of groupoids.

As to the relation of $\pi(E)$ to $\pi(M)$, we may consider components of M separately, and choose base points in them. If $q = 1$, there are two cases: if ξ is trivial, E has two components E_1, E_2, each homotopy equivalent to M (this is the case $w^\xi = 1$); if ξ is not trivial, E is connected, and is (homotopy equivalent to) a double covering of M, with $\pi_1(E) = \mathrm{Ker}(w^\xi)$. If $q = 2$, we have an exact homotopy sequence

$$\pi_2(M) \xrightarrow{\partial} \pi_1(S^1) \to \pi_1(E) \to \pi_1(M) \to 1\,;$$

the elements $x \in \pi_1(M)$ act on $\pi_1(S^1) \cong \mathbf{Z}$ as automorphisms via $w^\xi(x) = \pm 1$. Some care is needed to see that the second homotopy group is not really relevant.

PROPOSITION 11.4. *Suppose there exists a $(q-1)$-spherical fibration ($q = 1, 2$) whose projection induces the homomorphism $f : A \to B$ of fundamental groups and with first Stiefel class $w : B \to \{\pm 1\}$. Then there exists a universal $P : X \to Y$ as above, so that given any $(q-1)$-spherical fibration $p : E \to M$, and isomorphisms $h_1 : \pi_1(E) \to A$, $h_2 : \pi_1(M) \to B$ with $fh_1 = h_2\pi_1(p)$ and $w(p) = wh_2$, there exists a unique homotopy class of fibrewise maps of p to P inducing the stated isomorphisms.*

Proof In the case $q = 1$, it suffices to take $Y = K(B, 1)$ and X the indicated double covering. We may act similarly if $q = 2$ when Ker f is infinite cyclic: the characteristic class in $H^2(B; \mathbf{Z})$ (with coefficients twisted by w) of the extension determines a unique bundle over $K(B, 1)$ with fibre S^1. The difficult case is that in which Ker f has finite order n, say (it is of course cyclic). Note that the case $n = 1$ was the one covered by the argument in (11.3).

We may thus suppose given an exact sequence

$$0 \to \mathbf{Z} \xrightarrow{n} \mathbf{Z} \to A \xrightarrow{f} B \to 1$$

of groups. Let us first suppose w trivial. Then the above induces exact sequences of spaces (in the sense of Spanier [S13])

$$0 \to \mathbf{Z} \xrightarrow{n} \mathbf{Z} \to \mathbf{Z}_n \to S^1 \xrightarrow{n} S^1 \to K(\mathbf{Z}_n, 1) \to K(\mathbf{Z}, 2) \to K(\mathbf{Z}, 2) \to K(\mathbf{Z}_n, 2)\,,$$
$$0 \to \mathbf{Z}_n \to A \xrightarrow{f} B \to K(\mathbf{Z}_n, 1) \to K(A, 1) \to K(B, 1) \to K(\mathbf{Z}_n, 2)\,.$$

Let Y be the pullback of $K(\mathbf{Z}, 2) \to K(\mathbf{Z}_n, 2)$ and $K(B, 1) \to K(\mathbf{Z}_n, 2)$. We then obtain four exact sequences of spaces, forming the diagram

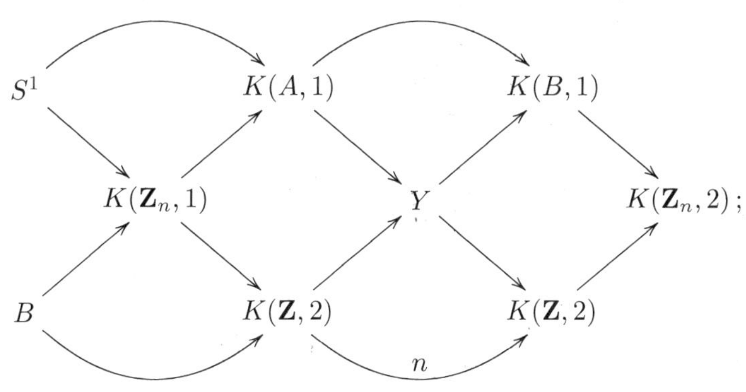

we claim that the fibration $S^1 \to K(A, 1) \to Y$ has the desired universal property. Note that Y itself is not a $K(\pi, 1)$; indeed, the other fibration $K(\mathbf{Z}, 2) \to Y \to K(B, 1)$ is precisely its Postnikov decomposition.

Given $p : E \to M$ as above, the bundle has a classifying map $M \to K(\mathbf{Z}, 2)$ and h_2 induces a map $M \to K(B, 1)$. We must show that these give homotopic maps $M \to K(\mathbf{Z}_n, 2)$; it will then follow that they have a common lift $M \to Y$. Now the cohomology class of the map $K(B, 1) \to K(\mathbf{Z}_n, 2)$ is, by construction, the characteristic class in $H^2(B;\mathbf{Z}_n)$ of the extension A of $\mathbf{Z}_n$ by B. But we assume that the fibration induced by $M \to K(\mathbf{Z},2)$ has fundamental group A, hence also that induced by $M \to K(\mathbf{Z}_n,2)$. Thus the two maps to $K(\mathbf{Z}_n,2)$ are homotopic.

The homotopy class of the lift $M \to Y$ depends on the choice of an explicit homotopy $M \times I \to K(\mathbf{Z}_n,2)$: two such differ by an element of $H^1(M;\mathbf{Z}_n) \cong H^1(A;\mathbf{Z}_n)$. We claim that this choice is fixed for us by h_1. For if we choose a homotopy carelessly, the bundle map covering our map $M \to Y$ induces h_1' making the diagram

$$\begin{array}{ccccccccc}
1 & \longrightarrow & \mathbf{Z}_n & \longrightarrow & \pi_1(E) & \xrightarrow{p} & \pi_1(M) & \longrightarrow & 1 \\
\| & & \| & & \downarrow h_1' & & \downarrow h_2 & & \| \\
1 & \longrightarrow & \mathbf{Z}_n & \xrightarrow{i} & A & \longrightarrow & B & \longrightarrow & 1
\end{array}$$

commute. Thus $h_1' h_1^{-1} = idp$ for some homomorphism $d : \pi_1(M) \to \mathbf{Z}_n$, determining an element of $H^1(M;\mathbf{Z}_n)$. It is now clear that by changing the homotopy in $K(\mathbf{Z}_n,2)$ using this element, we can make $h_1' = h_1$.

The above argument assumed $w^\xi = 1$. However, if we use cohomology classes rather than maps, we find that much the same holds in the general case. The maps $K(B,1) \to K(\mathbf{Z}_n,2)$ and $Y \to K(\mathbf{Z},2)$ must be replaced by cohomology classes twisted by w^ξ; similarly in the last part of the argument the classes in $H^1(M;\mathbf{Z}_n)$ are so twisted. The construction of Y is changed: in fact, if $\varepsilon \in H^2(B;\mathbf{Z}_n)$ is the class of the extension, and β_n the Bockstein belonging to the sequence

$$0 \to \mathbf{Z} \xrightarrow{n} \mathbf{Z} \to \mathbf{Z}_n \to 0\,,$$

we take $\beta_n(\varepsilon)$ as the k-invariant of Y: this has order (at most) n. There are then exact sequences of groups (twisted by w^ξ)

$$\begin{array}{ccccccc}
0 \longrightarrow & H^2(B;\mathbf{Z}) & \longrightarrow & H^2(Y;\mathbf{Z}) & \longrightarrow & \mathbf{Z} & \xrightarrow{\beta_n(\varepsilon)} H^3(B;\mathbf{Z}) \\
 & \downarrow & & \downarrow & & \downarrow & \qquad\downarrow \\
0 \longrightarrow & H^2(B;\mathbf{Z}_n) & \longrightarrow & H^2(Y;\mathbf{Z}_n) & \longrightarrow & \mathbf{Z}_n & \longrightarrow H^3(B;\mathbf{Z}_n) \\
 & \downarrow \beta_n & & \downarrow \beta_n & & & \\
 & H^3(B;\mathbf{Z}) & \longrightarrow & H^3(Y;\mathbf{Z}) & & &
\end{array}$$

The class $\varepsilon \in H^2(B;\mathbf{Z}_n)$ maps to zero in $H^3(Y;\mathbf{Z})$ since Y has k-invariant $\beta_n(\varepsilon)$, hence its image in $H^2(Y;\mathbf{Z}_n)$ lifts to a class in $H^2(Y;\mathbf{Z})$. Choose this class to have image $n \in \mathbf{Z}$ (possible since the image in $\mathbf{Z}_n$ is zero), and denote it by κ: this we take as the class of our 1-spherical fibration; it is determined only modulo $nH^2(B;\mathbf{Z})$, but as any two choices are equivalent, we fix one.

Now given M as above, since we have a circle fibration inducing the given fundamental groups, the image of ε in $H^2(M;\mathbf{Z}_n)$ is in the image of $H^2(M;\mathbf{Z})$ and hence annihilated by β_n. Thus the map $M \to K(B,1)$ lifts to a map $M \to Y$, and two lifts differ by an element of $H^2\big(M;\pi_2(X)\big)$. Now the class induced from κ differs from the class of the circle bundle over M by an element of $\mathrm{Ker}\big(H^2(M;\mathbf{Z}) \to H^2(M;\mathbf{Z}_n)\big) = nH^2(M;\mathbf{Z})$. Thus we can re-choose the lift $M \to Y$ so that κ induces the class of the circle bundle over M and this lift is determined uniquely modulo

$$\mathrm{Ker}\big(H^2(M;\mathbf{Z}) \to H^2(M;\mathbf{Z})\big) \;=\; \mathrm{Im}\, H^1(M;\mathbf{Z}_n)\,.$$

The rest of the argument is the same as in the simpler case first discussed.

Remark. The last version of the proof shows that the only condition to be satisfied by the extension

$$1 \to \mathbf{Z}_n \to A \to B \to 1$$

in order to be realised by circle fibrations is that the action of B on the normal subgroup $\mathbf{Z}_n$ must be induced from an action on $\mathbf{Z}$. Similarly, the fibration $S^1 \to X \to Y$ has the universal property in question if, and only if, X is an Eilenberg-MacLane space.

The above results extend immediately to the disconnected case on taking disjoint unions.

For the next results, we suppose given a pushout diagram Φ of groupoids

$$\Phi\ :\quad \begin{array}{ccc} A & \longrightarrow & C \\ \downarrow & & \downarrow \\ B & \longrightarrow & D \end{array}$$

together with a universal $(q-1)$-spherical fibration $p : X \to Y$ inducing $A \to B^*$. Let Z be a $K(C,1)$ meeting the mapping cylinder M_Y of p in X; then $M_Y \cup Z$ has fundamental groupoid D. We will write $K(\Phi)$ for the triad $(M_Y \cup Z; M_Y, Z; X)$. A Φ-*object* shall consist of:

a simple Poincaré pair (N^n, M) and a manifold pair (W^{n+q}, V);

a simple Poincaré embedding $(N, M) \to (W, V)$;

a 'smoothing' of the embedding $M \to V$;

and a map of triads

$$\left(\begin{array}{ccc} M^{n-1} & \longrightarrow & V^{n+q-1} \\ \downarrow & & \downarrow \\ N^n & \longrightarrow & W^{n+q} \end{array}\right) \longrightarrow \left(\begin{array}{ccc} X & \longrightarrow & Z \\ \downarrow & K(\Phi) & \downarrow \\ M_Y & \longrightarrow & M_Y \cup Z \end{array}\right)$$

**Here $q = 1$ or 2. Note that Φ includes an implicit choice of orientation character $w_\Phi = w(Y \cup Z) : D = \pi(Y \cup Z) \to \{\pm 1\}$.*

compatible with $w(Y \cup Z)$. The object will be called *restricted* if the map of triads induces isomorphisms of fundamental groupoids.

As in §9 there exist natural definitions of cobordism of Φ-objects, and of restricted cobordism, leading to a cobordism group $LS_n^1(\Phi)$ depending functorially on Φ, a restricted cobordism set $LS_n^2(\Phi)$, with preferred base element, and a forgetful map $LS_n^2(\Phi) \to LS_n^1(\Phi)$. These results, analogous to (9.1) and (9.2), are trivial, and we leave the proofs to the reader. More interesting is the analogue to (9.3), which is still an easy corollary of (3.3), using the methods of (11.3):

If the above Φ-object is restricted, it is cobordant to zero in the restricted sense if and only if the given smoothing of the Poincaré embedding $M \to V$ can be extended to a smoothing of the Poincaré embedding $(N, M) \to (W, V)$; provided $n \geqslant 5$.

It follows at once from this that we have a good obstruction theory for our smoothing problem, with obstruction groups the $LS_n^1(\Phi)$, provided we prove the analogue of (9.4), viz.

THEOREM 11.5. *Let $n \geqslant 4$, and assume the groupoids of Φ of finite type. Then the map $LS_n^2(\Phi) \to LS_n^1(\Phi)$ is bijective.*

This result is not trivial, and we give the proof of surjectivity; injectivity follows as usual by presenting the problem as a relative case of surjectivity.

Proof Suppose given a Φ-object, in the notation above. First consider the map $N \to Y$: exactly as in (9.4) we can perform surgery, leaving the boundary (M) fixed to make this 2-connected. This gives a simple Poincaré cobordism of N to N', say, with a $(q-1)$-spherical fibration induced over it by the map to Y. Now (as in the proof of (11.3)) we attach the mapping cylinder of this fibration to $\big(M(p) \cup C\big) \times I$ along $M(p) \times 1$, to obtain a (simple) Poincaré cobordism of the larger Poincaré complex, say to $M(p') \cup C'$. Next, apply (9.4) again to perform surgery on C' to make the induced map into Z 2-connected, and to leave the boundary (V) alone.

We have thus constructed a cobordism of simple Poincaré embeddings, starting with the given embedding, and such that the fundamental groupoids are as desired at the end. It remains to show that the cobordism of $M(p) \cup C$ is simply homotopy equivalent to a manifold cobordism of W; the desired result clearly follows from this. Now our cobordism was constructed by adding 1-handles and 2-handles only: for this one needs a (Poincaré or manifold) embedding of S^0 or S^1, with a trivialisation of the normal spherical fibration, block bundle, or vector bundle. But since O_n, $\widetilde{PL}_n$ and G_n all have the same 1-type, a fibre homotopy trivialisation induces a unique bundle trivialisation, and now since the dimension is at least 5 it follows (e.g. by (11.3)) that a Poincaré embedding determines a manifold embedding for S^0 or S^1. This shows that the Poincaré surgeries can be smoothed in turn to give a manifold cobordism of W, and thus completes the proof of the theorem.

The computation of these groups is now established by

THEOREM 11.6. *There exists an exact sequence, natural for maps of triads Φ as above,*

$$\cdots \to L_{n+1}(B) \xrightarrow{p} L_{n+q+1}(C \to D) \xrightarrow{q} LS_n(\Phi) \xrightarrow{r} L_n(B) \to \dots$$

Proof We first define the homomorphisms. For p, take an object over Y representing $\chi \in L_{n+1}(B)$: this is possible by (9.4.1). Over Y there is a universal $(q-1)$-spherical fibration, or equivalently, bundle or block bundle. Taking the induced (D^q, S^{q-1})-bundle over the object gives a relative object defining an element of $L_{n+q+1}(A \to B)$: our construction is evidently additive, cobordism-invariant, and natural, so gives a natural homomorphism*

$$p_0 \ : \ L_{n+1}(B) \to L_{n+q+1}(A \to B).$$

Define p by composing p_0 with the map $L_{n+q+1}(A \to B) \to L_{n+q+1}(C \to D)$ induced by Φ (regarded as a map of pairs of groupoids).

We define r by taking the surgery obstruction (relative to ∂M) for the tangential structure induced on M. Now $pr = 0$, for given a Φ-object representing χ, let $N' \subset W$ (with $\partial N' = M$) be obtained by transversality as in (11.3). Then $r(\chi)$ is the surgery obstruction for $N' \to N$, and $p_0 r(\chi)$ for the induced map of D^2-bundles. But using the fact that $N \subset W \subset W \times I$, we see that the surgery problem for D^2-bundles is induced from one for triads, with fundamental groupoid Φ (cf. (11.3) again). The exact sequence

$$L_{n+q+2}(\Phi) \to L_{n+q+1}(A \to B) \to L_{n+q+1}(C \to D)$$

now implies that $pr(\chi) = 0$.

Conversely, suppose given an object $N' \to N$ etc. representing an element of Ker p (throughout this paragraph, all manifolds will have a boundary, suppressed in the notation, which is to be fixed). The D^q-bundle defined (as we have seen) by the map $A \to B$ gives a relative object $(M', E') \to (M, E)$ with surgery obstruction on the kernel of the second, hence in the image of the first map above. So we can embed this in a more relative object $(W'; M', C'; E') \to (W; M, C; E)$. Now since Φ is a pushout, $\pi(M \cup C) \cong \pi(W)$. Thus, if M' and C' are amalgamated, we can do surgery (by (3.3)) to obtain a simple homotopy equivalence (of pairs) $(W'', V'') \to (W, M \cup C)$. Now N is Poincaré embedded in $M \cup C$; hence, using this equivalence, in V''. Our constructions (plus using transversality on the surgery we did) show that the induced tangential structure on N is that defined by the object $N' \to N$, which is thus in Im r, as required.

The definition of q needs some care, as although we have already observed that if the induced tangential structure on N comes from a manifold structure, we

* *Warning: the copy of the ring $B = \mathbf{Z}[\pi(Y)]$ in $L_{n+q+1}(A \to B)$ has the w_B-twisted involution, with $w_B : B \to D \to \{\pm 1\}$ the restriction of the orientation character w_Φ : $D \to \{\pm 1\}$ (see previous footnote), while the copy of B in $L_{n+1}(B)$ has the $w(p)w_B$-twisted involution, with $w(p)$ the orientation character of the $(q-1)$-spherical fibration $p : X \to Y$.*

have an obstruction in $L_{n+q+1}(C \to D)$ to obtaining N as a smooth submanifold; this obstruction need not be determined by the smoothing of N, essentially since the map $L_{n+1}(B) \to \mathscr{S}(N)$ is not usually injective.

The only way round this seems to be to give a new interpretation of $L_{n+q+1}(C \to D)$ as a cobordism group, using yet another definition of 'object'. We will attempt to treat this informally. Suppose given a Poincaré embedding $(\xi, C, E, h) : M^n \to V^{n+q}$, with fundamental groupoids as above. Using transversality as in (11.3) we construct a submanifold $M^m \subset V^{n+q}$. Now again as in (11.3) suppose we have a cobordism L of M to M'; then we can attach a disc (block) bundle A over L to $V \times I$ to form a manifold W, which we regard as a manifold triad. In fact we generalise this by permitting M and V to have boundaries, but assume the Poincaré embedding induced by a given embedding on the boundary, and L a product 'along the boundary', and then W a manifold tetrad. Given all this data, we have a surgery obstruction in $L_{n+q+1}(C \to D)$; this obstruction is unchanged if everything is altered by a cobordism, and conversely we claim: if two different set-ups have the same surgery obstruction, they are cobordant. This is the crucial point to verify: once $L_{n+q+1}(C \to D)$ is identified as the cobordism set, we can define q by ignoring all but the given Poincaré embedding of manifolds, with boundary induced by a manifold embedding. Exactness of (q, r) will then be immediate. Since the exactness of (p, q) was proved in all essentials in the discussion preceding (11.4), the proof of (11.6) will then be complete.

Assume then $(\xi_i, C_i, E_i, h_i) : M_i^n \to V_i^{n+q}$ $(i = 1, 2)$ Poincaré embeddings of manifolds; that transversality gives $M_i' \subset V_i^{n+q}$, and we have cobordisms L_i of M_i to M_i' such that the two eventual surgery obstructions coincide. We first construct a cobordism of M_1 to M_2, mapping into Y (the Y of (11.4)): as no restriction is yet imposed on the boundary, we can for instance take $M_1 \times I \cup M_2 \times I$. Now perform surgery to make the map 2-connected, thus obtaining a cobordism N, say. Glue the D^q-bundle induced over N to $(V_1 \cup V_2) \times I$, to obtain a cobordism of Poincaré embeddings, as in the figure

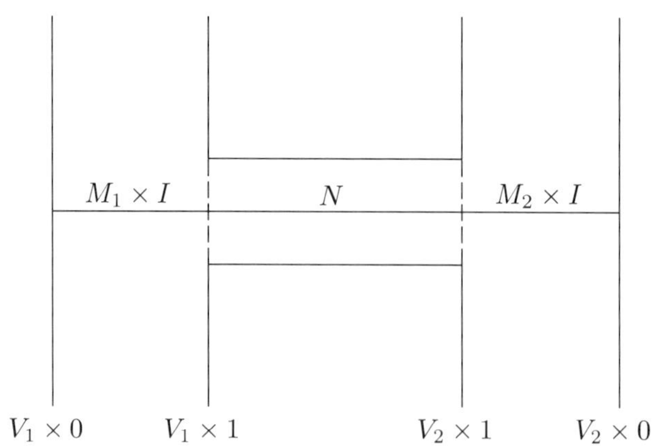

We now construct a manifold mapping into the above: in fact, we just glue together W_1, W_2, and the same disc bundle over N in the natural way. As usual (we have suppressed details above, in the interests of brevity and ease of reading) we have corresponding bundles and stable trivialisations. Our map is the identity on $(V_1 \times 0) \cup (V_2 \times 0)$, and on $\partial V_i \times I$. And our hypothesis implies that the surgery obstruction for our map induces zero in $L_{n+q+1}(C \to D)$, since we have glued, with opposite orientations, two problems with the same obstruction. It follows that we can find a cobordism to the empty set, whose fundamental groupoid is $C \to D$.

In the cobordism of the figure above to $\emptyset$, we simply take the induced boundary cobordism, and glue to the original figure: the result, reinterpreted in the obvious way, is a cobordism of the two original Poincaré embeddings. Since Φ is a pushout, it has the same fundamental groupoid as the support of the cobordism. Thus we can perform surgery (*not* keeping the disc bundle over N fixed) to obtain a simple homotopy equivalence. We do not use this to change the object, but can now regard the Poincaré embedding as going into a manifold W, a cobordism of V_1 to V_2. By transversality using the Poincaré embedding, we can find a cobordism $N' \subset W$ of M_1' to M_2'. It remains to find a cobordism of L_1 to L_2 which gives $N \cup N'$ on the boundary. But $L_1 \cup N \cup L_2$ is embedded in the manifold first constructed ($W_1 \cup W_2 \cup$ disc bundle over N), and hence in the manifold from which W was obtained by surgery. The desired cobordism is thus found by using transversality on the trace of this surgery.

This completes the construction of the desired cobordism and, with it, that of (11.6).

COROLLARY 11.6.1. *Multiplication by $P_2(\mathbf{C})$ induces isomorphisms for $n \geqslant 5$*

$$LS_n(\Phi) \to LS_{n+4}(\Phi)\,.$$

This follows by the naturality of the sequence of (11.6) and the Five Lemma, using (9.10).

We conclude this chapter with an analogue of (10.4) which will be used in the next chapter.

THEOREM 11.7. *Let $M^m \subset V^{m+q}$ be a (smooth or PL) submanifold, $q = 1$ or 2, and let Φ be the diagram of fundamental groupoids of the corresponding Poincaré embedding. Suppose $m \geqslant 5$, and let $x \in LS_{m+1}(\Phi)$. Then there exists a smoothing $M' \subset V$ of the same simple Poincaré embedding such that the obstruction to extending $M \times 0 \cup M' \times 1$ to a smoothing $N \subset V \times I$ of the product embedding is precisely x.*

Proof First consider $r(x) \in L_{m+1}(\pi(M))$: by (10.4) we can find a cobordism N of M to M', retracting to M, so that $M' \to M$ is a simple homotopy equivalence and the surgery obstruction for $N \to M \times I$ is $r(x)$. Form the bundle B with fibre D^q over N induced by the retraction from the normal disc bundle A of M in V: then the surgery obstruction for $B \to A \times I$ is $p_0 r(x)$, with the notation of (11.6). Attach B along $A \times 1$ to $V \times I$ to give W

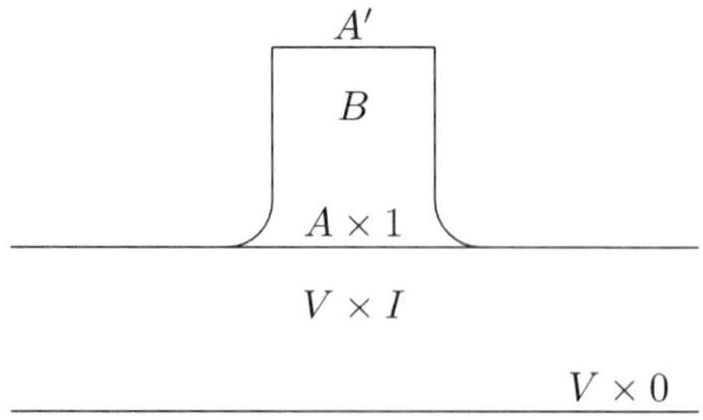

and round corners as indicated in the figure. There is an obvious retraction of W on $V \times I$; we wish to do surgery keeping $V \times 0$ and A' fixed. By naturality, the surgery obstruction is $ip_0 r(x) = pr(x) = 0$, so surgery can be performed to obtain a simple homotopy equivalence. W has then been converted to an s-cobordism; by the s-cobordism theorem the upper end is diffeomorphic to V, and we have $M' \subset A' \subset V$. If y is the obstruction to extending $M \times 0 \cup M' \times 1$ to a smoothing of the product Poincaré embedding, we evidently have $r(u) = r(x)$, so $x = y + q(z)$ for some z.

Using the additive property of surgery obstructions, we now see that it is enough to prove the result with the additional assumption $x = q(z)$. For this case we apply (10.4) to say that there is a cobordism of $V \times I$, keeping $(V \times 0) \cup A$ fixed, and retracting on $V \times I$, such that the upper end is mapped by a simple homotopy equivalence, and the surgery obstruction (rel $V \times 0 \cup A$) for the whole is z. Now by the same argument as above, the upper end of this cobordism is an s-cobordism of $V \times 0$, so can again be identified with $V \times I$, and $V \times 1$ contains $A \supset M$: we claim that this gives the desired embedding of M in V. Indeed, the calculation of the obstruction to extending the two embeddings of M in V to an embedding of $M \times I$ in $V \times I$ amounts to reversing the constructions above, which yields $q(z)$.

Remark. It is possible to obtain a more precise result, analogous to (10.4); however, we do not need this, and the detailed statement would have to be very technical.

The LS-groups and the surgery exact sequence. The surgery exact sequence of §10 and the codimension q surgery exact sequence of §11 are related as follows. The algebraic surgery exact sequence for a space V with a decomposition $V = C \cup E(p)$ and the exact sequence for $LS_*(\Phi)$ of Theorem 11.6 (with p the projection of a $(q-1)$-spherical fibration ξ over a subspace $M \subset V$ and Φ the pushout square of fundamental groups) are related by a natural transformation

$$\begin{array}{ccccccccc}
\cdots \to & L_{m+q+1}(V) & \to & \mathscr{S}_{m+q+1}(V) & \to & H_{m+q}(V;\mathbf{L}_\bullet) & \to & L_{m+q}(V) & \to \cdots \\
& \downarrow & & \downarrow & & \downarrow & & \downarrow & \\
\cdots \to & L_{m+q+1}(V,C) & \to & LS_m(\Phi) & \to & L_m(M) & \to & L_{m+q}(V,C) & \to \cdots
\end{array}$$

with

$$L_*(V) = L_*(\pi(V)) \ , \ L_*(M) = L_*(\pi(M)) \ ,$$
$$L_*(V,C) = L_*(\pi(C) \to \pi(V))$$

and

$$H_{m+q}(V;\mathbf{L}_\bullet) \to H_{m+q}(V,C;\mathbf{L}_\bullet) \cong H_m(M;\mathbf{L}_\bullet) \to L_m(M)\,.$$

If V^{m+q} is an $(m+q)$-dimensional manifold and $M^m \subset V$ is a codimension q submanifold, the structure invariant $s(f) \in \mathscr{S}_{m+q+1}(V) = \mathscr{S}_{TOP}(V)$ of a simple homotopy equivalence $f : W \to V$ from an $(m+q)$-dimensional manifold W has image the splitting obstruction $s_M(f) \in LS_m(\Phi)$ ([R7, 7.2]).

Codimension 2 and homology surgery. A codimension 2 manifold embedding $M^m \subset V^{m+2}$ is a generalised knot. The general philosophy for applying surgery theory to codimension 2 embeddings is to classify the homotopy types of manifolds within a homology type. For example, the complement of a knot $k : S^1 \subset S^3$ has the homology type of S^1, by Alexander duality; by Dehn's lemma k is unknotted if and only if the complement has the homotopy type of S^1. Motivated by the classification of high-dimensional $\mathbf{Z}_2$-invariant knots (using the invariant of Browder and Livesay [B29]) López de Medrano [L22] posed the problem of deciding if a normal map is normal cobordant to a homology equivalence. Cappell and Shaneson ([C9], et.al.) went on to develop homology surgery theory and many of its applications to codimension 2 embeddings. In this theory the surgery obstruction groups $L_*(\pi)$ for homotopy equivalences of manifolds with fundamental group π are replaced by the algebraic Γ-groups $\Gamma_*(\mathscr{F})$ for Λ-coefficient homology equivalences of manifolds with fundamental group π, with $\mathscr{F} : \mathbf{Z}[\pi] \to \Lambda$ a morphism of rings with involution. See Ranicki [R7, 7.8], [R13] for accounts of the surgery method in high-dimensional knot theory, including the analogues $\Gamma S_*(\Phi)$ of the codimension 2 LS-groups $LS_*(\Phi)$. The codimension 2 LS-groups $LS_*(\Phi)$ with $\pi_1(M) \cong \pi_1(V)$ are identified in [R7, 7.8.12] with the algebraic L-groups of $\mathbf{Z}[\pi_1(C)]$ with an antistructure, analogous to the identification in Theorem 12.9 of the LS-groups $LS_*(\Phi) = LN_*(\pi_1(C) \to \pi_1(V))$ in the norientable codimension 1 case with $\pi_1(M) \cong \pi_1(V)$ with the algebraic L-groups of $\mathbf{Z}[\pi_1(C)]$ with an antistructure. See Levine and Orr [L15] for a survey of the applications of surgery theory to knots and links.

12. Submanifolds : Other Techniques

First-time readers may omit this chapter, proceeding directly to §13.

Our discussion in §11 gave a reasonably complete picture of the obstructions encountered in smoothing Poincaré embeddings in codimensions 1 and 2, in terms of the groups $L_m(\pi)$. It follows that any other account, on comparison with this one, will yield information about these groups. Our present intention is to survey all known techniques for the submanifold problem, and to see what information can be deduced.

The only direct attack on the smoothing problem in the case $q = 2$ known to me is that of Casson [C10]. The results of this paper are all contained in our (11.3), and it seems probable that the methods would yield no more than that result. Thus we consider only the case $q = 1$.

We divide cases when M has codimension one in V into three types, in increasing order of difficulty.

(A) M separates V, i.e. C is disconnected.

(B) M is 2-sided in V (i.e. ξ is trivial) but does not separate it.

(C) M is 1-sided in V.

The three cases are illustrated by the following diagrams

(A)

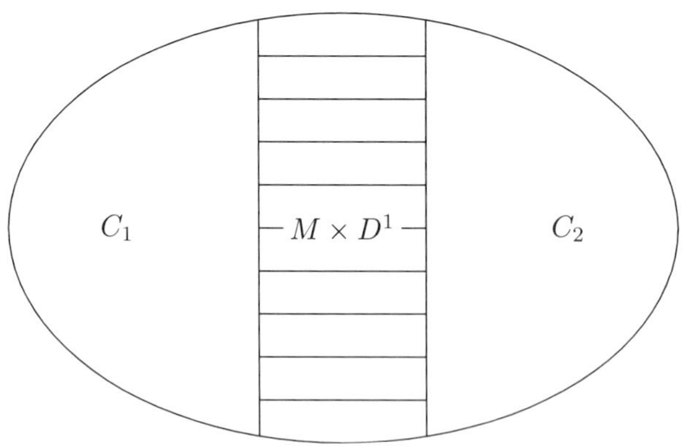

(B)

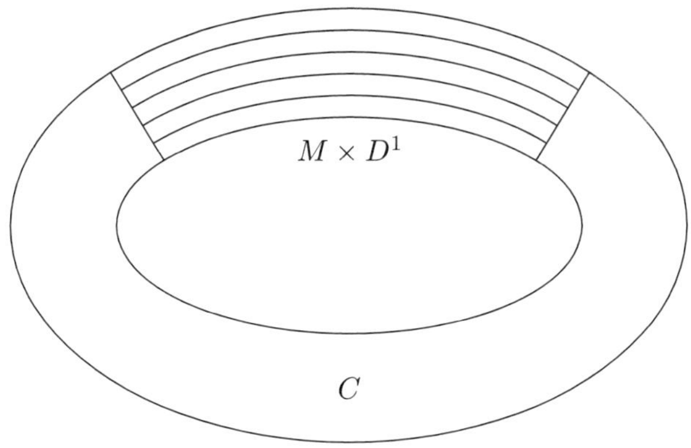

(C)

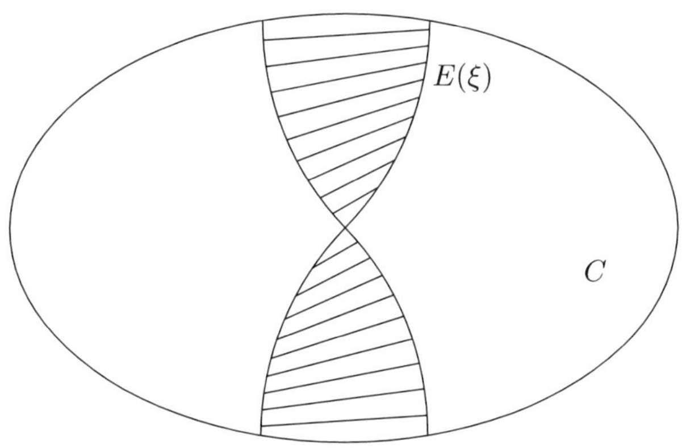

Here is a brief summary of the algebraic expressions of the codimension 1 splitting obstruction groups $LS_*(\Phi)$ in the three cases.

(A)+(B) In 12.4.1 the LS-groups are identified with the triad L-groups

$$LS_*(\Phi) \;=\; L_{*+2}(\Phi)\,.$$

In case (A) there is defined a pushout square of fundamental groups

$$\Phi\ :\quad \begin{array}{ccc} \pi_1(M) & \xrightarrow{i_1} & \pi_1(C_1) \\ {\scriptstyle i_2}\downarrow & & \downarrow \\ \pi_1(C_2) & \longrightarrow & \pi_1(V) \end{array}$$

by van Kampen's theorem, with C_1, C_2 *the two components of* C *(assuming* M, V *connected) and* $i_1 : \pi_1(M) \to \pi_1(C_1)$, $i_2 : \pi_1(M) \to \pi_1(C_2)$ *the morphisms induced by the inclusions* $i_1 : M \to C_1$, $i_2 : M \to C_2$. *The fundamental group of* V *is the amalgamated free product*

$$\pi_1(V) \;=\; \pi_1(C_1) *_{\pi_1(M)} \pi_1(C_2)$$

and the codimension 1 splitting groups LS-groups $LS_*(\Phi)$ *are the obstructions to the existence of a Mayer-Vietoris exact sequence in the L-groups*

$$\begin{aligned}\cdots \to L_{n+1}(\pi_1(V)) \xrightarrow{\partial} L_n(\pi_1(M)) \xrightarrow{(i_1\ i_2)} L_n(\pi_1(C_1)) \oplus L_n(\pi_1(C_2))&\\ \to L_n(\pi_1(V)) \xrightarrow{\partial} L_{n-1}(\pi_1(M)) \to \cdots&.\end{aligned}$$

For injective $\pi_1(M) \to \pi_1(V)$ *(or equivalently injective* $i_1 : \pi_1(M) \to \pi_1(C_1)$, $i_2 : \pi_1(M) \to \pi_1(C_2)$*) Waldhausen [W3] obtained an exact sequence in algebraic K-theory*

$$\begin{aligned}\cdots \to Wh_2(\pi_1(V)) &\xrightarrow{\partial} Wh(\pi_1(M)) \oplus \widetilde{\mathrm{Nil}}_2(\Phi)\\ &\xrightarrow{(i_1\ i_2)\oplus 0} Wh(\pi_1(C_1)) \oplus Wh(\pi_1(C_2))\\ &\to Wh(\pi_1(V)) \xrightarrow{\partial} \widetilde{K}_0(\mathbf{Z}[\pi_1(M)]) \oplus \widetilde{\mathrm{Nil}}_1(\Phi) \to \dots\end{aligned}$$

with $Wh_*(\pi_1(V)) \to \widetilde{\mathrm{Nil}}_*(\Phi)$ *split surjections. Cappell [C4] defined algebraic* UNil*-groups* $\mathrm{UNil}_*(\Phi)$ *for injective* $\pi_1(M) \to \pi_1(V)$, *such that*

$$\begin{aligned}LS_n(\Phi) \;=&\\ \widehat{H}^{n+1}\Big(\mathbf{Z}_2; \mathrm{Ker}\big((i_1\ i_2) : Wh(\pi_1(M)) &\to Wh(\pi_1(C_1)) \oplus Wh(\pi_1(C_2))\big)\Big)\\ &\oplus \mathrm{UNil}_{n+2}(\Phi)\ ,\end{aligned}$$

and used the realisation theorems (5.8), (6.5) *to obtain geometrically an exact sequence*

$$\begin{aligned}\dots \to L_{n+1}(\pi_1(V)) &\xrightarrow{\partial} L'_n(\pi_1(M)) \oplus \mathrm{UNil}_{n+1}(\Phi)\\ &\xrightarrow{(i_1\ i_2)\oplus 0} L_n(\pi_1(C_1)) \oplus L_n(\pi_1(C_2)) \to L_n(\pi_1(V))\\ &\xrightarrow{\partial} L'_{n-1}(\pi_1(M)) \oplus \mathrm{UNil}_n(\Phi) \to \dots\end{aligned}$$

with $L_*(\pi_1(V)) \to \mathrm{UNil}_*(\Phi)$ *split surjections and* $L'_*(\pi_1(M))$ *the intermediate L-groups (see §17D) associated to the $*$-invariant subgroup*

$$\begin{aligned}&\mathrm{Im}\Big(Wh_2(\pi_1(V)) \to Wh(\pi_1(M))\Big)\\ &= \;\mathrm{Ker}\Big((i_1\ i_2) : Wh(\pi_1(M)) \to Wh(\pi_1(C_1)) \oplus Wh(\pi_1(C_2))\Big) \subseteq Wh(\pi_1(M))\,.\end{aligned}$$

In case (B) the fundamental group of V is the HNN extension

$$\pi_1(V) = \pi_1(C) *_{\pi_1(M)} \{t\}$$

determined by the morphisms $i_1, i_2 : \pi_1(M) \to \pi_1(C)$ induced by the two inclusions $i_1, i_2 : M \to C$, and the codimension 1 splitting groups $LS_*(\Phi) = L_{*+2}(\Phi)$ are the obstructions to the existence of a Mayer-Vietoris exact sequence in the L-groups

$$\cdots \to L_{n+1}(\pi_1(V)) \xrightarrow{\partial} L_n(\pi_1(M)) \xrightarrow{i_1 - i_2} L_n(\pi_1(C)) \to L_n(\pi_1(V)) \xrightarrow{\partial} L_{n-1}(\pi_1(M)) \to \cdots .$$

For injective $\pi_1(M) \to \pi_1(V)$ (or equivalently injective $i_1, i_2 : \pi_1(M) \to \pi_1(C)$) there is an exact sequence in algebraic K-theory

$$\cdots \to Wh_2(\pi_1(V)) \xrightarrow{\partial} Wh(\pi_1(M)) \oplus \widetilde{\mathrm{Nil}}_2(\Phi) \xrightarrow{(i_1-i_2)\oplus 0} Wh(\pi_1(C)) \to Wh(\pi_1(V)) \xrightarrow{\partial} \widetilde{K}_0(\mathbf{Z}[\pi_1(M)]) \oplus \widetilde{\mathrm{Nil}}_1(\Phi) \to \cdots$$

([W3], generalising the splitting theorem of Bass, Heller and Swan [B9] for the Whitehead group of a polynomial extension). The algebraic UNil*-groups* $\mathrm{UNil}_*(\Phi)$ *of [C4] are such that*

$$LS_n(\Phi) = \widehat{H}^{n+1}\Big(\mathbf{Z}_2; \mathrm{Ker}(i_1 - i_2 : Wh(\pi_1(M)) \to Wh(\pi_1(C)))\Big) \oplus \mathrm{UNil}_{n+2}(\Phi)$$

and there is defined an exact sequence

$$\cdots \to L_{n+1}(\pi_1(V)) \xrightarrow{\partial} L'_n(\pi_1(M)) \oplus \mathrm{UNil}_{n+1}(\Phi) \xrightarrow{(i_1-i_2)\oplus 0} L_n(\pi_1(C)) \to L_n(\pi_1(V)) \xrightarrow{\partial} L'_{n-1}(\pi_1(M)) \oplus \mathrm{UNil}_n(\Phi) \to \cdots$$

with $L_*(\pi_1(V)) \to \mathrm{UNil}_*(\Phi)$ split surjections and $L'_*(\pi_1(M))$ the intermediate L-groups associated to the $*$-invariant subgroup

$$\mathrm{Im}\Big(Wh_2(\pi_1(V)) \to Wh(\pi_1(M))\Big) = \mathrm{Ker}\Big(i_1 - i_2 : Wh(\pi_1(M)) \to Wh(\pi_1(C))\Big) \subseteq Wh(\pi_1(M)) .$$

(C) If $\pi_1(M) \to \pi_1(V)$ is an isomorphism then $\pi_1(C) = \pi' \to \pi_1(V) = \pi$ is the injection of a subgroup of index 2, and the codimension 1 splitting groups $LS_*(\Phi) = LN_*(\pi' \to \pi)$ are identified in §12C with the algebraic L-groups of $\mathbf{Z}[\pi]$ with an antistructure.

See Ranicki [R7, 7.6], [R8], [R9, §23], [R10] for a further discussion of the development of codimension 1 splitting obstruction theory since the first edition of the book. In particular, the combination of [R7, pp. 666–685] and [R10, §8] gives a purely algebraic treatment of the Cappell UNil*-groups using chain complexes with Poincaré duality.*

12A. Separating Submanifolds

We begin our discussion of type (A) with the simplest case, when no surgery obstructions are encountered. This case was first discussed by Browder [B18], and his results improved by Wagoner [W1] and Sullivan (unpublished), but all assuming all spaces involved to be 1-connected.

THEOREM 12.1. *Let* (Y_1, X) *and* (Y_2, X) *be simple Poincaré pairs, with*

$$Y_1 \cap Y_2 = X \ , \ Y_1 \cup Y_2 = Y \ ,$$

and $i_* : \pi(X) \cong \pi(Y_1)$*. Let* V^v *be a closed (smooth or PL) manifold,* $v \geqslant 6$*, and* $\phi : V \to Y$ *a simple homotopy equivalence. Then there exists a submanifold* V_1^v *with boundary* M^{v-1} *of* V*, such that if* $V_2 = V - Int\ V_1$*,* ϕ *is homotopic to a map* ϕ^1 *which induces a simple homotopy equivalence of triads* $(V; V_1, V_2; M) \to (Y; Y_1, Y_2; X)$*.*

Proof Replacing Y_1, Y_2 by mapping cylinders if necessary, we may assume that X has product neighbourhoods $X \times [-1, 0]$, $X \times [0, 1]$ in them; in particular, we have a collapsing map $\pi : Y \to \Sigma X \to [-1, 1]$. Make $\pi \circ \phi$ transverse to 0, moving only points in $\phi^{-1}\big(\pi^{-1}(-\frac{1}{2}, \frac{1}{2})\big)$: then we can modify ϕ correspondingly (keeping the projection on X constant where it is defined), and so suppose ϕ such that $\pi \circ \phi$ is transverse to 0. Set $M' = \phi^{-1}(X)$, $V_i' = \phi^{-1}(Y_i)$ $(i = 1, 2)$.

Choose a homotopy inverse ϕ^{-1} to ϕ. Choose also an inverse to the tangent bundle τ_V – i.e. a bundle ν_V and trivialisation F of $\tau_V \oplus \nu_V$. Set $\nu = (\phi^{-1})^* \nu_V$: then we can identify $\nu_V = \phi^* \nu$. Now apply (3.3) to

$$(\phi \,|\, V_1') : (V_1', M') \to (Y_1, X)$$

with $\nu \,|\, Y_1$ and $F \,|\, V_1'$. Thus there is a cobordism

$$\psi : (Q, P) \to (Y_1, X)$$

of this pair to a pair (V_1, M) mapped by a simple homotopy equivalence, and also a stable trivialisation F' of $\tau_Q \oplus \psi^* \nu$ extending F.

Now construct a manifold W by glueing $V \times I$ to Q along $V_i' \times 1$. The maps $(\phi \times id)$ and ψ combine to give a map $\psi' : W \to Y \times I$; F and F' combine to give a stable trivialisation F'' of $\tau_W \oplus \psi'^* \nu$.

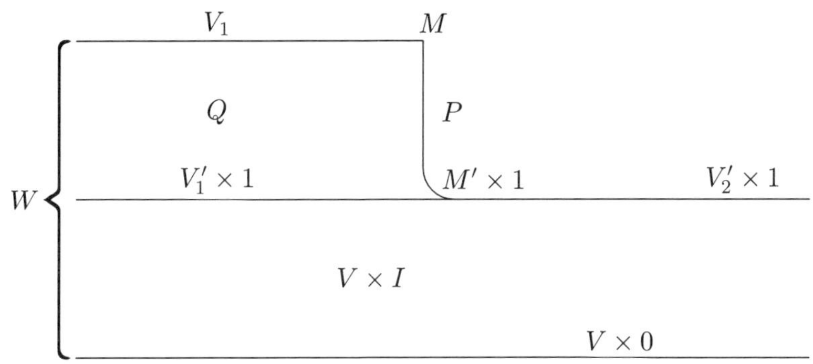

In the smooth case, we must round the corner at $M' \times 1$. Now (as in (11.3)) we define triads by

$$X\{1,2\} = W\,, \qquad X\{1\} = (V \times 0) \cup V_1\,, \qquad X\{2\} = (V_2' \times 1) \cup P\,,$$

$$Z\{1,2\} = Y \times I\,, \quad Z\{1\} = (Y \times 0) \cup (Y_1 \times 1)\,, \quad Z\{2\} = (Y_2 \times 1)\,.$$

Then ψ' induces a map of degree 1 of triads $X \to Z$ and a simple homotopy equivalence $\partial_2 X \to \partial_2 Z$; we constructed ν, F'' above. To apply (3.3), we need only check that $\pi(Y_2 \times 1) \to \pi(Y \times I)$ is an isomorphism. But by hypothesis $\pi(X) \to \pi(Y_1)$ is an isomorphism, and we have $Y_1 \cap Y_2 = X$, $Y_1 \cup Y_2 = Y$, so by van Kampen's theorem $\pi(Y_2) \to \pi(Y)$ is an isomorphism as required. Thus by (3.3) we can perform surgery (leaving $\partial_2 X$ fixed) to obtain a simple homotopy equivalence. Thus (as in (11.3)) the resulting manifold is an s-cobordism of V, hence is diffeomorphic to $V \times I$. But the upper end is split into M_1 and $M_2 = X'\{2\}$, which are mapped by simple homotopy equivalences, as desired.

Like the earlier embedding theorems, (12.1) can easily be relativised – one must start with Poincaré $(n+1)$-ads Y_1, Y_2, a simple homotopy equivalence $X = \partial_n Y_1 \to \partial_n Y_2$ (which, using a mapping cylinder, we take as the identity, and then assume $|Y_1| \cap |Y_2| = |X|$) and a manifold structure on the glued n-ad Y. If inclusion induces an isomorphism $\pi(X) \to \pi(\delta_n Y_1)$, and if $\dim X \geqslant 5$, we can proceed. Of course the result has a relative form: given manifold structures on $\partial_{n-1} Y_i$ which fit on $\partial_{n-1} X$, we can extend them. The proof extends in the same way as that of (11.3). As for (11.3.3) we deduce uniqueness up to concordance.

One can also combine (12.1) with (11.3.4) to obtain a new proof of (11.3) (this presupposes an alternative proof of (11.3.4)), and this method was at one time used by Browder [B18]. However, his later and more direct argument seems preferable.

We now drop the hypothesis that $\pi(X) \to \pi(Y_1)$ is an isomorphism in the above. We find that an obstruction appears, but that it is expressible by our earlier theory.

THEOREM 12.2_A. *Let (Y_1, X) and (Y_2, X) be simple Poincaré pairs with*

$$Y_1 \cap Y_2 = X \ , \ Y_1 \cup Y_2 = Y \ .$$

Let V^v be a closed (smooth or PL) manifold, $v \geqslant 6$, and $\phi : V \to Y$ a simple homotopy equivalence. Then there is a single, well-defined obstruction to smoothing X to a locally flat submanifold of V, and it lies in $L_{v+1}(\pi(Z))$, where Z is the triad $(Y; Y_1, Y_2; X)$.

Proof We proceed as in the proof of (12.1) to define V_i', M', ν' and F. We now have a Poincaré tetrad W:

$$|W| = Y \times I, \quad |\partial_2 W| = Y \times 0, \quad |\partial_1 W| = Y_1 \times 1, \quad |\partial_3 W| = Y_2 \times 1,$$

a corresponding manifold tetrad U, and a map (induced by ϕ) $\psi : U \to W$ of tetrads inducing a simple homotopy equivalence $\partial_2 U \to \partial_2 W$. The only choice in defining this was the application of transversality. Hence the bordism class of (U, ψ, ν, F) is uniquely determined. Thus the obstruction to performing surgery rel $\partial_2 U$ to obtain a simple homotopy equivalence is well-defined, and lies in $L_{v+1}(\pi(\delta_2 U)) \cong L_{v+1}(\pi(Z))$, since $Z \times 1 \subset \delta_2 U$ is a homotopy equivalence. If surgery can be performed, then as in (12.1) we can smooth X to a locally flat submanifold M of V.

This is not yet adequate for the applications to be made later of this idea. We must first relativise, and then show that each element of the suggested obstruction group does occur as an obstruction for the relativised problem. Relativising is perfectly straightforward, and the only extension we need is when V has a boundary, already split into two parts, and we seek to extend the splitting to V in a particular way.

THEOREM 12.2_B. *Let V_1, V_2 be manifold triads with $M = \partial_2 V_1 = \partial_2 V_2$, and V^v the manifold pair formed by glueing them. Let π be the fundamental groupoid of the triad $(V; V_1, V_2)$, $x \in L_{v+2}(\pi)$, and $v \geqslant 6$. Then there exists a new smoothing $M' \subset V$ of the Poincaré embedding defined by $M \subset V$ (relative to $\partial M \subset \partial V$) such that the obstruction to extending $(M \times 0) \cup (\partial M \times I) \cup (M' \times 1) \subset \partial(V \times I)$ to a smoothing of the product Poincaré embedding $M \times I \subset V \times I$ is x.*

Proof Define a tetrad U (as in (12.12_A)) by

$$|U| = V \times I, \quad |\partial_3 U| = V \times 0 \cup \partial V \times I, \quad |\partial_2 U| = V_1 \times 1, \quad |\partial_1 U| = V_2 \times 1.$$

Thus dim $|U| = v+1$, and $x \in L_{v+2}(\pi(\delta_3 U))$. Regard $(I, \partial I)$ as a pair, and form the product pentad $U \times I$. By (12.1), there exists a degree 1 map of pentads $\phi : W \to U \times I$, inducing simple homotopy equivalences on ∂_3 and ∂_4 (indeed, the identity on $U \times 0$) and having surgery obstruction x.

Now amalgamate $\partial_1(U \times I)$ and $\partial_2(U \times I)$, obtaining $V \times 1 \times I$; correspondingly with W, to obtain a map $\phi : aW \to a(U \times I) = aU \times I$ of amalgamated tetrads. By (3.3) we can perform surgery (relative to ∂_3 and ∂_4) to obtain a simple homotopy equivalence of tetrads. The surgery gives a cobordism mapped into $V \times I \times I$, with first face mapped into $V \times 1 \times I$. Make this map transverse to $M \times 1 \times I$ (leaving it fixed on W): thus we obtain a cobordism of pentads. Replacing W (if necessary) by the other end of the cobordism, which of course has the same surgery obstruction x, we see that we may assume that $\phi : W \to U \times I$ is such that the induced map of the amalgamated tetrads $aW \to aU \times I$ is a simple homotopy equivalence.

It follows that $|W|$ is an s-cobordism, relative to the boundary, of $|U| = V \times I$, hence we can identify it with $V \times I \times I$, and $|\partial_1 W \cup \partial_2 W|$ with $V \times 1 \times I$. Thus $W\{3,4\}$ is identified with a submanifold N of $V \times 1 \times I$, with boundary $(M \times 0) \cup (\partial M \times I) \cup (M' \times I)$ (for, by construction, all is canonical on $V \times 0 \cup \partial V \times I$). The induced simple homotopy equivalence of the subdivided $V \times 1$ shows that M' and M smooth the same Poincaré embedding. But our construction shows that the obstruction to extending $(M \times 0) \cup (\partial M \times I) \cup (M' \times 1)$ to a smoothing of the product Poincaré embedding is precisely x.

Pentads occurred here in discussing a non-relativised theorem; this shows that the discussion of n-ads in general is avoidable only at the cost of extra complications in cases like this. Of course, (12.2) can be systematically relativised too, but we will not need this.

Let Φ be a pushout diagram of finitely presented groups

$$\Phi : \quad \begin{array}{ccc} B & \xrightarrow{b_1} & C_1 \\ {\scriptstyle b_2}\downarrow & & \downarrow{\scriptstyle c_1} \\ C_2 & \xrightarrow{c_2} & D \end{array}$$

and write Ψ for the groupoid triad

$$\Psi : \quad \begin{array}{ccc} B \cup B & \xrightarrow{b_1 \cup b_2} & C_1 \cup C_2 \\ {\scriptstyle 1 \cup 1}\downarrow & & \downarrow{\scriptstyle c_1 \cup c_2} \\ B & \xrightarrow{c_1 b_1} & D \end{array}$$

deduced from it. Comparing the discussion above with (11.5), it is clear that $LS_n(\Psi) \cong L_{n+2}(\Phi)$. Moreover, we can identify $L_n(B) = L_{n+1}(B \cup B \to B)$ in a natural way, and (11.6) is then simply one of the exact sequences of the triad Ψ. Thus in the case when M separates V, our theory of (11.5)–(11.6) reduces entirely to the former one.

We can combine (12.2) with the theory of §11 in a different way.* For given a simple Poincaré embedding $(\xi, C, E, h) : M^m \to V^{m+q}$, with V a manifold, we

* *The discussion includes non-separating submanifolds. In particular, the definition of $LN_n(\phi)$ further below includes the non-separating case.*

apply (12.2) to $h : V \to M(\xi) \cup C$. We then find an obstruction to obtaining a corresponding splitting of V. If this obstruction vanishes, we split V: one part, say N, is simply homotopy equivalent to M, and we can then treat separately the problem of embedding M in N (cf. (11.3.4)). This leads to the following. Set $A = \pi(E)$, $B = \pi(M)$, $C = \pi(D)$, $D = \pi(V)$ as before, and

$$\Phi : \begin{array}{ccc} A & \longrightarrow & C \\ \downarrow & & \downarrow \\ B & \longrightarrow & D \end{array} \qquad \Psi = s_2\partial_2\Phi : \begin{array}{ccc} A & \longrightarrow & A \\ \downarrow & & \downarrow \\ B & \longrightarrow & B \end{array}$$

PROPOSITION 12.3. *We have a commutative diagram of exact sequences**

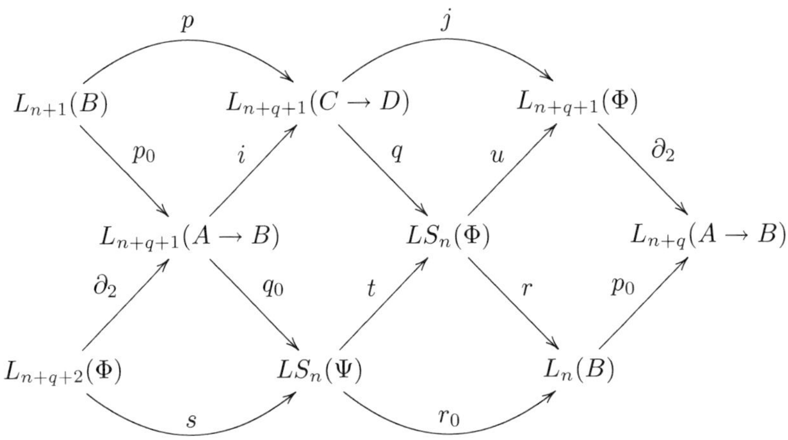

Proof Theorem (11.6) gives exactness of (p, q, r), also of (p_0, q_0, r_0), which is a special case; (3.2) gives a third sequence (i, j, ∂_2). We now define $s = q_0\partial_2$, t is induced by a map of triads, and u is induced, as in the description above, by taking the obstruction to splitting V as indicated by the Poincaré embedding. Triangles involving p and s are commutative by definition; $rt = r_0$ by naturality. The definition of q shows that $j = uq$. We have $qi = tq_0$ by naturality; finally, $\partial_2 u(x) = p_0 r(x)$ since if x is the class of a Poincaré embedding $(\xi, C, E, h) : M \to V$, and the splitting $V = N \cup C'$ is obtained by transversality, each of $\partial_2 u(x)$ and $p_o r(x)$ represents the surgery obstruction for $(N, \partial N) \to (M(\xi), E)$.

Since our diagram is commutative and three sequences are exact, by [W16] exactness of the fourth sequence follows if $ut = 0$. But this follows, since $L_{n+q+1}(\Psi) = 0$, from the commutative diagram

$$\begin{array}{ccc} LS_n(\Psi) & \longrightarrow & LS_n(\Phi) \\ \downarrow u_0 & & \downarrow u \\ L_{n+q+1}(\Psi) & \longrightarrow & L_{n+q+1}(\Phi)\,. \end{array}$$

* *The diagram is defined for all $q \geqslant 1$, but for $q \geqslant 3$ it collapses, since in that case $A = C$, $B = D$, $L_*(\Phi) = 0$, $LS_*(\Psi) = LS_*(\Phi)$.*

This fourth sequence, hinted at in the discussion preceding the lemma, shows how to 'localise' the problem of smoothing an embedding by looking at $LS_n(\Psi)$, which depends only on $A \to B$; namely, first (assuming the obstruction in $L_{n+q+1}(\Phi)$ vanishes), we embed in V^{m+1} a manifold N^{m+1} with relative boundary E', such that $\big(M(\xi), E\big) \to (N, E')$ is a simple homotopy equivalence; then try to embed M in N. If $q = 1$ and ξ is trivial, $M(\xi) = M \times I$, $E = M \times \partial I$, so N is an s-cobordism, and we can certainly embed M.

Let us abbreviate our notation for these more important $LS_n(\Psi)$, and write

$$LN_n(\phi) \;=\; LS_n(s_2\phi)\,.$$

The preceding remark now leads to

LEMMA 12.4. $LN_n(B \cup B \to B) = 0$.

A direct proof from the definition can easily be given using (12.1). Substituting in (12.3), we find that

COROLLARY 12.4.1. *Suppose* $q = 1$ *and* ξ *trivial. Then*

$$u\,:\,LS_n(\Phi) \cong L_{n+2}(\Phi)\,.$$

Notice particularly that this result does not assume that V separates M. In the case where it does, comparison with our earlier result yields an isomorphism (also deducible *a priori*) of the L-groups of the triads

$$\begin{array}{ccc} B & \longrightarrow & C_1 \\ \downarrow & & \downarrow \\ C_2 & \longrightarrow & D \end{array} \qquad \text{and} \qquad \begin{array}{ccc} B \cup B & \longrightarrow & C_1 \cup C_2 \\ \downarrow & & \downarrow \\ B & \longrightarrow & D \end{array}$$

Our method of proof of (12.1) and (12.2) differs from the original method of Browder (which appeared first in [B18]). Thus in spite of our success, there is still room for hope that the original method, with suitable modification, may lead to a theorem different from (though not, of course, contradicting) the above. In particular, we make the

CONJECTURE. *Let*

$$\Phi\ :\quad \begin{array}{ccc} A & \longrightarrow & C \\ \downarrow & & \downarrow \\ B & \longrightarrow & D \end{array}$$

be a pushout diagram of groups, with all maps injective. Then $L_n(\Phi) \equiv 0$.

A weaker conjecture is that $L_n(\Phi)$ is a subquotient of the Whitehead group $Wh(A)$: cf. (12.5) below. It is easy to give examples to show that some condition such as injectivity is necessary; for an example, let $A \to C$ be $\times 2 : \mathbf{Z} \to \mathbf{Z}$ and B (hence also D) trivial.

Although $(\times 2)$ does *not* induce $\times 2 : L_i(\mathbf{Z}) \to L_i(\mathbf{Z})$ for all i, it does so for $L_{4k+1}(\mathbf{Z}) \cong \mathbf{Z}$, and so is certainly not an isomorphism.

For recent results see §17E.

Cappell [C2], [C3], [C4], [C5] showed that these conjectures are false in general, using the groups $\mathrm{UNil}_*(\Phi)$ *of unitary nilpotent hermitian forms. The* UNil-*groups were defined for the pushout square* Φ *of any amalgamated free product*

$$\pi_1(V) \;=\; \pi_1(C_1) *_{\pi_1(M)} \pi_1(C_2)$$

with $\pi_1(M) \to \pi_1(C_1)$, $\pi_1(M) \to \pi_1(C_2)$ *injective. (See the notes at the end of the previous section of §12 for some of the properties of the* UNil-*groups). Specifically, for any* $k \geq 1$ *let*

$$M_0^{4k} \;=\; S^{4k} \subset V_0^{4k+1} \;=\; P_{4k+1}(\mathbf{R}) \# P_{4k+1}(\mathbf{R}) ,$$

so that

$$\pi_1(M_0) \;=\; \{1\} \subset \pi_1(V_0) \;=\; D_\infty \;=\; \mathbf{Z}_2 * \mathbf{Z}_2$$

with D_∞ *the infinite dihedral group. The* UNil-*groups of the corresponding pushout square of groups*

$$\Phi_0 \;:\; \begin{array}{ccc} \{1\} & \longrightarrow & \mathbf{Z}_2 \\ \big\downarrow & & \big\downarrow \\ \mathbf{Z}_2 & \longrightarrow & D_\infty \end{array}$$

were shown to be non-zero, with

$$\bigoplus_\infty \mathbf{Z}_2 \subseteq \mathrm{UNil}_{4k+2}(\Phi_0) .$$

Every non-zero element $x \neq 0 \in \mathrm{UNil}_{4k+2}(\Phi_0)$ *was realized as the splitting obstruction of a simple homotopy equivalence of* $(4k+1)$*-dimensional manifolds* $f : W^{4k+1} \to V_0$

$$s_M(f) \;=\; x \neq 0 \in LS_{4k}(\Phi_0) \;=\; L_{4k+2}(\Phi_0) \;=\; \mathrm{UNil}_{4k+2}(\Phi_0) .$$

The UNil-*groups were shown to be 2-primary in general*

$$\mathrm{UNil}_*(\Phi)[\tfrac{1}{2}] \;=\; 0 ,$$

with $\mathrm{UNil}_*(\Phi) = 0$ *for amalgamated free products which are 'square root closed', i.e. such that*

$$\{g_i \in \pi_1(C_i) : (g_i)^2 \in \pi_1(M)\} \;=\; \pi_1(M) \quad (i = 1, 2) .$$

Similarly in the two-sided case (B) considered in the next section.

12B. Two-sided Submanifolds

We next consider two-sided but non-separating manifolds. The only such case which has previously been studied by surgery is that which leads to the fibration theorem of Browder and Levine [B27] and the subsequent extension of this by Farrell (thesis, Yale University, 1967). I had originally planned an exposition along the lines of Farrell's thesis; then, the penultimate paragraph in his appendix hinted at a clearer result, which I proved by an extension of the same method. It appears, however, that the result can also be proved by a simple trick, which I introduced in [W12].

Suppose given a short exact sequence

$$(*) \qquad 1 \to \pi \xrightarrow{i} \Pi \xrightarrow{j} \mathbf{Z} \to 0$$

of finitely presented groups. Choose $g \in \Pi$ with $j(g) = 1$, and let conjugation by g induce the automorphism α of π. We write $Wh(\pi)$ for the Whitehead group of π, and $(Wh\,\pi)^\alpha$ for the subgroup of α_*-invariant elements. Now $\mathbf{Z}_2$ operates on $Wh(\pi)$ by taking conjugate (under $x \mapsto \overline{x}$) transposed matrices; α_* commutes with this action, so we can regard $(Wh\,\pi)^\alpha$ as a $\mathbf{Z}_2$-module.

We also need some notation for groupoids, since we cannot represent our pushout diagram below using only disjoint unions of groups. Denote by I the connected groupoid with two vertices 0, 1 and trivial vertex groups (this is the category with objects 0, 1 in which each $\mathrm{Hom}(A, B)$ contains just one morphism). We denote by Φ the diagram of groupoids below (which is a pushout) (cf. [B38]):

$$\begin{array}{ccc} \pi \cup \pi & \xrightarrow{(1,\alpha)} & \pi \\ \big\downarrow & & \big\downarrow \\ \pi \times I & \xrightarrow{\beta} & \Pi \end{array}$$

where β is defined by $\beta(1, 01) = g$ and $\beta(x, 00) = x$ for all $x \in \pi$.

THEOREM 12.5. *We have a natural isomorphism, for* $n \geqslant 6$,

$$LS_n(\Phi) \cong \widehat{H}^{n+1}\big(\mathbf{Z}_2; (Wh\,\pi)^\alpha\big)\,.$$

Proof It is easy to construct a smooth or PL fibration $L^{n-1} \subset U^n \to S^1$ whose induced exact sequence of fundamental groups coincides with $(*)$. Now by (11.7), for $n \geqslant 6$ we can find another smoothing $L' \subset U$ of the same simple Poincaré submanifold[†] such that the obstruction to extending $(L \times 0) \cup (L' \times 1)$

[†] *i.e. the same simple Poincaré embedding*

to a smoothing of the product Poincaré embedding is a prescribed element of $LS_n(\Phi)$.

Let $\widetilde{U}$ denote the covering space of U induced from the universal covering of S^1 by $\mathbf{R}$. The inclusion of L in U lifts to an inclusion in $\widetilde{U}$: more precisely, we can regard U as formed from $L \times I$ by attaching $L \times 1$ to $L \times 0$ by a homeomorphism G inducing α on $\pi_1(L)$. Then $\widetilde{U}$ is naturally identified with $L \times \mathbf{R}$; the group of covering transformations is generated by T with $T(x,t) = \big(G(x), t+1\big)$. Since $L' \subset U$ smooths the same Poincaré embedding, it also lifts to an embedding $h : L' \subset L \times \mathbf{R}$, which we may choose to lie to the right of $L \times 0$. Let W be the compact submanifold of $L \times \mathbf{R}$ bounded by $L \times 0$ and $h(L')$: it is well known that such a W is an h-cobordism. Let it have Whitehead torsion τ_W. We are now ready to play our trick.

Lemma 12.6. *There is an embedding $e : W \to U \times I$, with*

$$\partial e(W) \;=\; (L \times 0) \cup (L' \times 1)\,.$$

Proof We construct e as the projection of an embedding $\widetilde{e}$ in the universal cover $\widetilde{U} \times I \cong (L \times \mathbf{R}) \times I$, of $W' = W \cup (L' \times I)$. Define the first component of $\widetilde{e}\,|\,W$ to be the inclusion $W \subset L \times \mathbf{R}$. Let $p : W \to \mathbf{R}$ be the projection, and let K exceed the least upper bound of $p(W)$. Define the second component of $\widetilde{e}\,|\,W$ to send w to $p(w)/K$.

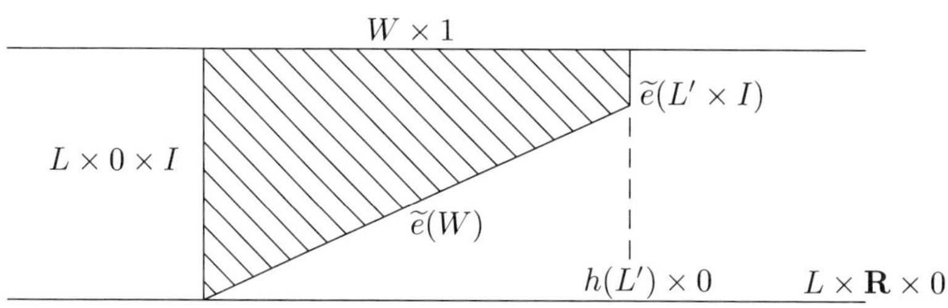

It is clear that W is embedded by $\widetilde{e}$ in $(L \times \mathbf{R}) \times I$, and moreover that the projection into $U \times I$ is still embedded. More, $\widetilde{e}(L') \subset h(L') \times I$. We now define $\widetilde{e}$ on $L' \times I$ as indicated in the diagram above: the image is the submanifold of $h(L') \times I$ bounded by $\widetilde{e}(L')$ and $h(L') \times 1$. (One could easily write down a formula for a topological embedding, but this would not be PL, and would need smoothing at the corner anyway). Again, it is clear that $\widetilde{e}$ induces, by projection, an embedding of $L' \times I$ in $U \times I$.

We must now verify that the projections of $\widetilde{e}(W)$ and $\widetilde{e}(L' \times I)$ intersect only along $\widetilde{e}(L' \times 0)$. Suppose that $\widetilde{e}(w)$ and $\widetilde{e}(m,t)$ have the same projection. Then, for some $i \in \mathbf{Z}$, $\widetilde{e}(M,t) = (T^i \times 1)\widetilde{e}(w)$, and so $h(m) = T^i(w)$. Now for $j > 0$, $T^j\big(h(L')\big)$ is connected and disjoint from $h(L')$ (for L' is embedded in U), so it lies completely to the right of W. Hence $i \geqslant 0$. But $p(m) = i + p(w)$, so if $i > 0$, the second coordinate of $\widetilde{e}(m,t)$ is $\geqslant p(m)/K > p(w)/K$, the second coordinate of $(T^i \times 1)\widetilde{e}(w)$. Thus overlap only occurs if $i = 0$, and along $\widetilde{e}(L' \times 0)$. Our embedding is already PL (in the PL case); in the smooth case, one must round the corner (e.g. by the Cairns-Hirsch theorem) to define e.

This direct construction, replacing an existence proof by surgery, shortens our proof considerably. It is clear that W smooths the product Poincaré embedding. It may not quite be true that W extends the given smoothing by $(L \times 0) \cup (L' \cup 1)$: the inclusion in $U \times I$ need not be homotopic (rel ∂W) to the given map. But this is easily corrected by isotoping L round U and extending to an isotopy of $U \times I$. Now the only thing which hinders us from smoothing the product simple Poincaré embedding is the torsion τ_W. Hence τ_W determines the obstruction in $LS_n(\Phi)$. If we glue two copies of $U \times I$ end to end, both surgery obstructions and Whitehead torsions are additive. It follows that we have an isomorphism from some subquotient of $Wh(\pi)$ onto $LS_n(\Phi)$.

Since the natural map $L^{n-1} \to L'^{n-1}$ is a simple homotopy equivalence, the torsion τ_W must satisfy $\tau_W^* = (-1)^{n+1}\tau_W$. Further, when U is cut along L we obtain an s-cobordism. The same must hold true for L'. But the new cobordism is obtained from the old by adding W at one end and removing it at the other. Also, the fundamental group is mapped by the identity at one end and by α at the other. It follows that $\alpha_*(\tau_W) = \tau_W$. Conversely, let τ satisfy

$$\alpha_*(\tau) = \tau\,, \qquad \tau^* = (-1)^{n+1}\tau\,. \tag{1}$$

Split a collar neighbourhood $L \times I$ of L in U as a product of h-cobordisms with torsions $\pm\tau$, with common boundary L' (for $n \geqslant 6$, this is known to be possible: see e.g. [M14]). Then the duality condition on τ implies that the natural map $L \to L'$ is a simple homotopy equivalence; the first condition deals with the simple homotopy type of the complement, and so L' and L define the same simple Poincaré embedding. Thus the relevant subgroup of $Wh(\pi)$ is precisely that defined by (1).

Next, let τ define zero in $LS_n(\Phi)$. Then, with the notations above, we can extend $(L \times 0) \cup (L' \times 1)$ to a smoothing $X \subset U \times I$ of the product Poincaré embedding. Lift X to $\widetilde{U} \times I$: the result may meet $\widetilde{e}\big(W \cup (L' \times I)\big)$ but is isotopic (rel ∂X) to an embedding meeting it only in ∂X: this is easily seen by engulfing, or by finding an h-cobordism of $L \times 0$ to $L' \times 1$ in $\widetilde{U} \times I$ disjoint (except at the ends) from both the above – e.g. a translate of X, together with pieces in $\widetilde{U} \times \partial I$ – and then noting that this is h-cobordant rel the boundary both to X and to $\operatorname{Im}\widetilde{e}$, so that we can isotope either into a neighbourhood of it disjoint from the other. The region between X and $\operatorname{Im}\widetilde{e}$ is now an h-cobordism, so is determined by its torsion.

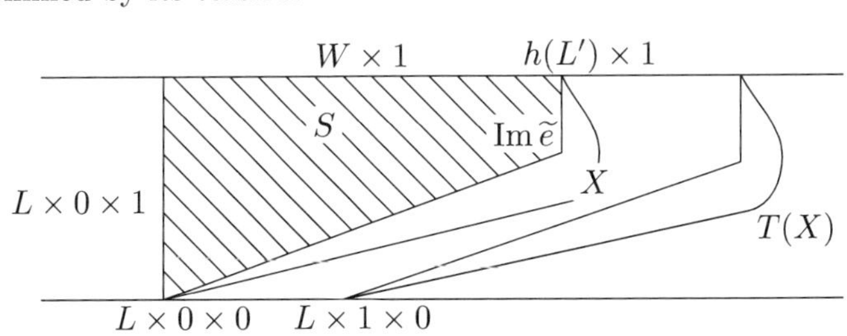

Thus we can embed X in a small neighbourhood of $\operatorname{Im}\widetilde{e}$ in $U \times I$, by embedding in this an h-cobordism with the given torsion. This embedding also smooths the product simple Poincaré embedding.

Now the shaded region S in the figure evidently collapses onto $W \times 1$, and the torsion of $L \times 1 \subset W \times 1$ is τ, hence so is the torsion of $L \times I \subset S$. Moving one to the right, the inclusion of $L \times I \times I$ in the larger region, hence also that of $L \times I$, has torsion $\alpha_* \tau$. Thus the h-cobordism of $\widetilde{e}(W)$ has torsion $\alpha_* \tau - \tau = 0$. If we now translate $\widetilde{e}(W)$ to X by a (small) h-cobordism with torsion σ, the corresponding h-cobordism of X to itself will (similarly) have torsion $\alpha_* \sigma - \sigma$. Now we need $\tau = \sigma + (-1)^{n+1}\sigma^*$ for X to be an s-cobordism, and $\alpha_*\sigma = \sigma$ for X to smooth the product Poincaré embedding. These conditions are both necessary and sufficient. Since, for a $\mathbf{Z}_2$-module A on which $\mathbf{Z}_2$ operates by $a \mapsto a^*$, one can identify $\widehat{H}^{n+1}(\mathbf{Z}_2; A)$ with

$$\frac{\{a \in A : a^* = (-1)^{n+1}a\}}{\{b^* + (-1)^{n+1}b \, : b \in A\}},$$

the theorem now follows.

COROLLARY 12.5.1. *Suppose given a simple Poincaré embedding of M^n in a manifold V^{n+1}, with M, V and the complement C connected and such that each inclusion $i_0, i_1 : M \to C$ induces an isomorphism of π_1, and ∂M is embedded in ∂V, and $n \geqslant 5$. Then we can smooth it to obtain a manifold embedding, provided an obstruction $\in \widehat{H}^{n+1}\big(\mathbf{Z}_2; Wh(\pi_1(M))^\alpha\big)$ vanishes, $\alpha = (i_{1*})^{-1}(i_{0*})$.*

This is merely a restatement, in geometrical terms, of the theorem. Now we will use it more algebraically. First we apply (12.3). The corresponding $LS_n(\Psi)$ vanish by (12.4). We thus obtain isomorphisms (or directly by (12.4.1))

$$L_{n+2}(\Phi) \cong LS_n(\Phi) \cong \widehat{H}^{n+1}\big(\mathbf{Z}_2; (Wh\,\pi)^\alpha\big).$$

THEOREM 12.6. *There is a commutative exact diagram*

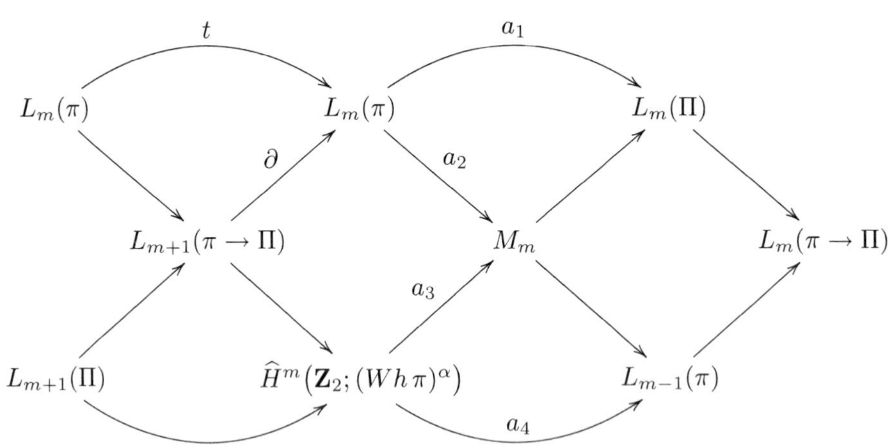

and the map $t = 1 - w(g)\alpha_*$.

Proof The sequence containing a_1 is the exact sequence of $\pi \to \Pi$. The sequence containing a_4 is one of the (isomorphic) sequences of the triad Φ, on identifying $L_{m+1}(\pi \cup \pi \to \pi)$ with $L_m(\pi)$ (essentially the diagonal subgroup of $L_m(\pi) \times L_m(\pi)$) in the obvious way. To define M_m, and obtain the sequence with a_3, we convert $\{\Pi\}$ into a triad $\sigma_1\sigma_1(\Pi)$, which maps naturally into Φ, and take the exact sequence of this map. Thus we define M as the L-group of

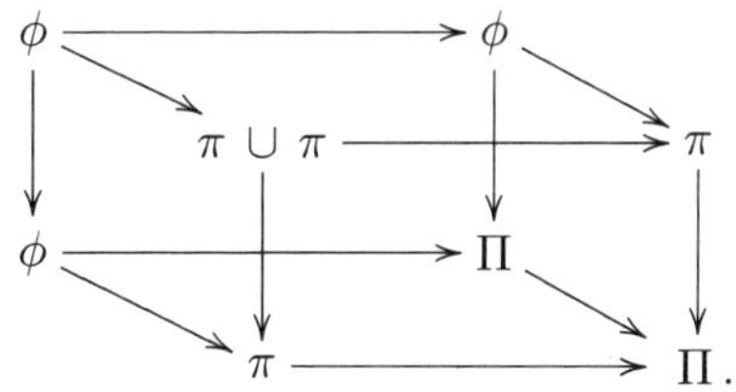

Now another exact sequence containing M is induced by the map of triads

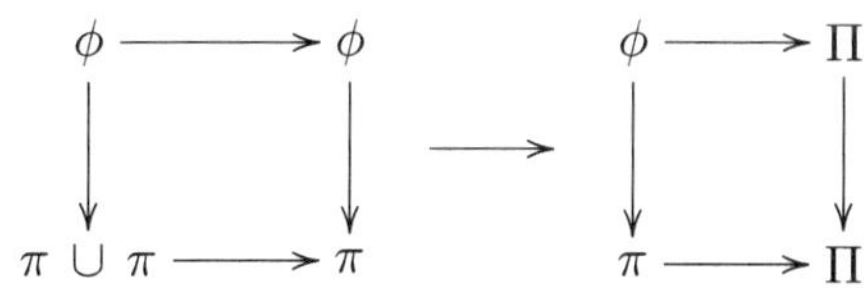

But each of these has the same L-groups as π (we can remove the top row of the first, and the right column of the second). We have now defined all our sequences: the tedious but trivial verifications of commutativity we leave to the reader (some care is needed in making the correct identifications).

To compute the composite t, we use the commutative diagram

$$\begin{array}{ccc} L_m(\pi) \cong L_{m+1}\big((1,1) : \pi \cup \pi \to \pi\big) & \xrightarrow{\partial} & L_m(\pi \cup \pi) \\ \big\downarrow & & \big\downarrow (1,\alpha)_* \\ L_{m+1}(\pi \to \Pi) & \xrightarrow{\partial} & L_m(\pi)\,. \end{array}$$

Thus $L_m(\pi)$ is identified with $\{x, -x\} \in L_m(\pi \cup \pi) \cong L_m(\pi) \times L_m(\pi)$.

Now $\{x, -x\}$ appears to be mapped into $x - \alpha_* x$. However, our map is not induced by a map of groups, and it is seen on inspection that if $w = -1$, the orientation of a fundamental class is changed. Thus the correct formula is $x - w(g)\alpha_*(x)$, as asserted.

We will find this result very useful in the next chapter.

Codimension 1 splitting in the non-separating two-sided case (B).
Let V be an $(n+1)$-dimensional manifold with a codimension 1 submanifold $M^n \subset V^{n+1}$ of type (B), with trivial normal bundle $M \times [0,1] \subset V$ and complement $C = \text{cl.}(V - M \times [0,1])$, such that V, M, C are connected. By definition, a homotopy equivalence $f : W \to V$ from an $(n+1)$-dimensional manifold W

h-splits at $M \subset V$ if f is h-cobordant to a map which is transverse regular at M with the restrictions $f| : f^{-1}(M) \to M$, $f| : f^{-1}(C) \to C$ homotopy equivalences. In the special case when the inclusions $i_0, i_1 : M \to C$ induce isomorphisms $i_{0*}, i_{1*} : \pi_1(M) \cong \pi_1(C)$ let $\alpha = (i_{1*})^{-1} i_{0*} : \pi_1(M) = \pi \to \pi$, so that

$$\pi_1(V) \;=\; \Pi \;=\; \pi \times_\alpha \mathbf{Z}$$

and as above there is an exact sequence

$$1 \to \pi \xrightarrow{i} \Pi \xrightarrow{j} \mathbf{Z} \to 0\,.$$

Farrell and Hsiang [F4] proved that in this case a homotopy equivalence $f : W \to V$ from an $(n+1)$-dimensional manifold W is such that $\tau(f) \in \operatorname{Im}(i : Wh(\pi) \to Wh(\Pi))$ if (and for $n \geqslant 5$ only if) f h-splits at $M \subset V$. The Whitehead groups of π, Π fit into the exact sequence of Bass [B7, Chapter XII] and Farrell and Hsiang [F3]

$$\begin{aligned}\dots \to Wh_2(\pi) &\xrightarrow{i} Wh_2(\Pi) \xrightarrow{\partial} Wh(\pi) \oplus \widetilde{\mathrm{Nil}}_1(\mathbf{Z}[\pi]) \oplus \widetilde{\mathrm{Nil}}_1(\mathbf{Z}[\pi])\\ &\xrightarrow{(1-\alpha)\oplus 0 \oplus 0} Wh(\pi) \xrightarrow{i} Wh(\Pi) \xrightarrow{\partial} \widetilde{K}_0(\mathbf{Z}[\pi]) \oplus \widetilde{\mathrm{Nil}}_0(\mathbf{Z}[\pi]) \oplus \widetilde{\mathrm{Nil}}_0(\mathbf{Z}[\pi])\\ &\xrightarrow{(1-\alpha)\oplus 0 \oplus 0} \widetilde{K}_0(\mathbf{Z}[\pi]) \xrightarrow{i} \widetilde{K}_0(\mathbf{Z}[\Pi]) \to \dots\end{aligned}$$

extended to the left to include the Wh_2-groups. The geometrically defined isomorphism of Theorem 12.5 is given algebraically by

$$LS_n(\Phi) \xrightarrow{\cong} \widehat{H}^{n+1}(\mathbf{Z}_2; (Wh\,\pi)^\alpha)\ ;\ s_M(f) \mapsto \partial\tau_2(f)\ ,$$

with $s_M(f) \in LS_n(\Phi)$ the splitting obstruction of a simple homotopy equivalence $f : W \to V$ and

$$\tau_2(f) \;=\; (-1)^n \tau_2(f)^* \in \operatorname{Ker}(\partial : Wh_2(\Pi) \to Wh(\pi))$$

the Wh_2-invariant of f. The groups $L_*(\pi \to \Pi)$, M_* in Theorem 12.6 are given by the intermediate L-groups of π decorated by $Wh(\pi)^\alpha \subseteq Wh(\pi)$

$$L_{m+1}(\pi \to \Pi) \;=\; L_m^{Wh(\pi)^\alpha}(\pi)$$

and the intermediate L-groups of Π decorated by $\operatorname{Im}(i : Wh_2(\pi) \to Wh_2(\Pi)) \subseteq Wh_2(\Pi)$

$$M_m \;=\; L_m^{Wh_2(\pi)}(\Pi)$$

– see §17D for intermediate L-theory, and Hsiang and Sharpe [H25], Weiss and Williams [W43] for the Wh_2-variant of surgery obstruction theory. See Novikov [N8], and Ranicki [R2], [R3] for the algebraic L-theory of the twisted Laurent polynomial extension $\mathbf{Z}[\Pi] = \mathbf{Z}[\pi]_\alpha[z, z^{-1}]$, generalising the algebraic K-theory methods and results of [B7], [F3]. Cappell [C4], [C5] has also defined UNil*-groups for type (B) codimension 1 splitting, assuming that the morphisms $i_{0*}, i_{1*} : \pi_1(M) \to \pi_1(C)$ are injections – see the notes at the ends of the previous two sections of §12. The* UNil*-groups vanish in the case $\Pi = \pi \times_\alpha \mathbf{Z}$ ([C5, Chapter VI].)*

12C. One-sided Submanifolds

Finally we consider the problem of 1-sided submanifolds. This seems to be less tractable than the other cases considered above – not only does the general case lead to complicated formulae, but even those special cases where the result can be neatly expressed do not seem to admit a direct geometrical argument. We follow the general lines of the treatment by Browder and Livesay [B29] of a special case of the problem.

In the notation introduced earlier in this chapter, our problem is as follows. Let $\pi' \subset \pi$ be a subgroup of index 2. Compute $LN_n(\pi' \to \pi)$. Compared with the solution in (12.4) for two-sided submanifolds, the results in this case are complicated, and we will have no further computations of groups LS_n in this chapter. We end up with a group whose definition resembles that of $L_n(\pi')$, and in one case coincides with it.

We will adopt the following notation. V^v is a compact manifold with corner dividing the boundary into two parts $\partial_0 V$, $\partial_1 V$; the latter is the total space of a (smooth or PL) bundle with fibre D^1. We are given a double covering $\widetilde{V}$ of V, trivial on $\partial_0 V$, which is an s-cobordism between the two copies of $\partial_0 V$. Let $\partial M \subset \partial_1 V$ be the zero cross-section: we wish to extend this to a submanifold of V whose inclusion is a simple homotopy equivalence. We will write T for the nontrivial covering transformation of $\widetilde{V}$ over V.

As a first step, let W be the mapping cylinder of the projection $\widetilde{V} \to V$, alias the associated bundle with fibre $D^1 = [-1, 1]$. There is a section over $\partial_0 V$ of the double covering (with fibre $\{-1, 1\}$): extend to a section $V \to W$. Over $\partial_1 V$ there is a natural way to do this, so that this section is transverse to the zero section, and meets it in $\partial M \times 0$. Now deform the section (rel $\partial_1 V$) to be transverse to the zero section, and denote by M the submanifold of V which is the locus of zeros of the section.

Since $\widetilde{V}$ is an s-cobordism, $(\widetilde{V}, \partial_1 \widetilde{V})$ is simply homotopy equivalent to $(\partial_0 V, \partial\partial_0 V)$, and so is a simple Poincaré pair (of formal dimension $v-1$). Hence also $(V, \partial_1 V)$ is a simple Poincaré pair. The homology class of $(M, \partial M)$ is dual to the cohomology class (rel $\partial_0 V$) of the covering; it follows that $(M, \partial M) \to (V, \partial_1 V)$ is a map of degree 1. Perhaps it is only really clear that the degree is odd, but we will discuss this point more fully below.

It will be seen that $\partial_1 V$ plays no essential rôle above, nor will it in the arguments below. To economise notation, we set $\partial_1 V = \emptyset$, $\partial V = \partial_0 V$, on the understanding that all would be equally well justified if $\partial_1 V$ were to be reintroduced.

Now we can regard the identity map of V as corresponding to a simple Poincaré embedding, which we seek to smooth. Our technique is to perform surgery on the manifold M we have just constructed, or rather, equivariant surgery on its double covering. We want $\widetilde{M} \subset \widetilde{V}$ to be a homotopy equivalence, and so must kill the relative homotopy groups. The pattern of the argument closely resembles that of §5 and §6, but many details are different.

LEMMA 12.7. *If $v > 2k$, we can do surgery on M inside V to make the inclusion $M \to V$ k-connected.*

Proof We proceed by induction on k, treating the cases $k \leqslant 2$ individually. We first describe what we mean by 'surgery on M inside V'. Suppose given an embedding

$$f : (D^r, \partial D^r) \times D^{v-r} \to (V, M)$$

with $f^{-1}(M) = \partial D^r \times D^{v-r}$. Then we say that the manifold

$$M' \;=\; M - f(\partial D^r \times D^{v-r}) \cup f(D^r \times \partial D^{v-r})$$

is obtained from M by surgery, using f. It is easy to construct, using f, a cobordism from $M \times 0$ to $M' \times 1$ lying in $V \times I$;

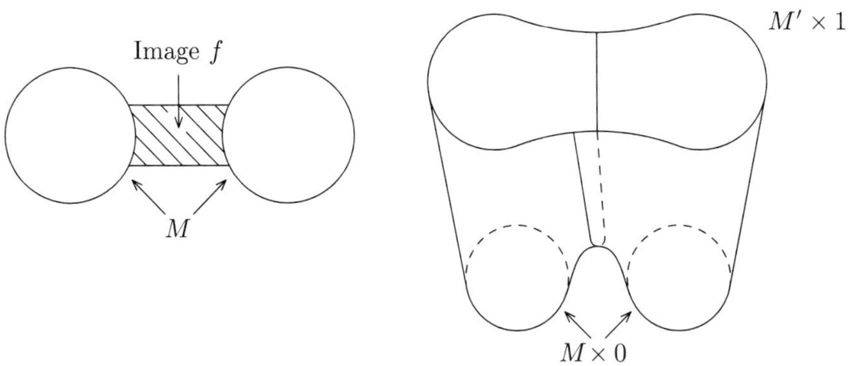

in fact, cobordism is the equivalence relation which interests us. It can be shown, as in the previously discussed cases, that cobordism (as equivalence relation) within V is generated by surgeries inside V, together with isotopies. We will not need this result, but use below a 'surgery' of a somewhat different form to this, which nevertheless gives a cobordism. (Even this can easily be avoided). In the smooth case, the above description should be completed by rounding the corner which appears at $f(\partial D^r \times \partial D^{v-r})$: this is easily accomplished using a model. We observe that it suffices to give an embedding $\overline{f} : (D^r, \partial D^r) \to (V, M)$, since f can then be constructed by taking a suitable (tubular or regular) neighbourhood.

We may suppose V connected; it is then clear that $M \subset V$ is 0-connected ($M \neq \emptyset$ as its mod 2 homology class is nonzero). Now recall that M was the locus of zeros of a section, transverse to the zero section, of the arc bundle over V associated to the double covering $\widetilde{V}$. Lifting this to $\widetilde{V}$ by the projection, $\widetilde{M}$ is the locus of zeros of a section s, transverse to the zero section, of the bundle $\widetilde{V} \times [-1, 1] \to \widetilde{V}$. Define A^+ (resp. A^-) to be the set of $P \in \widetilde{V}$ such that $s(P)$

has second coordinate $\geqslant 0$ (resp. $\leqslant 0$). Then $\widetilde{V} = A^+ \cup A^-$ and $\widetilde{M} = A^+ \cap A^-$; also $T(A^+) = A^-$.

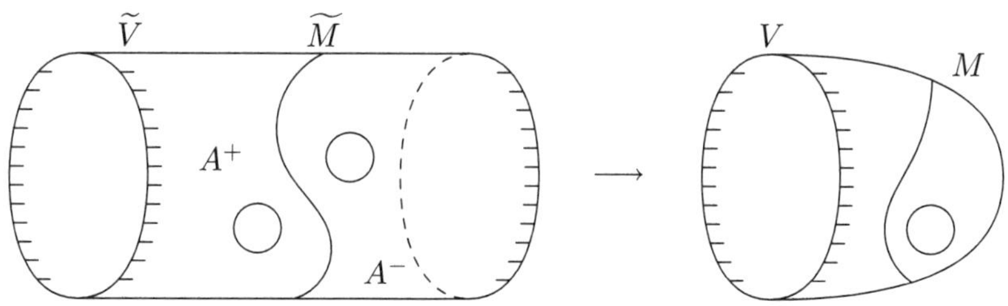

It is now clear in what sense the inclusion $M \subset V$ has degree 1: we must orient $\widetilde{M}$ as $\partial_*[A^+]$, where A^+ inherits its orientation (possibly with twisted coefficients) from $\widetilde{V}$.

Now $\widetilde{V}$ is an s-cobordism; let α be an arc joining the ends, and disjoint from its image by T. If now α meets more than one component of $\widetilde{M}$, some component arc β of $\alpha \cap A^+$ or $\alpha \cap A^-$ must have its ends on different components of $\widetilde{M}$. We thicken β to obtain an embedding $(D^1, \partial D^1) \times D^{v-1} \to (\widetilde{V}, \widetilde{M})$; if the transverse disc D^{v-1} is small enough, this projects to an embedding in V, which we use to perform surgery. This decreases the number of components of $\widetilde{M}$ by one (at least); by induction we may suppose $\widetilde{M}$ connected.

Since $i : M \to V$ has degree 1, the induced map $i_* : \pi_1(M) \to \pi_1(V)$ is surjective (else one could lift i to some covering space of V). Hence i is 1-connected. We next wish to make $\pi_1(i)$ bijective: now by (1.2) we can do this by adding a finite set of 2-handles, whose classes are given by maps $f_i : (D^2, \partial D^2) \to (V, M)$.

We are treating the case $k = 2$, so our hypothesis is $v \geqslant 5$, so these maps can be assumed to be disjoint embeddings, moreover, we may take them to meet M transversely. Choose lifts $\widetilde{f}_i : D^2 \to \widetilde{V}$: then $f_i^{-1}(M)$ gives a finite set of disjoint simple closed curves in D^2, separating D^2 into $R^+ = \widetilde{f}_i^{-1}(A^+)$ and $R^- = \widetilde{f}_i^{-1}(A^-)$. We choose the $\widetilde{f}_i$ so that the region meeting ∂D^2 is a component of R^+.

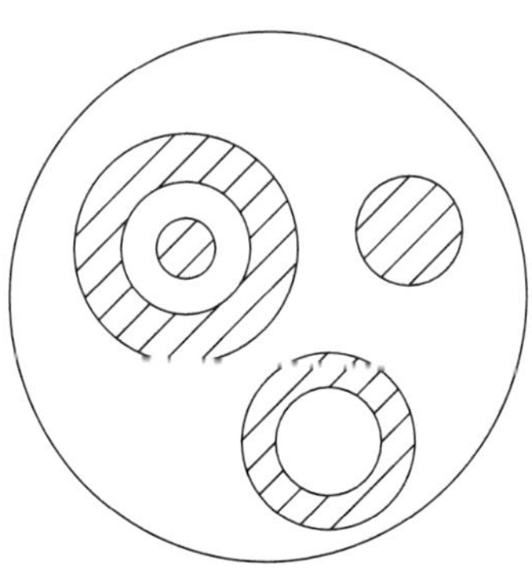

R^- is shaded.

Now thicken f_i to an embedding $F_i : D^2 \times D^{v-2} \to V$ in such a way that $F_i^{-1}(M) = f_i^{-1}(M) \times D^{v-2}$. We delete, for each i, $M \cap \operatorname{Im} f_i$ from M, and replace it by

$$(R^- \times \partial D^{v-2}) \cup \left(S^1 \times (D^{v-2} - \tfrac{1}{2}D^{v-2})\right) \cup \left(D^2 \times \partial(\tfrac{1}{2}D^{v-2})\right),$$

where $\frac{1}{2}D^{v-2}$ denotes the disc concentric with D^{v-2} with radius $\frac{1}{2}$. The subset deleted has skeleton a collection of circles, of codimension $\geqslant 3$, so this deletion does not affect the fundamental group. The attachments add various relations between the generators; in particular, we have killed the desired elements. This argument is a modification of the original one of Browder [B17] (he used only ordinary surgeries). We may thus suppose that i_* induces an isomorphism of fundamental groups. Now by (2.2) the induced map

$$\pi_2(M) \cong H_2(M; \Lambda) \to H_2(V; \Lambda) \cong \pi_2(V)$$

is surjective, so i is 2-connected.

Finally we treat the case $k > 2$, supposing inductively that i is already $(k-1)$-connected. Then van Kampen's theorem implies that

$$\pi_1(\widetilde{M}) \cong \pi_1(A^+) \cong \pi_1(A^-) \cong \pi_1(\widetilde{V}) = \pi' .$$

The next remark is trivial enough, but it (and extensions) will be used so frequently in the sequel that it is useful to have a reference.

LEMMA 12.8. *In the situation above, we have isomorphisms for all r*

$$K_{r-1}(M; \Lambda) = H_r(V, M; \Lambda) = H_r(\widetilde{V}, \widetilde{M}; \Lambda') = H_r(A^+, \widetilde{M}; \Lambda') \oplus H_r(A^-, \widetilde{M}; \Lambda') .$$

Moreover,

$$K_{r-1}(A^+; \Lambda') = H_r(\widetilde{V}, A^+; \Lambda') \ = \ H_r(A^-, \widetilde{M}; \Lambda') .$$

If, further, $i : M \to V$ is $(r-1)$-connected, then

$$\pi_r(i) \cong K_{r-1}(M; \Lambda) \ ,$$

and

$$\pi_r(A^+, \widetilde{M}) \ = \ H_r(A^+, \widetilde{M}; \Lambda') .$$

The proof is immediate (given what we have already done in §2). We return to the proof of (12.7). Taking $r = k$ in the lemma, we have

$$\pi_k(i) = \pi_k(A^+, \widetilde{M}) \oplus \pi_k(A^-, \widetilde{M}) ,$$

and the two summands are interchanged by T. Now $\pi_k(i)$ is finitely generated over Λ by (1.2) (or rather, by its proof), hence over Λ', hence so is $\pi_k(A^+, \widetilde{M})$. We choose a finite set of generators, and represent by maps $f_i : (D^k, \partial D^k) \to (A^+, \widetilde{M})$. Since $v > 2k$ we may suppose by general position that these, and their projections in V, are disjoint embeddings. We now use them to perform surgery. Since we are below the middle dimension, this has the effect of killing the classes of the ∂f_i in $K_{k-1}(M; \Lambda)$. But we chose these to be Λ'-generators of $K_k(A^+, \widetilde{M}; \Lambda')$ and hence (T interchanges the summands) Λ-generators of $H_k(\widetilde{V}, \widetilde{M}; \Lambda')$, so we have killed the whole group, and made i k-connected. This proves the lemma.

We must now consider what happens in the middle dimensions; here, the even and odd dimensional cases have to be considered separately. First let $v = 2k+1$, so that $\dim M = 2k$ is even. Then the only remaining nonvanishing homology or cohomology group of $(A^+, \widetilde{M})$, with any coefficient module, is the $(k+1)$st. By Lemma 2.3, $H^+ = H_{k+1}(A^+, \widetilde{M}; \Lambda')$ is a finitely generated projective Λ'-module, which is stably free and s-based. Now by performing surgery on further $(k-1)$-spheres in M, we add free modules to this (cf. (5.5)), so may suppose it free and based. We write also $H^- = H_{k+1}(A^-, \widetilde{M}; \Lambda')$ and

$$H = H^+ \oplus H^- = H_{k+1}(\widetilde{V}, \widetilde{M}; \Lambda') = H_{k+1}(V, M; \Lambda) .$$

First consider the simple hermitian form (on H) over Λ', arising from intersections and self-intersections in $\widetilde{M}$: then H^+ and H^- are complementary lagrangians. That they are lagrangians follows from (5.7) (take $N = A^+$). Next we consider the simple hermitian form over Λ, using intersections in M. Choose $t \in (\pi - \pi')$. Then for $x, y \in H^+$ we have $\lambda(x, y) = t\lambda_0(x, y)$, say, where $\lambda_0(x, y) \in \Lambda'$.

Similarly, $\mu(x) = t\mu_0(x)$, where μ_0 lies in a certain quotient group of Λ' (to be described below). Notice that since $H^+ = H^- t$, the maps λ_0 and μ_0 entirely determine λ and μ.

We next determine the relations satisfied by λ_0 and μ_0. Notice that we are working entirely over Λ': however, some account needs to be taken of t, and we write

$$g_0 = t^2, \quad g^\alpha = t^{-1}gt\,(g \in \pi'), \quad \text{and} \quad \overline{g}^\alpha = w(g)t^{-1}g^{-1}t :$$

then
$$w(g_0) = 1 \quad \text{and} \quad w(g^\alpha) = w(g).$$

(Q1) $\lambda_0(x, y)$ is Λ'-linear in y (this is clear).

(Q2)
$$\begin{aligned}\lambda_0(y,x) &= t^{-1}\lambda(y,x) = (-1)^k t^{-1}\overline{\lambda(x,y)} \\ &= (-1)^k t^{-1}\overline{t\lambda_0(x,y)} = (-1)^k t^{-1}\overline{\lambda_0(x,y)}\,\overline{t} \\ &= (-1)^k w(t)\overline{\lambda_0(x,y)}^\alpha g_0^{-1}.\end{aligned}$$

(Q3) $\mu_0(x+y) - \mu_0(x) - \mu_0(y) = \lambda_0(x,y)$ (clear)

(Q4)
$$\begin{aligned}\lambda_0(x,x) &= t^{-1}\lambda(x,x) = t^{-1}\big(\mu(x) + (-1)^k\overline{\mu(x)}\big) \\ &= t^{-1}\big(t\mu_0(x) + (-1)^k w(t)\overline{\mu_0(x)}t^{-1}\big) \\ &= \mu_0(x) + (-1)^k w(t)\overline{\mu_0(x)}^\alpha g_0^{-1}.\end{aligned}$$

(Q5)
$$\begin{aligned}\mu_0(xr) &= t^{-1}\overline{r}\mu(x)r \\ &= t^{-1}\overline{r}t\mu_0(x)r = \overline{r}^\alpha\mu_0(x)r.\end{aligned}$$

and we should add that (by a similar calculation) μ_0 lies in the quotient of Λ' by the group of elements of the form

$$y - (-1)^k w(t)\overline{y}^\alpha g_0^{-1}.$$

Also note that duality in M at the chain level extends to a simple equivalence $C^*(V,M) \to C_*(V,M)$. In the double covers, this splits as a direct sum of two equivalences over Λ', the first being $C^*(A^+,\widetilde{M}) \to C_*(A^-,\widetilde{M})$. This is a simple equivalence, being the composite of Lefschetz duality $C^*(A^+,\widetilde{M}) \to C_*(A^+,\partial V)$, a dimension shifting isomorphism (recall that $\widetilde{V}$ is an s-cobordism) $C_*(A^+,\partial V) \to C_*(\widetilde{V},A^+)$, and excision. In our case, the chain complex is simply equivalent to the $(k+1)$st homology group (with appropriate basis), and the above coincides with the duality given by λ_0, composed with multiplication by t_0. It follows that the 'hermitian' form λ_0 is simple.

In our case this makes better algebraic sense: if $t^2 = g_0 = 1$, we have precisely a simple $(-1)^k w(t)$-hermitian form with respect to the anti-involution

$$x \mapsto \overline{x}^{\alpha} = t^{-1}\overline{x}t$$

of the ring Λ'. If, in particular, t is central these coincide with the forms used above to define $L_0(\pi')$ and $L_2(\pi')$ $\big($precisely, $L_{2k+1-w(t)}(\pi')\big)$.

We can now enunciate the main result of this part of the chapter.

THEOREM 12.9. *The surgery obstruction groups $LN_n(\pi' \to \pi)$, for $n \geqslant 5$, are determined from the class of simple 'hermitian' forms just described by the same algebraic process as we used in §5 for n even or in §6 for n odd (in this case we will assume the existence of $t \in \pi - \pi'$ of order 2).*

One interesting case is when π' is abelian, and α takes each element to its inverse: here our hermitian forms specialise to be orthogonal or symplectic.

For the particular case when t is central we have the

COROLLARY 12.9.1.
$$LN_n(\pi' \to \pi' \times \mathbf{Z}_2^-) \cong L_n(\pi'),$$
$$LN_n(\pi' \to \pi' \times \mathbf{Z}_2^+) \cong L_{n+2}(\pi').$$

Here, $\mathbf{Z}_2^+$ and $\mathbf{Z}_2^-$ denote groups of order 2 with w trivial resp. nontrivial: not that $\pi(V) = \pi' \times \mathbf{Z}_2^-$ implies $\pi(M) = \pi' \times \mathbf{Z}_2^+$ since the normal bundle is non-trivial. We now deduce

COROLLARY 12.9.2. *We have an isomorphism of exact sequences*

$$\begin{array}{ccccccc}
L_{n+1}(\pi' \times \mathbf{Z}_2^+) & \xrightarrow{p_0} & L_{n+2}(\pi' \to \pi' \times \mathbf{Z}_2^-) & \xrightarrow{q_0} & LN_n(\pi' \to \pi' \times \mathbf{Z}_2^-) & \xrightarrow{r_0} & L_n(\pi' \times \mathbf{Z}_2^+) \\
\downarrow{\scriptstyle s} & & \downarrow & & \downarrow{\scriptstyle b} & & \downarrow{\scriptstyle s} \\
L_{n+1}(\pi' \times \mathbf{Z}_2^+) & \longrightarrow & L_{n+1}(\pi' \to \pi' \times \mathbf{Z}_2^+) & \longrightarrow & L_n(\pi') & \longrightarrow & L_n(\pi' \times \mathbf{Z}_2^+)
\end{array}$$

where the upper sequence of maps comes from (11.6), the lower sequence from (3.2), b is the isomorphism of (12.9.1), and s is induced algebraically by the ring automorphism sending the nontrivial element T of $\mathbf{Z}_2$ to $-T$.

Commutativity of the third square follows at once from the definitions of s, b and r_0. Since j is injective (the retraction $\pi' \times \mathbf{Z}_2^+ \to \pi'$ induces a left inverse for it), so is r_0, and we have an induced isomorphism of their cokernels.

We now begin the proof of (12.9), for the case when n is even. We must show that the surgery obstruction groups $LN_n(\pi' \to \pi)$ are obtained from the forms above by the same process as in §5. We can use the same definitions of hyperbolic form and lagrangian as there, and hence define a Grothendieck group $\mathscr{G}$ of forms modulo hyperbolic forms. We have shown that a representative surgery problem for an element of $LN_n(\pi' \to \pi)$ defines (after suitable preliminary surgeries) an element of $\mathscr{G}$: we thus have a correspondence between $LN_n(\pi' \to \pi)$ and $\mathscr{G}$ which is clearly additive. Next we prove that this is a well-defined map $LN_n(\pi' \to \pi) \to \mathscr{G}$.

Since the correspondence is additive, it suffices to show that if the geometrical obstruction to a problem vanishes, then we have (at least stably) a hyperbolic form. So we can assume there exists a cobordism $N \subset V \times I$ of $M \times 0$ to $M' \times 1$, with $M' \subset V$ a simple homotopy equivalence. By Lemma (12.7), we can perform surgery on N in $V \times I$ to make the inclusion map k-connected. It then follows that N has a handle decomposition based on M' with only k-handles and $(k+1)$-handles: we wish to get rid of the latter, by killing $H_k(N, M; \Lambda)$. Now as in (12.8), if we write B^+, B^- for the parts into which $\widetilde{N}$ separates $\widetilde{V} \times I$, with $A^+ \subset \partial B^+$, we have

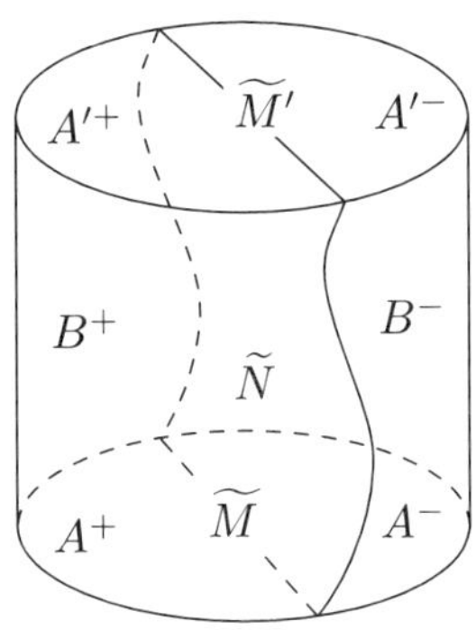

$$\begin{aligned} H_k(N, M; \Lambda) &= H_k(\widetilde{N}, \widetilde{M}; \Lambda') = H_{k+1}(\widetilde{V} \times 1, \widetilde{V} \times 0 \cup \widetilde{N}; \Lambda') \\ &= H_{k+1}(B^+, A^+ \cup \widetilde{N}; \Lambda') \oplus H_{k+1}(B^-, A^- \cup \widetilde{N}; \Lambda') \\ &= K^+ \oplus K^-, \text{ say.} \end{aligned}$$

If we write also Φ^+ for the triad $(B^+; A^+, \widetilde{N})$, then $K^+ = \pi_{k+1}(\Phi^+)$ by Namioka's relativisation [N1] of the Hurewicz theorem.

Since $K^+ \oplus K^-$ is finitely generated over Λ', so is K^+; select a finite set of generators, (which then also generate $H_k(N, M; \Lambda)$ over Λ), and represent by maps $f_j : (D^{k+1}, D^k_+, D^k_-, S^{k-1}) \to \Phi^+$. If these are put in general position, they have only isolated intersections and self-intersections, which are easily disposed of by our usual method; hence we may suppose them disjoint embeddings. We now thicken the images and perform corresponding surgery (in $\widetilde{V} \times I$) on N relative to M. By (1.4), this kills $H_k(N, M; \Lambda)$. Since the $f_j(S^{k-1})$ are necessarily trivial in M, it is easily verified (as in §5) that the effect on $H_{k+1}(A^+, \widetilde{M}; \Lambda')$ is to add a hyperbolic form. But for the new M, $\partial_* H_{k+2}(B^+, A^+ \cup \widetilde{N}; \Lambda')$ is a lagrangian in $H_{k+1}(A^+, \widetilde{M}; \Lambda')$. We will not write out the proof in detail: it is closely analogous to that of (5.7). Thus our original form was stably hyperbolic. Hence our correspondence is in fact a homomorphism.

We next show that it is injective, in other words, that if we have a hyperbolic form, surgery is possible. The formulae above show that if $\{e_1, \dots, e_r\}$ is a preferred Λ'-base of a lagrangian of (λ_0, μ_0), then the same elements of $H^+ \subset H$ form a preferred Λ-base of a lagrangian for (λ, μ). Hence if we can perform surgery on these classes, we do indeed obtain a simple homotopy equivalence $M \to V$. Now represent these classes by maps $\widetilde{f}_i : (D^{k+1}, S^k) \to (A^{+2k+1}, \widetilde{M}^{2k})$. We wish to arrange that not only these, but also the projected maps f_i into (V, M), are disjoint embeddings – for then we can use them to perform surgery on M to obtain the desired conclusion.

It will suffice, in fact, to make $f_1 : (D^{k+1}, S^k) \to (V, M)$ an embedding, for we can then perform surgery using this, and proceed by induction. Consider $\widetilde{f}_1$: since $(A^+, \widetilde{M})$ is k-connected, and $k \geqslant 3$, a theorem of Hudson [H28] allows us to suppose $\widetilde{f}_1$ an embedding. Now if $T : \widetilde{V} \to \widetilde{V}$ is the covering transformation, since $T\widetilde{f}_1$ maps D^{k+1} into A^-, its image can only overlap that of $\widetilde{f}_1$ on $\widetilde{M}$, corresponding to a double point of $\widetilde{f}_1 : S^k \to M$. But since the class e_1 of f_1 has $\mu(e_1) = 0$, the last part of (5.2) shows us how to remove such double points, by 'pushing' an arc α of S^k across a disc in M. We can now extend this push to an isotopy of D^{k+1} which introduces no new singularities, as follows. Choose a coordinate neighbourhood of the disc in M across which the push is to be made (viz. the product of the new model used in [M13, 6.6] and an arc normal to M). We may suppose

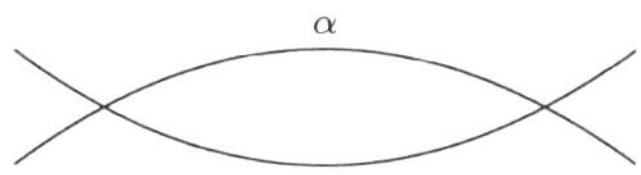

that a neighbourhood N_α of α in D^{k+1} corresponds to the product $(N_\alpha \cap M) \times I$. If now h_t is the isotopy of $N_\alpha \cap M$, we extend it by

$$h_t(P, u) = h_{t(1-u)}(P)$$

to the disc. Clearly this extension has the desired properties: in the PL case, we cannot use the same equations, but it is not difficult to arrange details to give the same geometrical picture. (Less attractively, we can use the above to obtain a smooth embedding of the model situation, and approximate this by a PL embedding, using the techniques of [L3]). This concludes the proof that if we have a hyperbolic form, we can do surgery: injectivity of the map $LN_n(\pi' \to \pi) \to \mathscr{G}$ follows.

It remains to show the map surjective – i.e. to realise any simple hermitian form by a surgery problem. We can do this following the techniques of (5.8) and (6.5). Namely, let M_0^{2k-1} ($k \geqslant 3$) have fundamental group π; let V_0^{2k} be the bundle over it with fibre $[-1, 1]$ corresponding to the subgroup $\pi' \subset \pi$, and identify $M_0 \subset V_0$ with the zero cross-section. Let $\widetilde{M}_0 \subset \widetilde{V}_0$ be the coverings with fundamental group π.

LEMMA 12.10. *Given any simple 'hermitian' form satisfying* (Q1)–(Q5), *there exists* $N^{2k} \subset V_0 \times I$ *with* $M_0 = N \cap (V_0 \times 0)$, *such that* $M_1 = N \cap (V_0 \times 1) \subset V_0 \times 1$ *smooths the same simple Poincaré embedding, and such that* $N \subset V_0 \times I$ *is* k*-connected, and determines the given simple hermitian form.*

Proof Suppose the form has rank $2r$. We first construct N. We can write down a simple hermitian form over Λ, related to the given form as above. Now by (5.8) we can construct a cobordism N of M (satisfying various ancillary conditions) such that $K_k(N; \Lambda)$ defines precisely this simple hermitian form. To be explicit, if our given form is on the free Λ'-module H^+ with basis $\{e_i\}$, and has $\lambda_0(e_i, e_j) = a_{ij}$, $\mu_0(e_0) = b_i$, this form is on the free Λ-module $H \supset H^+$ with the same basis, and has $\lambda(e_i, e_j) = ta_{ij}$, $\mu(e_i) = tb_i$. Thus H^+ is a lagrangian in $K_k(\widetilde{N}; \Lambda')$. We attach $(k+1)$-handles to $\widetilde{N} \times I$, along $\widetilde{N} \times 1$, by embeddings representing the classes e_i: call the result A^+. We let A^- be a homeomorphic copy of A^+, attached to it along $\widetilde{N} \times 0$ by the map corresponding to the covering transformation of $\widetilde{N}$ over N: set $\widetilde{W} = A^+ \cup A^-$. This construction is due to López de Medrano [L20].

The homeomorphism of A^+ on A^- induces a fixed point free involution of $\widetilde{W}$ (since it is free on $\widetilde{N}$): let the orbit space be $W \supset N$. All will be proved if we can show that W is an s-cobordism of V_0, and hence identify it with $V_0 \times I$. Now up to simple homotopy, N was formed from M by attaching a certain number of k-cells, and W from $V_0 \cup N$ by attaching the same number of $(k+1)$-cells. It will suffice, then, to show that the matrix of incidence numbers (over Λ) of these cells is (stably) a product of elementary matrices. In fact we can take it to be the identity, since the $\{e_i\}$ were a preferred base of the given form, and hence can be taken to be the classes of the attached k-cells; and the $(k+1)$-cells were attached with boundary classes precisely $\{e_i\}$.

We now turn to the proof of (12.9) in the case when n is odd. We must consider the case $v = 2k+2$, and $\dim M = (2k+1)$ odd. Again by (12.7) we can assume $M \subset V$ k-connected, and hence $\widetilde{M} \subset A^+$ also k-connected. Select a finite set of Λ'-generators for $K_{k+1}(A^+, \widetilde{M}; \Lambda')$: we claim that these can be represented by disjoint framed embeddings $\widetilde{f}_i : (D^{k+1}, S^k) \times D^{k+1} \to (A^+, \widetilde{M})$. Indeed, we could take generic maps, and "push the singularities away across the boundary" since $(A^+, \widetilde{M})$ is 1-connected; or (perhaps better) appeal to the theorem of Hudson [H28] which formalises this idea. We can also suppose, by a general position argument, that the $\widetilde{f}_i(S^k \times D^{k+1})$ are disjoint from their images by the covering transformation, or equivalently that they project to disjoint embeddings f_i in M.

Delete from A^+ the interiors of the images of these embeddings forming A_0^+, say: let W^+ denote the union of the images, $U^+ = W^+ \cap \widetilde{M}$, $X^+ = W^+ \cap A_0^+$, so $U^+ = \bigcup \widetilde{f}_i(S^k \times D^{k+1})$ and $X^+ = \bigcup \widetilde{f}_i(D^{k+1} \times S^k)$; similarly with $-$ for $+$. Let $M_0 = M - \bigcup \widetilde{f}_i(S^k \times \text{Int } D^{k+1})$, with double covering $\widetilde{M}_0 = \widetilde{M}_0 - \text{Int }(U^+ \cup U^-)$. Similarly we define $U \subset M$ and $X \subset W \subset V$, with $\widetilde{U} = U^+ \cup U^-$, etc.

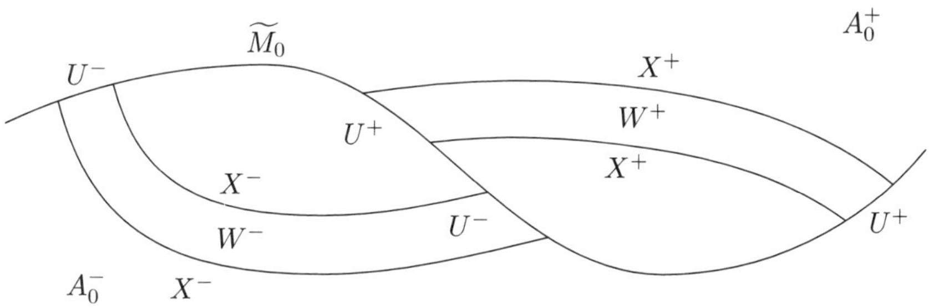

We have repeated on M the construction of §6, which led us to consider a diagram (1), and the pair of lagrangians $H_{k+1}(U, \partial U; \Lambda)$ and $K_{k+1}(M_0, \partial U; \Lambda)$ in $H_k(\partial U; \Lambda)$. We now show – as in (12.8) – that this splits as a direct sum.

LEMMA 12.11. *The diagram* (1) *of* §6, *viz.*

$$\begin{array}{ccccccccc}
0 & & K_{k+1}(M, M_0) = K_{k+1}(U, \partial U) & & & & K_k(M_0) & & 0 \\
& \searrow & \nearrow & \searrow & & \nearrow & & \searrow \nearrow & \\
& K_{k+1}(M) & & & K_k(\partial U) & & & K_k(M) & \\
& \nearrow & \searrow & \nearrow & & \searrow & & \nearrow \searrow & \\
0 & & K_{k+1}(M, U) = K_{k+1}(M_0, \partial U) & & & & K_k(U) & & 0
\end{array} \qquad (1)$$

with coefficients Λ *throughout, splits over* Λ' *as the direct sum of two subdiagrams interchanged by* T, *of which the first is*

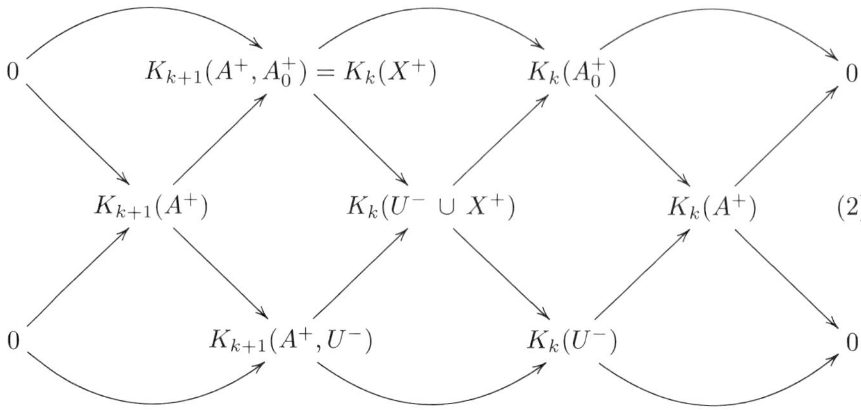

(2)

with coefficients Λ' *throughout.*

Proof The splitting of $K_*(M)$ is provided by (12.8). Also, $\widetilde{U}$ is the disjoint union $U^+ \cup U^-$. It is now easy to see that the lower sequence splits as asserted. For the rest, we first observe that each component of W (or $\widetilde{W}$, W^+, W^-) is contractible, and that

$$K_k(\partial U; \Lambda) \;=\; K_k(U; \Lambda) \oplus K_k(X; \Lambda)\ ,$$

admitting a further split into components with affixes $^{\pm}$. We also have isomorphisms

$$K_k(X; \Lambda) \cong K_{k+1}(U, \partial U; \Lambda) \cong K_{k+1}(M, M_0; \Lambda)\,,$$
$$K_k(X^+; \Lambda') \cong K_{k+1}(W^+, X^+; \Lambda') \cong K_{k+1}(A^+, A_0^+; \Lambda')\,,$$

using the explicit definitions of W, X, U and excision isomorphisms. Next, since the complement of $A_0^+ \cup A_0^-$ is Int W, a union of contractible sets, we have $K_i(A_0^+ \cup A_0^-) = 0$ for if $i = k$, $k+1$; now as $A_0^+ \cap A_0^- = \widetilde{M_0}$, the splitting of $K_k(M_0; \Lambda)$ follows using the Mayer-Vietoris sequence.

Since our isomorphisms are induced by inclusions and boundary maps, it is easy to check that the obvious maps in (2) are indeed induced by splitting diagram (1), and hence that (2) commutes.

We can now parallel the discussion of §6. First observe that $K_k(U^- \cup X^+)$, as submodule of $K_k(\partial U)$, can be treated like the H^+ arising in the even dimensional case above. It admits a nonsingular form satisfying (Q1)–(Q5), which is even a hyperbolic form for – as is readily seen – $K_k(X^+)$ is a lagrangian. In fact, the splitting above, with the knowledge that $K_{k+1}(M_0, \partial U)$ is a lagrangian in §6, shows that $K_{k+1}(A^+, U^-)$ is another lagrangian here†. Retaining the notation of the even-dimensional case, we now *assume* $t^2 = 1$. It should not be difficult to develop an analogous theory without this hypothesis, but since we have no applications in mind for this case, we forbear from doing so.

†Strictly speaking, we must also check preferred bases over Λ'. This needs a new argument which is, however, not essentially different from the one in (5.7).

The point of the assumption is that we now have a simple hermitian or skew-hermitian form in the original sense, so can use the general algebraic results of §6, in particular (6.2) and (6.3): we also use the same notation.

Now the choice of the embedding of W^+ determines a coset $TU_r(\Lambda')\alpha \subset SU_r(\Lambda')$: this is immediate. Next we will show that the embedding $M \subset V$ determines a stable double coset $TU(\Lambda')\alpha TU(\Lambda') \subset S'U(\Lambda')$. The effect of a surgery, deleting a component of U, say U^α from M and replacing it by X^α, changes A^+ to $A^+ - W^{+\alpha} \cup W^{-\alpha}$, so the old $W^{+\alpha}$ now lies in A^-, so $U^{+\alpha}$, $X^{+\alpha}$ are renamed $X^{-\alpha}$, $U^{-\alpha}$; similarly if we interchange $\pm$. Thus our kernel $K_k(U^- \cup X^+)$ is unchanged, but our standard base is altered by σ (in the position α). So when we include surgery in the equivalence relation, only the class of α in the quotient group $SU(\Lambda')/RU(\Lambda')$ is determined; conversely, if this class is zero, surgery is possible.

In order to complete the proof that the group $LN_n(\pi' \to \pi)$ in question is isomorphic to $SU(\Lambda')/RU(\Lambda')$, it remains to show first, that the element we have constructed in this group is invariant under cobordism (not just the surgeries above), and second, that any element of this group can appear as an obstruction (cf. (6.5)).

Our next task is to investigate the effect on α of changing the choice of the submanifold W of V. The argument closely parallels (and uses) that of §6. Now W^+ is the image of a set of disjoint embeddings $f_i : D^{k+1} \times (D^{k+1}, S^k) \to (A^+, \widetilde{M})$, which determine homology classes in $H_{k+1}(A^+, \widetilde{M}; \Lambda')$. Given another such set of embeddings g_i, determining the same homology classes, we claim that there are disjoint isotopies H_i of f_i to g_i. For as $k \geqslant 2$, $(A^+, \widetilde{M})$ is 2-connected, and the assertion follows from [H28]. However, the projected embeddings in V need not be disjointly isotopic, as the projections of the H_i need not be disjoint embeddings in $V \times I$: H_i may meet TH_j. This shows that the ρ_{ij} of §6 have values in $t\Lambda' \subset \Lambda$; and correspondingly for the ν_i. We claim that they can take arbitrary values subject to this restriction: indeed, we need only subject the spheres $f_i(0 \times S^k) \subset M$ to regular homotopies constructed in §6, and note that these lift to disjoint isotopies in $\widetilde{M}$, which we can then extend to $D^{k+1} \times D^{k+1}$. Now the effect of this on α was computed in §6; translating back to the present context, it means that α is multiplied on the right by an arbitrary element of $UU_r(\Lambda')$.

To investigate the effect of changing the homology classes, we refer again to §6, noting the single difference that only Λ' (and so π') may be used for coefficients, not the whole of Λ. Now the argument of §6 shows that such changes will stabilise α, and that they multiply it on the right by elements of $TU_r(\Lambda')$. To show that we can get arbitrary such elements, it is only necessary to check the construction for (T4) (the others are trivial). Here, we have two embeddings $f_1, f_2 : (D^{k+1}, S^k) \times D^{k+1} \to (A^+, \widetilde{M})$, and join them by an arc in $\widetilde{M}$ which is then thickened to obtain a disc in A^+ meeting $\widetilde{M}$ in a trivial handlebody. Our construction then consisted of performing the diffeomorphism of [W8, p. 272] on the handlebody; now this diffeomorphism was constructed as the restriction of a rotation of euclidean space, and so certainly extends to our disc.

Next, we show the cobordism invariance of the class of α modulo $RU(\Lambda')$: we cannot here use the argument of (6.1) since we have not shown that cobordism (as equivalence relation) is generated by surgeries. Since we evidently have an additive correspondence between the groups $LN_{2k+1}(\pi' \to \pi)$ and $SU(\Lambda')/RU(\Lambda')$ it is sufficient to assume M cobordant to an M' with $i' : M' \subset V$ a simple homotopy equivalence, and then show that $\alpha \in RU(\Lambda')$. Let $N \subset V \times I$ be a cobordism; by (12.7) we can suppose this inclusion $(k+1)$-connected, so that we have the exact sequence

$$0 \to K_{k+1}(M;\Lambda) \to K_{k+1}(N;\Lambda) \to K_{k+1}(N,M;\Lambda) \to K_k(M;\Lambda) \to 0$$

in which, by (2.3), the two middle terms are stably free and s-based (and may, as usual, be assumed free and based). We claim that – as in (12.8) and (12.11) – this sequence splits over Λ' as the direct sum of two sequences which are interchanged by T. For introduce B^+, B^- as in the proof of the corresponding assertion in the even dimensional case; then one summand is the sequence

$$0 \to K_{k+1}(A^+;\Lambda') \to K_{k+1}(B^+;\Lambda') \to K_{k+1}(B^+,A^+;\Lambda') \to K_k(A^+;\Lambda') \to 0\,.$$

Again, as usual, we may suppose

$$H_{k+2}(B^-, A^- \cup \widetilde{N};\Lambda') \;=\; H_{k+1}(B^+, A^+;\Lambda')$$

free and based over Λ'. Now consider the triad $\Phi = (B^-; A^-, \widetilde{N})$. We have $A^- \cap \widetilde{N} = \widetilde{M}$, and $(\widetilde{N}, \widetilde{M})$ is k-connected, $(A^-, \widetilde{M})$ also. By Namioka's relativisation [N1] of the Hurewicz theorem, we have $\pi_{k+2}\Phi \cong H_{k+2}(B^-, A^- \cup N;\Lambda')$. We choose a preferred basis, and represent by maps

$$f_i : (D^{k+2}; D^{k+1}_-, D^{k+1}_+; S^k) \to (B^-; A^-, \widetilde{N}; \widetilde{M})\,.$$

By Hudson's embedding theorem [H28], these maps may be supposed disjoint embeddings (we apply the theorem first to the restriction to the union of the D^{k+1}_+, then, keeping that fixed, to the rest). We can thicken (multiplying by D^{k+1}), again obtaining embeddings. Since, by the sequences above, the $f_i(S^k)$ generate the summand $K_k(A^+;\Lambda')$ of $K_k(M;\Lambda)$, we can set

$$W^- \;=\; \bigcup f_i(D^{k+1}_- \times D^{k+1})\;,$$

and correspondingly for all the others introduced above.

Now observe that if we perform surgery on M using the embeddings U we obtain precisely the cobordism N of M to a simple homotopy equivalence. Thus the lagrangians $K_{k+1}(M_0, \partial U;\Lambda)$ and $K_{k+1}(X, \partial U;\Lambda) \cong K_k(U;\Lambda)$ of $K_k(\partial U;\Lambda)$ are complementary. Using the splitting (12.11), it follows that $K_{k+1}(A^+, U^-;\Lambda')$ and $K_{k+1}(X^-, \partial U^-;\Lambda') \cong K_k(U^-;\Lambda')$ are additively complementary in $K_k(U^- \cup X^+;\Lambda')$.

Now our construction has identified the free based modules $K_{k+1}(B^+, A^+; \Lambda')$ in (3) and $K_k(U^-)$ in (2). Comparing (3) with the lower sequence in (2), we see that the induced isomorphism of $K_{k+1}(B^+; \Lambda')$ on $K_{k+1}(A^+, U^-; \Lambda')$ also preserves preferred bases; and these bases are dual to those in the preceding sentence. It follows that $K_{k+1}(A^+, U^-; \Lambda')$ and $K_k(U^-; \Lambda')$ are complementary even when we take preferred bases into account. Hence the stable class of α lies in $RU(\Lambda')$, as asserted (cf. end of §6).

Finally, we produce a construction analogous to that of (12.10) for the odd dimensional case: this will complete the proof of (12.9).

LEMMA 12.12. *Let M_0^{2k} ($k \geqslant 3$) have fundamental group π: let V_0^{2k+1} be the bundle over it with fibre $[-1, 1]$ associated to the subgroup $\pi' \subset \pi$; let $\alpha \in SU_r(\Lambda')$. Then there exists $M^{2k+1} \subset V_0 \times I$ with $\partial M = M_0 \times 0 \cup M_1 \times 1$, where $M_1 \subset V$ smooths the same simple Poincaré embedding as M_0, and $W \subset \widetilde{V}_0 \times I$ which, with the notation above, determines α.*

Proof First embed r disjoint copies of (D^{k+1}, D^k_+) in (V_0, M_0), and perform surgeries using the D^k_- on M_0 inside V_0: this surgery determines a cobordism $N_0 \subset V_0 \times [0, \frac{1}{2}]$ of $M_0 \times 0$ to $M_2 \times \frac{1}{2}$, say; M_2 is the connected sum of M_0 with r copies of $S^k \times S^k$. The surgery gave (we see by explicit calculation) a lagrangian $K_{k+1}(B_0^+, A_2^+; \Lambda')$ in $K_k(A_2^+; \Lambda')$. Applying α gives a new lagrangian. Represent its basis elements by framed embeddings $(D^{k+1}, S^k) \to (A_2^-, \widetilde{M}_2)$ and again perform surgeries to obtain a cobordism $N_1 \subset V_0 \times [\frac{1}{2}, 1]$ of $M_2 \times \frac{1}{2}$ to $M_1 \times 1$. It is clear that M_0, M_1 smooth the same simple Poincaré embedding. We must show that α determines the surgery obstruction for $N = N_0 \cup N_1$. This follows on general principles by an argument along the lines of (7.2), which we leave to the reader.

However, we can also argue directly, as in (6.5). For each copy of D^{k+1} we started with, we have a nontrivial sphere $S^k \subset N_0$; let U be the union of neighbourhoods of these, and obtain W from discs spanning them in $D^{k+1} \times D^k \times I$ (these may be constructed explicitly). We can then identify $K_k(A_2^+; \Lambda')$ with $K_k(U^- \cup X^+)$, and $K_{k+1}(B_0^+, A_2^+; \Lambda')$ corresponds to $K_k(X^+)$. But $K_{k+1}(N, U)$ is freely spanned by the classes of the attached $(k+1)$-cells; correspondingly for the summand $K_{k+1}(B^+, U^-)$; and these were defined as the images by α of the basis of the first lagrangian. The result follows.

See Ranicki [R7, pp. 687–735],[R6] for a chain complex treatment of codimension 1 splitting obstruction theory in the one-sided case (C). See Hambleton, Taylor and Williams [H10] and Muranov [M23] for the application of the theory to the computation of the L-groups of finite groups.

Part 3

Calculations and Applications

13A. Calculations: Surgery Obstruction Groups

Up to now, the groups $L_m(\pi)$ have been defined and treated in a purely abstract fashion, and we have given no discussion of them, beyond mentioning that they are abelian. For applications, however, more precise information is necessary, and we now compute a few of the groups explicitly, and give some further calculation based on the results of §12. We will then discuss in the same cases how surgery obstructions are to be computed as elements of these groups.

THEOREM 13A.1. *Let π be of order 1 or 2. Then the groups $L_m(\pi)$ are given by the following table:*

		$m=0$	$m=1$	$m=2$	$m=3$
$\lvert\pi\rvert = 1$	$: \pi = 1$	$\mathbf{Z}\,(\sigma/8)$	0	$\mathbf{Z}_2(c)$	0
$\lvert\pi\rvert = 2,\ w = 1$	$: \pi = \mathbf{Z}_2^+$	$\mathbf{Z}\oplus\mathbf{Z}\,(\sigma/8, \widetilde{\sigma}/8)$	0	$\mathbf{Z}_2(c)$	$\mathbf{Z}_2(d)$
$\lvert\pi\rvert = 2,\ w = -1$	$: \pi = \mathbf{Z}_2^-$	$\mathbf{Z}_2(c)$	0	$\mathbf{Z}_2(c)$	0

Explanation. The first column defines the class of groups π; the second gives the shorthand notation which we will use for that class. The symbol in parentheses after each group denotes the isomorphism of $L_m(\pi)$ onto that group; in particular, c denotes the Kervaire-Arf invariant [W18, 4.7], σ the signature, and $\widetilde{\sigma}$ the signature of the double covering.

Proof In the case when m is even, the definition of the groups L_0 and L_2 is the same as that of the group $\mathscr{G}$ in [W18, §4], except for reference to a preferred class of bases. However, since $Wh(\pi)$ (the Whitehead group: see [M14]) is trivial for the groups in question, this makes no difference. The results for m even thus follow from [W18, 4.13–14]. In the case $|\pi| = 1$, the calculation is much older: the result is implicit in [M10], [K5].

When m is odd, we cannot use the formulation of [W18, §5] directly. However, it was shown there that if $\pi = 1$ or $\mathbf{Z}_2^-$, there is no obstruction to successful completion of surgery: the case $\pi = 1$ again goes back to [K5], and for $\mathbf{Z}_2^-$ see [W18, §6]. Since, by (6.5), every element of $L_m(\pi)$ occurs as obstruction to a surgery problem, we can conclude in these cases that $L_m(\pi)$ vanishes.

The remaining cases $\pi = \mathbf{Z}_2^+$, $m \equiv \pm 1 \pmod 4$ must be considered in more detail. Here we have two alternative methods: to follow up [W18, §6], or to use the results of (12.9). We prefer to do both, and check that they lead to the same result. For the former, we know that there is at most one obstruction, and that mod 2, to performing surgery. It turns out that this is genuine if $m \equiv 3 \pmod 4$ and bogus when $m \equiv 1$.

First let $m = 4k - 1$, $k \geqslant 2$. Then by [W18, (6.1)], given any corresponding surgery problem $\phi : M \to X$, we can do surgery to make ϕ $(2k - 1)$-connected, and $K_{2k-1}(M)$ finite. By [W18, (5.7)], the obstruction to performing further surgeries to obtain a homotopy equivalence lies in a certain group $\mathscr{G} \cong \mathbf{Z}_2$. It follows from the theory of §9 that we have an isomorphism of $L_3(\mathbf{Z}_2^-)$ on a subgroup of $\mathscr{G}$ (that the natural map respects group structure follows since both surgery obstructions add for disjoint unions). An example to show that the map is onto is easily found: since one appears naturally in (13A.9) below, we will not repeat it here.

For the case $m = 4k+1$, if $\phi : M \to X$ is a $2k$-connected map representing a surgery problem, we know by [W18, (6.5)] that if $G = K_{2k}(M)$ is finite, we can do surgery to kill it; and by [W18, (6.1)] that the parity of the rank of G is a surgery invariant, and that if this rank is even, we can do surgery to make G finite. It is enough, then, to assume this rank odd, and derive a contradiction. By the same result, we can do surgery to reduce rank G to 1. An examination of the proof of [W18, (6.2)] now shows that if $K_{2k}(M)$ contains an element x with $|x\Lambda|$ finite and > 2, we can (as there) do two surgeries with the net effect of reducing the order of the torsion subgroup. So we can assume there is no such x. Then it is easily seen that $K_{2k}(M)$ is a direct sum of cyclic modules generated by elements x_0, x_i such that the annihilator of x_0 is $\langle T-1\rangle$ or $\langle 2T-2\rangle$ and that of x_i is $\langle T-1, T+1\rangle$. If there are any such x_i ($\neq x_0$), then the Λ-module $K_{2k}(M)$ cannot be defined by a presentation with the same number of relations as of generators. But this, with [W18, (5.4) and (6.3)], leads to a contradiction. Thus $K_{2k}(M)$ admits a single generator. Now perform the construction of §6 above: we then have the hyperbolic form $\{e, f\}$, the lagrangian $\{e\}$, and another lagrangian, generated say by

$$e' = e(a+bT) + f(c-cT),$$

where $c = 1$ or 2 (recall that the annihilator of f modulo e' is the ideal determined above). Since e' generates an additive direct summand, we must have $\langle a+bT, c-cT\rangle = \langle 1\rangle$; in particular, a and b have opposite parity. Also,

$$0 = \mu(e') = (a+bT)(c-cT) = c(a-b)(1-T) \pmod 0,$$

so $c(a-b) = 0$. But $c \neq 0$, hence $a = b$. This gives a contradiction: the result is established

We now check our result by the second method alluded to above: here we take the $L_i(1)$ and $L_i(\mathbf{Z}_2^-)$ as given. First compute $L_m(1 \to \mathbf{Z}_2^-)$ by the exact sequence. We need to compute $L_0(1) \to L_0(\mathbf{Z}_2^-)$ and $L_2(1) \to L_2(\mathbf{Z}_2^-)$. The first is zero, since by [W18, (4.13, Complement)], c vanishes on $L_0(1)$. The second is an isomorphism, since c maps each group isomorphically to $\mathbf{Z}_2$. The exact sequence now shows

$$L_0(1 \to \mathbf{Z}_2^-) \cong L_0(\mathbf{Z}_2^-) \cong \mathbf{Z}_2, \qquad L_1(1 \to \mathbf{Z}_2^-) \cong L_0(1) \cong \mathbf{Z},$$
$$L_2(1 \to \mathbf{Z}_2^-) = 0 = L_3(1 \to \mathbf{Z}_2^-).$$

If we use the isomorphism $L_i(1 \to \mathbf{Z}_2^+) \cong L_{i+1}(1 \to \mathbf{Z}_2^-)$ of (12.9.2), and the known values of $L_i(1)$, the exact sequence for $1 \to \mathbf{Z}_2^+$ now reduces to

$$0 \to L_3(\mathbf{Z}_2^+) \to \mathbf{Z}_2 \to \mathbf{Z}_2 \to L_2(\mathbf{Z}_2^+) \to 0\,,$$
$$0 \to L_1(\mathbf{Z}_2^+) \to 0 \to \mathbf{Z} \to L_0(\mathbf{Z}_2^+) \to \mathbf{Z} \to 0\,.$$

Since 1 is a retract (preserving w) of $\mathbf{Z}_2^+$, $L_2(1) \cong \mathbf{Z}_2$ is a retract of $L_2(\mathbf{Z}_2^+)$, which thus cannot vanish. Hence the map $\mathbf{Z}_2 \to \mathbf{Z}_2$ above is zero. The values of the $L_i(\mathbf{Z}_2^+)$ (up to isomorphism) now all follow.

We next reconsider the multisignature as defined in [W18, 4.9], and obtain a new formulation of this invariant, which clarifies its functorial properties. Most of the arguments are taken from the (Liverpool) Ph.D. thesis of D. W. Lewis (see [L19]), where further discussion will be found. Several of the ideas have also been discovered by Ted Petrie [P3]. We assume throughout that π is a finite group.

The multisignature was defined as follows. Decompose the real group ring $\mathbf{R}[\pi]$ as a direct sum of matrix rings (there is one summand corresponding to each class of irreducible real representations of π). The group ring has an anti-involution $\sum_{g\in\pi} \lambda_g g \mapsto \sum_{g\in\pi} w(g)\lambda_g g^{-1}$; under this, the summands are preserved or interchanged in pairs. Now simple anti-involuted algebras were classified by Weil [W40]; we will refer instead, however, to [W22]. Here they are classified, according to the centre, into 5 types: $(\mathbf{C}+\mathbf{C}, s)$, $(\mathbf{C}, 1)$, $(\mathbf{C}, c)$, $(\mathbf{R}+\mathbf{R}, s)$ and $(\mathbf{R}, 1)$. In each type we have a 'Brauer' group, isomorphic respectively to 1, $\mathbf{Z}_2$, 1, $\mathbf{Z}_2$ or $\mathbf{Z}_2 \oplus \mathbf{Z}_2$; we label the elements of these groups, according to the simple real Lie group obtained from the anti-involuted algebra by setting $U(A,*) = \{a \in A : a^*a = 1\}$, as $GL(\mathbf{C})$; $O(\mathbf{C})$, $Sp(\mathbf{C})$; U; $GL(\mathbf{R})$, $GL(\mathbf{H})$; $O(\mathbf{R})$, $Sp(\mathbf{R})$, $O(\mathbf{H})$, $Sp(\mathbf{H})$. The isomorphism classification of the simple anti-involuted algebra is determined by: class in the Brauer group, dimension over $\mathbf{R}$, and – for the types $O(\mathbf{R})$, U and $Sp(\mathbf{H})$ only – a further invariant which can be regarded as a signature.

Next one must observe that the classification of (non-singular) hermitian or skew-hermitian forms over an anti-involuted algebra depends only on the class of this algebra in the Brauer group. Lewis proved this *ad hoc*, but it follows also from the general theory of Fröhlich and McEvett [F12]. Aside from the rank, the only invariant needed for classification is the signature, which appears only

for hermitian forms, for types U, $O(\mathbf{R})$, $Sp(\mathbf{H})$,

for skew-hermitian forms, for types U, $Sp(\mathbf{R})$, $O(\mathbf{H})$.

The multisignature is the collection of signatures arising from the relevant direct summands of $\mathbf{R}[\pi]$.

This account is valid for any semisimple algebra over $\mathbf{R}$ (and hence for ones over $\mathbf{Q}$); we now take stock of the simplifications arising from the fact that we have a group ring. Suppose first that we take the orientable case: $w = 1$. Here Lewis makes the simple but important observation that we have a positive anti-involution in the sense of Weil [W40]. For we see at once, taking the elements of π as a basis, that

$$\operatorname{trace}\Big(\sum_{g\in\pi}\lambda_g g\Big) = |\pi|\lambda_1$$

and so

$$\operatorname{trace}\Big\{\Big(\sum_{g\in\pi}\lambda_g g\Big)\Big(\sum_{g\in\pi}\lambda_g g^{-1}\Big)\Big\} = |\pi|\sum_{g\in\pi}\lambda_g^2 > 0$$

unless each λ_g is 0. Thus for hermitian forms we have a signature for each irreducible real representation; for skew-hermitian forms, only for those of type **C**.

We can now reformulate the multisignature in the orientable case in the style of Atiyah and Singer [A8, pp. 578–579]. From $\lambda : H \times H \to \mathbf{Z}[\pi]$ we pick out the constant term, giving a $(-1)^k$-symmetric bilinear map (for manifolds of dimension $2k$). Extend this to $H^c = H \otimes_{\mathbf{Z}} \mathbf{C}$ (using the same matrix on a base of $H \subset H^c$) as a $(-1)^k$-hermitian form ϕ in the ordinary sense. Thus

$$\phi(y,x) = (-1)^k\overline{\phi(x,y)} \quad , \quad \phi(xg,yg) = \phi(x,y)$$

(H^c inherits from H a π-module structure). Now choose a positive definite π-invariant hermitian form $\langle\,,\rangle$ on H^c: this is unique up to homotopy. Consider the eigenvalue problem:

$$\phi(x,y) = \beta\langle x,y\rangle \quad \text{for all } y\,.$$

The eigenvalues β all satisfy $\overline{\beta} = (-1)^k\beta$, as follows from the first property above; thus they are real (k even) or pure imaginary (k odd), nonzero since ϕ is nonsingular. Thus $H^c = H^c_+ \oplus H^c_-$, where H^c_ε ($\varepsilon = \pm 1$) is the sum of eigenspaces corresponding to positive multiples of ε or εi. It follows from the second property above that this decomposition is π-invariant.† Now take the element of $\mathbf{R}[\pi]$ given by the character of H^c_+ minus that of H^c_- as our new invariant.

It follows easily that the constructed character is real for k even and pure imaginary for k odd. Lewis shows that its components in terms of irreducible representations coincide (up to sign) with the multisignature as previously defined, except for a factor of 2 in the case of quaternionic representations (in fact for k even our character corresponds to a real representation). However, naturality properties are much easier to obtain with the representation form of the definition. In particular, χ is multiplicative for products. This is observed in [A8, p. 580]; see also [C16, p. 39], particularly for signs.

We turn now to the nonorientable case. First tensor the anti-involuted algebra by $(\mathbf{C}, c)$. The effect on types is that summands of type $(\mathbf{C}+\mathbf{C}, s)$ or $(\mathbf{C}, c)$ are twinned; those of type $(\mathbf{C}, 1)$ or $(\mathbf{R}+\mathbf{R}, s)$ become of type $(\mathbf{C}+\mathbf{C}, s)$; those of

†If this is carried through in the nonorientable case, it turns out that orientation-reversing elements of π interchange H^c_+ and H^c_-.

type $(\mathbf{R}, 1)$ become of type $(\mathbf{C}, c)$. Hence (checking cases) the number of resulting summands of type $(\mathbf{C}, c)$ is the *sum* of the numbers of signatures for hermitian and skew-hermitian forms. But for $\mathbf{C}[\pi]$ the direct summands correspond to irreducible complex representations of π; the anti-involution $\sum_{g\in\pi}\lambda_g g \mapsto \sum_{g\in\pi}\overline{\lambda}_g g^{-1}$ preserves each of these, as we have just seen, so the anti-involution $\sum_{g\in\pi}\lambda_g g \mapsto \sum_{g\in\pi} w(g)\overline{\lambda}_g g^{-1}$ interchanges the summands corresponding to characters χ and $w\cdot\chi$. Thus the summands of type $(\mathbf{C}, c)$ (i.e. the fixed ones) correspond to irreducible complex characters χ of G such that $\chi(g) = 0$ for all g with $w(g) = -1$.

Let $H = \operatorname{Ker} w$; let α be the class mod inner automorphisms of the automorphisms induced on H by conjugating by an element of $G - H$. Then α acts on the set of representations of H; if ψ is an irreducible representation of H and $\psi^\alpha \sim \psi$, ψ extends to a representation of G; otherwise the induced representation ψ^G is irreducible. The representations we want are those of the latter sort.

We now again reformulate the multisignature in representation-theoretic terms, this time in the nonorientable case. All we need do is take the multisignature of the orientation double covering (i.e. consider only the action of H). For consistent choice of signs, we suppose chosen a base point and an orientation at that point, which thus lifts to an orientation of the double covering. Now a covering transformation induces an automorphism of H of class α, and changes the orientation, so our character ψ of H must satisfy $\psi^\alpha = -\psi$. Using the above results it is not difficult to see that such characters specify precisely the multisignature as defined above.

We extend this character by 0 on $G - H$ so as to regard it as a function on G. It then satisfies the identity $\psi(x^{-1}yx) = w(x)\psi(y)$. We call such functions *twisted class functions* on G: note that (taking $x = y$) they all vanish outside H. The space (over $\mathbf{C}$) of such functions has the same dimension as the group (over $\mathbf{Z}$) of our characters, which we call twisted characters.

We can summarise much of the above in

THEOREM 13A.2. *The multisignature function on $L_{2n}(G)$ is equivalent to the representation defined above: in the orientable case it is a (virtual) representation of G with character real (n even) or pure imaginary (n odd); in the nonorientable case a representation (as above) of $H = \operatorname{Ker} w$ satisfying also $\psi^\alpha = -\psi$.*

We conclude our discussion by showing that the multisignature is not trivial. In fact we obtain the following, in slightly sharper form.

THEOREM 13A.3. *The cokernel of the multisignature map $L_{2k}(\pi) \to \alpha\mathbf{Z}$ is a finite 2-group.*

Proof The multisignature classified forms over $\mathbf{R}$: we must lift these back to forms over $\mathbf{Q}$ and, ultimately over $\mathbf{Z}$. To lift to $\mathbf{Q}$ we use Galois cohomology. Start with a hyperbolic form over $\mathbf{Z}$ of sufficiently large rank. Its algebraic automorphism group is of a type determined above (if the form is hermitian, take the type of the anti-involuted algebra; if skew-hermitian, take this and interchange "O" and "Sp"). For factors not of type $O(\mathbf{R})$, U, $Sp(\mathbf{H})$, do not twist. For these, take the connected and simply connected almost-simple group mapping to them: Spin, SU, $Sp(\mathbf{H})$. For these, according to Kneser [K15], Galois cohomology is determined by its behaviour at the infinite prime; the effect of twisting here is to introduce a signature, which can take any value divisible by 8, 4, 2 in the three cases. (The signature is congruent mod 2 to the rank; going from O, U to SO, SU imposes a condition on the determinant and hence congruence mod 4; similarly going from SO to Spin we must watch the Clifford algebra).

Twist the hyperbolic form over $\mathbf{Q}$ by the chosen cohomology class. By construction, this is trivial in all p-adic fields, so over each of these we still have a hyperbolic form: choose a standard basis at each, and intersect to get a lattice over $\mathbf{Z}$. As $\mathbf{Z}[\pi]$-module, this is flat and locally free, hence projective: call it P. We have yet to make P free, and to find a base with respect to which it is simple hermitian. We accomplish both these in one step, at the expense of doubling the chosen value of the multisignature.

Write the dual module of P as $P^* = \mathrm{Hom}_{\mathbf{Z}[\pi]}(P, \mathbf{Z}[\pi])$; let $e : P \to P^*$ be the adjoint isomorphism for the chosen nonsingular sesquilinear form; by this isomorphism, we can regard it as a form on P^*. Choose an inverse module Q to P and a basis for the free module $F = P \oplus Q$: invest $F^* = P^* \oplus Q^*$ with the dual basis. Now take the orthogonal direct sum of two copies of the original form and the hyperbolic form on Q: regard this as a form on

$$P \oplus P^* \oplus (Q \oplus Q^*) \cong F \oplus F^*$$

and give this the free basis above. Its adjoint is an isomorphism of $F \oplus F^*$ on $F^* \oplus F$, i.e. an automorphism of $F \oplus F^*$, which I claim is simple. For it is the direct sum of $(\pm)$ the identity maps of Q and Q^*, and the automorphism of $P \oplus P^*$ with matrix $\begin{pmatrix} 0 & \alpha \\ \alpha^{-1} & 0 \end{pmatrix}$, which is well known to be simple.

Finally note that since our form is trivial (by construction) over the 2-adic integers, there is no difficulty in finding a quadratic map μ with the desired properties.

As to further direct calculations, we will not give details here since they involve technical knowledge of algebraic number theory, and algebraic K-theory, and because results are being rapidly developed at the time of writing. We confine ourselves instead to a description of some results.

THEOREM 13A.4. *(i) For all finite π, the kernel of the multisignature is a finite 2-group.*
(ii) Let π be cyclic of odd order; write $L_{2k}(\pi) = L_{2k}(1) \oplus \widetilde{L}_{2k}(\pi)$: then multisignature maps $\widetilde{L}$ to characters trivial on 1. This last map has zero kernel, and the image consists of the characters (real or imaginary as appropriate) divisible

by 4.
(iii) For all finite π, *the transfer* $\tau : L_0(\pi) \to L_0(1)$ *is surjective.*

We will also need the classification of non-simple hermitian forms over $\mathbf{Z}\pi$ – i.e. all as above, except the requirement that the isomorphism $A\lambda$ be simple. Suppose π cyclic of odd order – hence in particular commutative. Then the determinant of $A\lambda$ with respect to a preferred basis is a unit D' of $\mathbf{Z}\pi$: if the form has rank $2r$ define the discriminant $D = (-1)^r D'$. Note that when π is trivial, D is always $+1$: thus in general D has augmentation $+1$. Also, since the form is hermitian, $\overline{D} = D$. Write U for the group of units D of $\mathbf{Z}\pi$ with $\varepsilon(D) = 1$ and $\overline{D} = D$. Let χ generate the (Pontrjagin) dual of G.

THEOREM 13A.5. *The form* ϕ *is determined up to stable isomorphism by* $D(\phi)$ *and the multisignature*

$$\sigma(\phi) \;=\; \sum_{0}^{(N-1)/2} 2\alpha_r\big(\chi^r + (-1)^k\chi^{-r}\big)$$

and, for k *odd, by* $c(\phi)$. *The only relation between these is that*

$$\operatorname{sign}\big(\chi^r(D)\big) \;=\; (-1)^{\alpha_r}\,.$$

Finally, I conjecture that if π is finite, $L_{2k+1}(\pi)$ is also finite. See §17E.

Next, the result of [W18, 4.11] can be somewhat improved.

PROPOSITION 13A.6. *Suppose all elements* $x \in \pi$ *with* $x^2 = 1$ *and* $w(x) = (-1)^{k-1}$ *commute (and so generate an elementary* 2*-group). Then any hyperbolic plane with zero Arf invariant is standard.*

Proof Let $\{e, f\}$ be a basis with $\lambda(e,e) = \lambda(f,f) = 0$, $\lambda(e,f) = 1$. Then $\mu(e) = \sum\{x : x \in I_1\}$ is a sum of certain elements $x \in \pi$ with $x^2 = 1$ and $w(x) = (-1)^{k-1}$, as above, each with multiplicity 1 (mod 2); similarly $\mu(f) = \sum\{x : x \in I_2\}$, say. Since we have zero Arf invariant, I_1 or I_2 – say I_1 – has even order. Now set

$$e' \;=\; e + \sum\{fx : x \in I_1\}\,.$$

Then $\lambda(e', f) = 1$, so it remains to compute

$$\mu(e') \;=\; \mu(e) + \sum\{x : x \in I_1\} + \big(\sum\{\overline{x} : x \in I_1\}\big)\mu(f)\big(\sum\{x : x \in I_1\}\big)\,.$$

The first two terms cancel and $\overline{x} = w(x)x^{-1} = (-1)^{k-1}x$. Now for $x \neq y \in I_1$, terms involving xzy and yzx above cancel, since all our elements of π commute; the remaining terms have the form $xzx = x^2z = z$, so the whole reduces to $|I_1|\mu(f)$, which vanishes mod 2 since $|I_1|$ is even. Now $\mu(e') = 0$, and the result follows easily.

Further calculation along these lines suggests that in the first non-abelian case (of order 6), new invariants analogous to c will be needed in order to deal with μ.

We next draw some deductions from the functorial character of the $L_m(\pi)$. If π is a retract of ρ (both inclusion and retraction commuting with w, of course), then $L_m(\pi)$ is a summand of $L_m(\rho)$. Thus $L_m(1)$ is a summand of any orientable $L_m(\rho)$. More generally, if $w(\rho) = 1$, $L_m(\pi)$ is a summand of $L_m(\pi \times \rho)$. In the nonorientable case, one naturally thinks of $\mathbf{Z}_2^-$. Here we can give a mild improvement of the periodicity theorem.

PROPOSITION 13A.7. *We have a natural isomorphism*

$$L_m(\pi \times \mathbf{Z}_2^-) \cong L_{m+2}(\pi \times \mathbf{Z}_2^-).$$

Remark. Ker $w \subset \pi \times \mathbf{Z}_2^- \to \pi$ is an isomorphism, and $\pi \times \mathbf{Z}_2^-$ is the direct product of the subgroups Ker w and $\mathbf{Z}_2^-$. Thus we may suppose $w(\pi) = 1$ without loss of generality.

Proof We note that the periodicity theorem was obtained by multiplying by $P_2(\mathbf{C})$, and $P_2(\mathbf{C}) \sim P_2(\mathbf{R}) \times P_2(\mathbf{R})$. Now consider

$$L_m(\pi \times \mathbf{Z}_2^-) \xrightarrow{\times P_2(\mathbf{R})} L_{m+2}(\pi \times \mathbf{Z}_2^- \times \mathbf{Z}_2^-) \to L_{m+2}(\pi \times \mathbf{Z}_2^-),$$

where the third map is induced by identifying the copies of $\mathbf{Z}_2^-$. Denote the composite by r. If we show that r^2 coincides with the periodicity isomorphism, it will follow that r is injective and surjective, and hence an isomorphism. Now consider the diagram

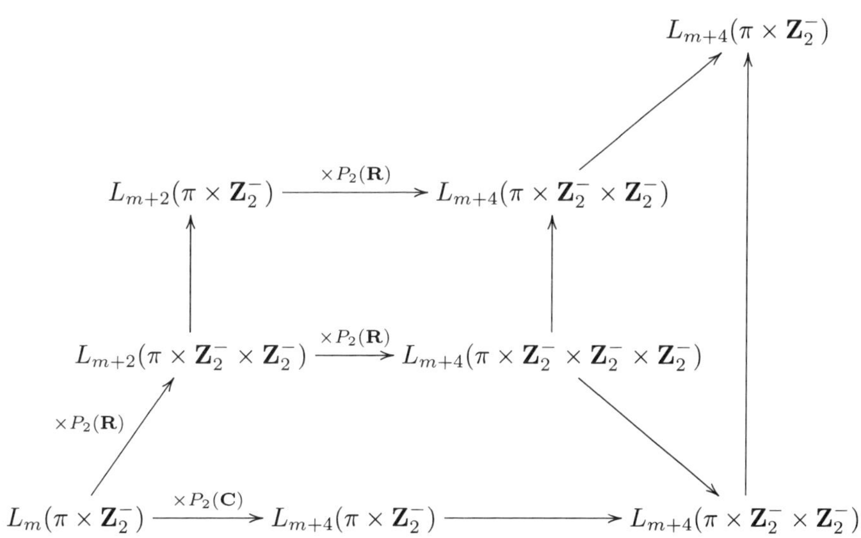

in which the diagonal map from $L_{m+4}(\pi \times \mathbf{Z}_2^- \times \mathbf{Z}_2^- \times \mathbf{Z}_2^-)$ is induced by identifying the last two copies of $\mathbf{Z}_2^-$, and the vertical map by identifying the first two copies. Then the two squares clearly commute, and the composite map $L_m(\pi \times \mathbf{Z}_2^-) \to L_{m+4}(\pi \times \mathbf{Z}_2^-)$ is just r^2. Since we can find a cobordism from $P_2(\mathbf{R}) \times P_2(\mathbf{R})$ to $P_2(\mathbf{C})$ which is (necessarily) nonorientable, and has fundamental group $\mathbf{Z}_2$ (e.g. by doing surgery on the classifying map of the tangent bundle of an arbitrary cobordism), the lower part of the diagram commutes. But the composite $L_{m+4}(\pi \times \mathbf{Z}_2^-) \to L_{m+4}(\pi \times \mathbf{Z}_2^-)$ is the identity. Thus r^2 coincides with multiplication by $P_2(\mathbf{C})$.

In the case $\pi = 1$, this naturally confirms part of (13A.1). The argument appears to admit generalisation, with ρ for $\pi \times \mathbf{Z}_2^-$, but needs an orientation-reversing central involution $x \in \rho$; and then $\rho = (\operatorname{Ker} w) \times \mathbf{Z}_2^-(x)$.

We next quote from §12

THEOREM 12.6. *Let $1 \to \pi \to \Pi \to \mathbf{Z} \to 0$ be an exact sequence of finitely presented groups; let $g \in \Pi$ map to $1 \in \mathbf{Z}$, and induce (by inner automorphism) the automorphism α of π. Then we have a commutative exact diagram:*

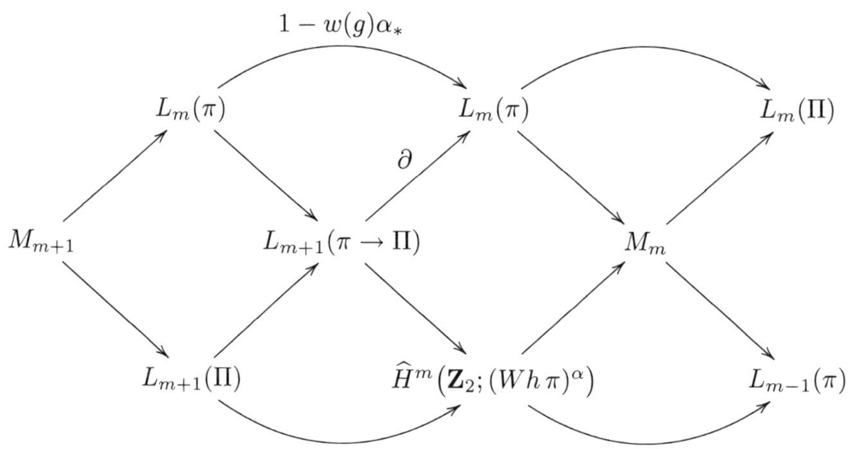

Thus if $\widehat{H}^m(\mathbf{Z}_2; (Wh\,\pi)^\alpha) = 0$ – e.g. if $Wh\,\pi = 0$ – we can identify $L_{m+1}(\pi \to \Pi)$ with $L_m(\pi)$ and ∂ with $1 - w(g)\alpha_*$. Note also that for $\Pi = \pi \times \mathbf{Z}^+$ we have splittings $L_{m+1}(\Pi) = L_{m+1}(\pi) \oplus L_{m+1}(\pi \to \Pi)$ and $M_{m+1} = L_{m+1}(\pi) \oplus L_m(\pi)$.

Applying this, we obtain

THEOREM 13A.8. *In the orientable case,*

$$L_m(\mathbf{Z}^r) \cong \bigoplus_{0 \leqslant i \leqslant r} \binom{r}{i} L_{m-i}(1)\,.$$

Further, we have the following table of $L_m(\pi)$:

π $\quad m$	0	1	2	3
$\mathbf{Z}$, orientable ($\mathbf{Z}^+$)	$\mathbf{Z}$	$\mathbf{Z}$	$\mathbf{Z}_2$	$\mathbf{Z}_2$
$\mathbf{Z}$, nonorientable ($\mathbf{Z}^-$)	$\mathbf{Z}_2$	0	$\mathbf{Z}_2$	$\mathbf{Z}_2$
$\mathbf{Z}^+ \oplus \mathbf{Z}_2^+$	$\mathbf{Z} \oplus \mathbf{Z} \oplus \mathbf{Z}_2$	$\mathbf{Z} \oplus \mathbf{Z}$	$\mathbf{Z}_2$	$\mathbf{Z}_2 \oplus \mathbf{Z}_2$
$\mathbf{Z}^+ \oplus \mathbf{Z}_2^-$ ($\cong \mathbf{Z}^- \oplus \mathbf{Z}_2^-$)	$\mathbf{Z}_2$	$\mathbf{Z}_2$	$\mathbf{Z}_2$	$\mathbf{Z}_2$
$\mathbf{Z}^- \oplus \mathbf{Z}_2^+$	$\mathbf{Z}_2 \oplus \mathbf{Z}_2 \oplus \mathbf{Z}_2$	0	$\mathbf{Z}_2$	$\mathbf{Z}_2 \oplus \mathbf{Z}_2$
$\mathbf{Z}^+ \oplus \mathbf{Z}^-$	$\mathbf{Z}_2 \oplus \mathbf{Z}_2$	$\mathbf{Z}_2$	$\mathbf{Z}_2$	$\mathbf{Z}_2 \oplus \mathbf{Z}_2$
$K = Gp\{x, y \mid y^{-1}xy = x^{-1}\}$, $w = 1$	$\mathbf{Z} \oplus \mathbf{Z}_2$	$\mathbf{Z} \oplus \mathbf{Z}_2$	$\mathbf{Z}_2$	$\mathbf{Z}_2 \oplus \mathbf{Z}_2$
K, $w(x) = 1$, $w(y) = -1$	$\mathbf{Z}_2 \oplus \mathbf{Z}_2$	$\mathbf{Z}$	$\mathbf{Z} \oplus \mathbf{Z}_2$	$\mathbf{Z}_2 \oplus \mathbf{Z}_2$

Note. The result for $\mathbf{Z}^r$ has been announced by J. Shaneson [S4], [S6].

The first result follows by induction on r. All the others are also direct applications, which we leave to the reader, noting that in each case the corresponding group $Wh(\pi) = 0$ (for $\pi = \mathbf{Z}_2$ or is free abelian). These examples show clearly that the preceding theorem is our most effective method yet for computing new groups $L_m(\pi)$. The fact that the groups $L_m(\pi)$ computed so far are all sums of copies of $\mathbf{Z}$ and $\mathbf{Z}_2$ is merely due to the fact that we have only treated simple examples. In this table, unlike (13A.1), we have not described the automorphisms; however, we will shortly be describing how to compute some obstructions. Here first is a final little calculation.

LEMMA 13A.9. *$L_3(\mathbf{Z}^+) \to L_3(\mathbf{Z}_2^+)$ is an isomorphism.*

The result will follow easily from the calculations of surgery obstructions in (13B.7) and (13B.8), but some readers may be interested to see some direct manipulations of matrices.

Proof Let e_1, f_1, e_2, f_2 be a standard basis for a skew-hermitian hyperbolic form over $\mathbf{Z}[\mathbf{Z}] = \mathbf{Z}[T, T^{-1}]$. Define

$$\begin{aligned} e_1'' &= e_1(1+T) + f_1 T + e_2(1+T) + f_2 \\ f_1'' &= e_1(1+T) + f_1(1+T) + e_2 T \\ e_2'' &= e_1 T + f_1(-1+T) + e_2(1+2T) + f_2(1+T) \\ f_2'' &= \qquad - f_1 \qquad\qquad + e_2(1+T) + f_2(1+T)\,. \end{aligned}$$

A little calculation shows that we have another standard basis, so we have indeed defined an automorphism. Now the matrix is also defined over the polynomial ring, so its image under the isomorphism $L_3(\mathbf{Z}^+) \to L_2(1)$ is represented by the cokernel of this map over $\mathbf{Z}[T]$. The following calculation is not really necessary for the proof, but seems worth inserting.

$$\begin{array}{rll} \text{Now write} & e_1' = e_1 + e_2 + f_2 & f_1' = e_1 + f_1 \\ & e_2' = e_1 + f_1 + e_2 & f_2' = e_2 + f_2\,; \\ \text{then} & e_1'' = e_1' + e_2'T & f_1'' = f_1' + e_2'T \\ & e_2'' = (e_1' - f_1') + (e_2' + f_2')T & f_2'' = (e_1' - f_1') + f_2'T \\ \text{and} & e_1 = e_1' - f_2' & f_1 = -e_1' + f_1' + f_2' \\ & e_2 = e_2' - f_1' & f_2 = -e_2' + f_1' + f_2' \end{array}$$

So e_1', f_1', e_2', f_2' span the whole module and e_1', f_1', $e_2'T$, $f_2'T$ form a basis for the image of the map over $\mathbf{Z}[T]$. Thus its cokernel admits the images of e_2', f_2' as a basis. But these give a hyperbolic plane, with $\mu(e_2') = \mu(f_2') = 1$, so the Arf invariant is 1. Thus we have the nonzero elements of $L_2(1)$, $L_3(\mathbf{Z}^+)$.

Now take the image in $L_3(\mathbf{Z}_2^+)$ by setting $T^2 = 1$. The surgery obstruction is represented by the pair of lagrangians $e_1 = f_2 = 0$ and $f_1'' = f_2'' = 0$. The quotient by these is generated by f_1 and e_2 subject to

$$f_1(1+T) + e_2T = 0$$
$$-f_1 + e_2(1+T) = 0$$

and so by e_2 with $e_2(2+3T) = 0$. Compute the surgery obstruction by the recipe at the end of [W18, §6]. The determinant of our module is $2 + 3T$, or, normalised, $-3 - 2T$. The obstruction is the class of -1 (mod 2), which is nontrivial. Thus the homomorphism is nonzero.

COROLLARY. *If N is even, $L_3(\mathbf{Z}_N^+)$ has $\mathbf{Z}_2$ as direct summand.*

For the map of the lemma then factorises as

$$L_3(\mathbf{Z}^+) \to L_3(\mathbf{Z}_N^+) \to L_3(\mathbf{Z}_2^+)\,.$$

The proof of (13A.1) illustrates well enough how to use diagram chasing to deduce relative L-groups from absolute ones. It seems worth, however, pointing out that some of the groups for submanifolds may be similarly obtained.

PROPOSITION 13A.10.

$$LN_n(A \cup A \to A) = 0$$
$$LN_n(A \to A \times \mathbf{Z}_2^-) \cong L_n(A)$$
$$LN_n(A \to A \times \mathbf{Z}_2^+) \cong L_{n+2}(A)$$
$$LN_n(A \times \mathbf{Z}^+ \to A) \cong \widehat{H}^{n+1}(\mathbf{Z}_2; Wh\,A)\,.$$

Proof The first three results are (12.4) and (12.9.1). For the fourth – which corresponds to codimension 2 – we first note that since A is a retract of $A \times \mathbf{Z}^+$, we can write

$$L_{n+1}(A \times \mathbf{Z}^+) \;=\; L_{n+1}(A) \oplus X_{n+1}(A)\,,$$

where the second summand* can be equivalently interpreted as

$$L_{n+1}(A \to A \times \mathbf{Z}^+) \quad \text{or} \quad L_{n+2}(A \times \mathbf{Z}^+ \to A)\,.$$

Now compare the exact sequences of (11.6) and (12.6):

$$\begin{array}{ccccccc}
L_{m+1}(A) & \xrightarrow{p} & L_{m+3}(A \times \mathbf{Z}^+ \to A) & \xrightarrow{q} & LN_m(A \times \mathbf{Z}^+ \to A) & \xrightarrow{r} & L_m(A) \\
\| & & \downarrow & & \downarrow t & & \| \\
L_{m+1}(A) & \longrightarrow & L_{m+2}(A \to A \times \mathbf{Z}^+) & \longrightarrow & \widehat{H}^{m+1}(\mathbf{Z}_2; Wh\, A) & \longrightarrow & L_m(A)\,.
\end{array}$$

The desired result will follow by the Five Lemma if we can define t to make the diagram commute. That such t exists can be seen geometrically by representing the element of LN_m as the obstruction to finding M^m in N^{m+2}, simply homotopy equivalent to N, and then taking the map $\partial N \to S^1$ induced by projecting the fundamental group on $\mathbf{Z}$, and taking the obstruction given by (12.5) to representing this as a fibre map.

Note. The reader may try computing $LN_n(\mathbf{Z} \xrightarrow{2} \mathbf{Z})$ $(q = 1)$ – the difficulty here is determining maps in exact sequences – and $LN_n(\mathbf{Z} \to \mathbf{Z}_m)$ $(q = 2)$, which I conjecture to be 0, and will mention further below in connection with fake lens spaces (see §17E).

There has been much progress in the computations of $L_*(\pi)$ for finite π since the first edition, starting with Wall's own sequence of papers [W28]–[W33], and continuing with the work (in alphabetic order) of Bak, Carlsson, Connolly, Hambleton, Kolster, Milgram, Pardon, Taylor, Williams and others. The computation techniques combine specific results in the classical theory of quadratic forms, algebraic number theory, algebraic groups and representation theory (e.g. the induction of Dress [D5]) with general results in algebraic K- and L-theory. See Ranicki [R8], [R9, Chapter 22] for some of the general algebraic L-theory results involving localisation and completion: for any group π the natural maps $L_*(\mathbf{Z}[\pi]) \to L_*(\mathbf{Q}[\pi])$ are isomorphisms modulo 8-torsion, and there is defined a Mayer-Vietoris sequence

$$\cdots \to L_n(\mathbf{Z}[\pi]) \to L_n(\widehat{\mathbf{Z}}[\pi]) \oplus L_n(\mathbf{Q}[\pi]) \to L'_n(\widehat{\mathbf{Q}}[\pi]) \to L_{n-1}(\mathbf{Z}[\pi]) \to \ldots$$

(with L'_* the appropriate intermediate L-groups) for the localisation-completion 'arithmetic square' of rings with involution

$$\begin{array}{ccc}
\mathbf{Z}[\pi] & \longrightarrow & \mathbf{Q}[\pi] \\
\downarrow & & \downarrow \\
\widehat{\mathbf{Z}}[\pi] & \longrightarrow & \widehat{\mathbf{Q}}[\pi]
\end{array}$$

* *In fact, $X_{n+1}(A) = L_n^h(A)$ is the free L-group of A – see §17D.*

with

$$\widehat{\mathbf{Z}} = \varprojlim_{m} \mathbf{Z}/m\mathbf{Z}$$

the profinite completion of $\mathbf{Z}$, *and* $\widehat{\mathbf{Q}}$ *the profinite completion of* $\mathbf{Q}$ (= *quotient field of* $\widehat{\mathbf{Z}}$). *See Hambleton and Taylor [H9] for a survey of the computations of* $L_*(\pi)$ *for finite* π.

See the notes at the end of §*15B for the L-theory of infinite groups.*

13B. Calculations: The Surgery Obstructions

This chapter is concerned with the multisignature part of the surgery obstruction with finite fundamental group, and with characteristic class formulae for the surgery obstructions of normal maps of closed manifolds.

For finite π the surgery obstruction of an even-dimensional normal map $\phi : M \to X$ is given modulo 2-primary torsion by the difference of the multisignatures of M and X, which for closed manifolds M, X are shown to be just the ordinary signatures.

Hirzebruch used multiplicative properties of bordism to construct the ℓ-genus $\ell \in H^{4*}(BO; \mathbf{Q})$, and to identify the signature of a $4k$-dimensional manifold M with the evaluation of $\ell(M) = (\tau_M)^*(\ell) \in H^{4*}(M; \mathbf{Q})$ on the fundamental class $[M] \in H_{4k}(M)$. Similarly, multiplicative properties of the surgery obstructions are used to identify the Kervaire invariant of a normal map of closed $(4k+2)$-dimensional manifolds with an evaluation of a universal class $\kappa \in H^{4*+2}(G/PL; \mathbf{Z}_2)$.

We come now to the second problem, that of calculating the surgery obstruction for a given problem. This is somewhat easier when the manifolds concerned have no boundary (in contrast to the general theory, where a boundary makes no difference). Also, integer obstructions are rather easier to compute than obstructions in finite groups; thus we begin by considering the multisignature.

If X^{2k} is an oriented Poincaré complex, $\pi_1(X) \to \pi$ a homomorphism*, then by considering the action of π on the induced regular covering $\widetilde{X}$ of X with group π, and in particular on $H_k(\widetilde{X})$, we obtain a character on $\sigma(\pi, X)$. Now let $\phi : M \to X$ have degree 1. Since (see (2.2)) $K_k(\widetilde{M})$ is orthogonal to $\phi^* H_k(\widetilde{X}) \subseteq H_k(\widetilde{M})$, the multisignature for the kernel – which is a surgery obstruction – is the difference of those for M and X. The rôle of this surgery obstruction is further clarified by

PROPOSITION 13B.1. (*see also* §14B). *Let M^{2k} be a closed PL manifold, $\pi_1(M) \to \pi$ compatible with $w_1(M)$. Then $\sigma(\pi, M)$ is a multiple of the character of the regular representation – i.e. it vanishes on elements $g \neq 1$ of π.*

An equivalent formulation is that all components of the multisignature are equal (up to signs depending on conventions). Unless k is even and $w = 1$, the result implies that σ is zero.

Proof For smooth manifolds, this can be deduced from the Atiyah-Bott fixed point theorem [A7]. However, the arguments necessary to obtain the PL case from this are strong enough to yield an independent proof, so we give this. It

* *π is finite here*

suffices to consider the orientable cases, as the rest follow by taking a double covering.

The characteristic map $M \to K(\pi,1)$ can be regarded as defining a bordism class. Now the signature is well-known to be a cobordism invariant, and an extension of the usual proof (cf. (5.7), which is another extension) shows that multisignature is an invariant of bordism class in $\Omega_{2k}^{PL}(K(\pi,1))$. This invariant is clearly additive, so defines a homomorphism

$$\Omega_{2k}^{PL}(K(\pi,1)) = \Omega_{2k}^{PL} \oplus \widetilde{\Omega}_{2k}^{PL}(K(\pi,1)) \to R(\pi)$$

where $R(\pi)$ is the character ring of π. Now $R(\pi)$ is torsion free. Since π is finite, $H_p(\pi;\Omega_{2k}^{PL})$ is finite also for $p > 0$, so by the spectral sequence of Ω_*^{PL} it follows that $\widetilde{\Omega}_{2k}^{PL}(K(\pi,1))$ is finite, hence is annihilated by the multisignature. But an element of the summand Ω_{2k}^{PL} is represented by a trivial map $M \to K(\pi,1)$, and so by a trivial covering $M \times \pi \to M$. However in this case $H_k(\widetilde{M}) = H_k(M) \otimes \mathbf{R}[\pi]$ with the hermitian form induced from $H_k(M)$, and the result follows easily.

The argument extends to the topological case, and will be further discussed in §14B.

It follows from this proposition that if X is a simple Poincaré complex, a necessary condition that X have the homotopy type of a closed PL manifold is that $\sigma(\pi, X)$ vanish for each $g \neq 1$ in π^*. And if X does satisfy this condition, and $\phi : M \to X$ represents a surgery problem, then the multisignature surgery obstruction will vanish if and only if we have equality of the usual signature. The reader should be warned not to discard the multisignature on account of this: indeed, the exact sequence (10.8) shows that most of it plays a non-trivial rôle in classifying odd-dimensional manifolds.

We obtain further information as follows. Observe first that the definition of $\sigma(\pi, X)$ extends without difficulty to any compact manifold X – or any Poincaré pair $(X, \partial X)$ – with $\pi_1(X) \to \pi$. The only small difference is that ϕ is no longer nonsingular, and our eigenvalue decomposition becomes $H^c = H_+^c \oplus H_0^c \oplus H_-^c$, where the middle term corresponds to the zero eigenvalue. This does not contribute to $\sigma(\pi, X)$, which is defined as before. The idea of using these signatures (at least, ignoring π) is due to S. P. Novikov. I am grateful to Dennis Sullivan for drawing my attention to it.

Observe again that in a surgery problem with boundary fixed, with a degree 1 map $M \to X$, the multisignature surgery obstruction is given by $\sigma(\pi, M) - \sigma(\pi, X)$. Some information can also be obtained when the boundary is not fixed.

**If X is a $4k$-dimensional Poincaré complex satisfying this condition the signature of a finite cover $\widetilde{X}$ of degree d must be $d\cdot$ signature(X). See Wall [W21] for the explicit construction of 4-dimensional simple Poincaré complexes X with cyclic fundamental group $\pi_1(X) = \pi = \mathbf{Z}_d$ (d prime), such that signature($\widetilde{X}$) $\neq d\cdot$ signature(X) for the universal cover $\widetilde{X}$. These X do not satisfy the condition, and are not homotopy equivalent to a PL (or TOP) manifolds. See also [R9, Chapter 22].*

Now let L^{2k-1} be a closed PL manifold, $\pi_1(L) \to \pi$. As before it defines a bordism class, and as in (13B.1) but using also finiteness of Ω_{2k-1}^{PL}, due to Williamson [W45], this class has finite order s, say. Write $sL = \partial V$ as singular manifold in $K(\pi, 1)$. We now define $\rho(\pi, L)$ as the restriction to $\pi - \{1\}$ of $s^{-1}\sigma(\pi, V)$. We must show that this is independent of the choices of s and V. Since $rsL = \partial(rV)$, and $\sigma(\pi, rV) = r\sigma(\pi, V)$, independence from s is clear. Now suppose also that $sL = \partial W$. Glueing along sL, we obtain a *closed* singular manifold $V - W$ in $K(\pi, 1)$. But by the proposition the restriction to $\pi - \{1\}$ of $\sigma(\pi, V - W)$ vanishes. The desired independence follows. A related argument has been independently discovered by Conner [C16].

We now summarise the basic properties of the invariant ρ.

THEOREM 13B.2. *Given a closed PL manifold L^{2k-1} and $\pi_1(L) \to \pi$, π finite, we can define an invariant $\rho(\pi, L) : \pi - \{1\} \to \mathbf{C}$. The class of $\rho(\pi, L)$ modulo (restricted) twisted characters is an invariant of bordism class in $\widetilde{\Omega}_{2k-1}^{PL}(K(\pi, 1))$, and defines a homomorphism of this group. For each k, the $\rho(\pi, L)$ form a group of functions on $\pi - \{1\}$, containing (for $k \geqslant 3$) the 16-fold multiples of (twisted) characters which are real (imaginary) for k even (odd) as a subgroup of finite index. Thus $\rho(\pi, L)$ is a twisted class function taking only real (imaginary) values for k even (odd).*

Proof We have defined ρ; it is clearly additive for disjoint unions. As to bordism invariance, let $sL = \partial V$ and $L - M = \partial W$. Glue s copies of W along L to V so that $\partial(V - sW) = sM$. By additivity, $\sigma(\pi, V - sW) = \sigma(\pi, V) - s\rho(\pi, W)$, hence $\rho(\pi, L) - \rho(\pi, M) = \sigma(\pi, W)$ comes from a (twisted) character. Realisability of 16-fold multiples follows from (13A.3) an (5.8); that these have finite index follows from the finiteness of the bordism group $\widetilde{\Omega}_{2k-1}^{PL}(K(\pi, 1))$. Finally, a class in Ω_{2k-1}^{PL} is represented by an L with trivial covering; then we can write $sL = \partial V$ with trivial covering, and as in (13B.1), $\sigma(\pi, V)$ must vanish on $\pi - \{1\}$.

It is perhaps appropriate to remark here that Andrew Casson had proved prior to the work of Kirby and Siebenmann that ρ is a topological invariant.

We now begin some calculations proper*. For M a closed PL manifold, $\mathscr{T}(M) \cong [M, G/PL]$ by (10.6), and we have the surgery obstruction map $\theta : \mathscr{T}(M) \to L_m(\pi(M))$. More generally, for M compact we have $\theta : \mathscr{T}(M, \partial M) \to L_m(\pi(M))$, and $\mathscr{T}(M, \partial M)$ is the group of homotopy classes of maps $(M, \partial M) \to (G/PL, *)$, which we will denote by $[(M, \partial M), (G/PL, *)]$. Following Sullivan [S22], we reinterpret this by varying M.

THEOREM 13B.3. *The surgery obstructions define a homomorphism*

$$\theta : \Omega_m^{PL}(K(\pi, 1) \times G/PL, K(\pi, 1) \times *) \to L_m(\pi) ,$$

natural in π.

Proof Represent a given bordism element by a triple (M, f, g) with M^m a compact PL manifold, $f : M \to K(\pi, 1)$ and $g : (M, \partial M) \to (G/PL, *)$. Then

*for arbitrary groups π

g determines $\{g\} \in \mathscr{T}(M, \partial M)$, $\theta\{g\} \in L_m\big(\pi(M)\big)$, and we set $\theta(M, f, g) = f_*\theta\{g\} \in L_m(\pi)$. The construction is evidently additive for disjoint unions, and if (M', f', g') represents the same bordism element there is a cobordism (W, f'', g'') between them (here, $\partial W = M \cup V \cup M'$ and $g'' : (W, V) \to (G/PL, *)$). Then g'' determines $\{g''\} \in \mathscr{T}(W, V)$ and

$$\theta\{g''\} \in L_{m+1}\big(\pi(M) \cup \pi(M') \to \pi(W)\big)$$

with
$$\partial\theta\{g''\} = (\theta\{g'\}, -\theta\{g\}).$$

By the exact sequence (3.1), $\theta\{g\}$ and $\theta\{g'\}$ have the same image in $L_m\big(\pi(W)\big)$, and hence *a fortiori* also (by f'') in $L_m(\pi)$. Thus the obstruction depends only on the bordism class, as asserted.

It follows from the theorem that the maps θ in question are determined if we know them for closed manifolds.

This result gives considerable simplification to the problem of computing surgery obstructions; note, however, that it is of no help for deciding whether a given Poincaré complex comes from a manifold or not*, but fits instead into the framework of the latter part of §10.

The rest of this chapter is devoted to the calculation of θ, in terms of characteristic classes, in some cases where we know the group $L_m(\pi)$. For the case $\pi = 1$ this was done by Sullivan [S22]. Two cases arise here: in the simpler, $4|m$ and $\frac{1}{8}\sigma : L_0(1) \to \mathbf{Z}$ is an isomorphism. Now the signature is relatively well understood, and we will merely quote Sullivan's result.

The maps $G/PL \to BPL \leftarrow BO$ induce homotopy and homology isomorphisms mod the class of finite groups, hence in particular rational cohomology isomorphisms. In $H^*(BO; \mathbf{Q})$ we have the Pontrjagin classes p_i and also the Hirzebruch classes ℓ_i which are expressed in terms of the p_i by certain known polynomials [H18]. Write $\ell(M) = \sum_1^\infty \ell_i(M)$ for the total Hirzebruch class of the tangent bundle of a PL manifold M. Write also $\ell_i(G/PL)$ for the corresponding classes in $H^{4*}(G/PL; \mathbf{Q})$ and $\ell(G/PL) = \sum_1^\infty \ell_i(G/PL)$. Then for $g : (M, \partial M) \to (G/PL, *)$ we have for $m = 4k$

$$\sigma\big(\theta(M, g)\big) \;=\; \ell(M) g^* \ell(G/PL)[M] \in L_{4k}(1) \;=\; \mathbf{Z}.$$

Sullivan [S22] also goes on to show that the classes $\frac{1}{8}\ell_i(G/PL)$ can be lifted to classes $\lambda_i(G/PL) \in H^{4*}(G/PL; \mathbf{Z}_{(\mathrm{odd})})$ with coefficient group the rationals with odd denominators†, but we will not need this result.

**An m-dimensional Poincaré complex is homotopy equivalent to a PL manifold if and only if the Spivak normal fibration is PL reducible and the surgery obstruction of any normal map $M \to X$ belongs to* $\mathrm{Im}(\theta) \subseteq L_m\big(\pi_1(X)\big)$.

†See Morgan and Sullivan [M20] for the definitive treatment of the classes $\lambda_i(G/PL)$, using the bordism of manifolds with singularities.

In the other case we have the isomorphism $c : L_2(1) \to \mathbf{Z}_2$, given by the Kervaire-Arf invariant. Sullivan has [S22] a formula for c on orientable manifolds, which we will discuss, and modify to cover the general case. First note that since (for any k) c gives an isomorphism of $L_{2k}(\mathbf{Z}_2^-)$ with $\mathbf{Z}_2$, we can identify c for a general $L_{2k}(\pi)$ as the map $L_{2k}(\pi) \to L_{2k}(\mathbf{Z}_2^-)$ induced by w (regarded as a homomorphism $\pi \to \mathbf{Z}_2^-$). Next, if $w_1(M)$ is the mod 2 reduction of an integral class x, evaluation on x defines a map $\pi(M) \to \mathbf{Z}^-$ through which w factors. But we have already seen that $L_{4k}(1) \to L_{4k}(\mathbf{Z}^-)$ is surjective and $L_{4k}(1) \to L_{4k}(\mathbf{Z}_2^-)$ is zero. Thus if M^{4k} has $w_1(M)$ coming from an integral class, c will vanish on $[M, G/PL]$.

The key to Sullivan's proof is the following product formula.

LEMMA 13B.4. *Let L, M be closed even-dimensional PL manifolds, and $g : M \to G/PL$. Then $c(L \times M, gp_2) = \chi(L)c(M, g)$.*

Here, c denotes the Kervaire-Arf invariant and χ the Euler characteristic. A word about the proof is in order, since I have not yet seen a complete proof, though the result has been known for at least three years. Sullivan's original proof used the work of Brown and Peterson [B33] relating c to spinor cobordism, together with a number of ingenious arguments which gradually generalised their [B35, 1.6 and A2] to the result above. A much simpler proof was announced by Rourke and Sullivan, and a sketch is given in [R20]. Finally, Browder claims that the result follows easily from his version [B22] of the Kervaire-Arf invariant.†

It seems that the result can be generalised: take any surgery problem and multiply by the closed manifold L so that the result has even dimension. Its Kervaire-Arf invariant c vanishes if $\dim L$ is odd; otherwise equals $\chi(L)$ times c of the original problem. Another generalisation (same references for proofs) is:

Let L, M be closed PL manifolds; $f : L \to G/PL$, $g : M \to G/PL$.

Then $$c(L \times M, fg) = \chi(L)c(M, g) + \chi(M)c(L, f).$$

We now give the proof of Sullivan's formula and of our generalisation of it: these also appear in [R20].

THEOREM 13B.5. *There is a unique class $\kappa = \sum_1^\infty \kappa_{2i}$, with $\kappa_{2i} \in H^{2i}(G/PL; \mathbf{Z}_2)$, such that for any closed PL manifold M^{2n} and $g : M \to G/PL$, we have*

$$c(M, g) = w(M) g^* \kappa[M].$$

The class κ is of the form $(1 + Sq^2 + Sq^2Sq^2)k$, where

$$k = \sum_{i=0}^{\infty} k_{4i+2},\ k_{4i+2} \in H^{4i+2}(G/PL; \mathbf{Z}_2).$$

Both k and κ are primitive.

†This is now available as [B24, Chapter III, §5].

Proof The bordism invariance of c shows that it defines homomorphisms $\mathcal{N}_{2n}^{PL}(G/PL) \to \mathbf{Z}_2$. Suppose inductively κ_{2i} defined for $i < n$ so that the formula holds in lower dimensions. We claim that

$$c(M, g) - \sum_{i=1}^{n-1} g^* \kappa_{2i} w_{2n-2i}[M]$$

vanishes on decomposable elements and hence (see treatment of $\mathcal{N}_*^{PL}$ in [B28]) factors through $H_{2n}(G/PL; \mathbf{Z}_2)$; thus defining (by duality) a unique $\kappa_{2n} \in H^{2n}(G/PL; \mathbf{Z}_2)$ such that the formula is valid for $2n$-manifolds.

Our claim is now verified by a simple calculation. Given closed manifolds A^{2k}, B^{2n-2k} and $g : A^{2k} \to G/PL$, we have

$$\begin{aligned} c(A \times B, gp_1) &= \chi(B) c(A, g) && \text{by the product formula} \\ &= w_{2n-2k}(B)[B] \sum_{i=1}^{k} g^* \kappa_{2i} \cdot w_{2k-2i}(A)[A] && \text{by inductive hypothesis} \\ &= (g^* \kappa \otimes 1) \cdot w(A \times B)[A \times B] && \text{by the Whitney sum formula,} \end{aligned}$$

so the formula holds in this case. If A and B both have odd dimension, each side of the equation is zero, so it is still true.

Next, primitivity of κ follows from the formula for $c(L \times M, fg)$. Now set $k = \sum_{i \geqslant 0} k_{2i} = (1 + Sq^2 + Sq^2 Sq^2 Sq^2)\kappa$. Then k is primitive, $\kappa = (1 + Sq^2 + Sq^2 Sq^2)k$, and it remains only to check that $k_{4i} = 0$. It suffices to show that each such class can be represented by a map of a manifold with $w_1(M)$ the mod 2 reduction of an integral class $x(M)$: $\mathbf{Z}_2$-manifolds in the terminology of [R20], [S22]. Recall the observation above that if M^{4k} is a $\mathbf{Z}_2$-manifold, then for any $f : M \to G/PL$ we have $c(M, f) = 0$.

LEMMA 13B.6. *Let X be a space, $\xi = \sum_{i \geqslant 0} \xi_i$ with $\xi \in H^{4i+2}(X; \mathbf{Z}_2)$. Then for any $\mathbf{Z}_2$-manifold M^{2n} and $f : M \to X$, we have*

$$\begin{aligned} w(M) f^*(\xi + Sq^2 \xi + Sq^2 Sq^2 \xi)[M] &= 0 && (n \text{ even}) \\ &= w(M) f^*(\xi)[M] && (n \text{ odd}) \end{aligned}$$

COROLLARY. *Assume (inductively) $k_{4i} = 0$ for $2i < n$. Then for any $\mathbf{Z}_2$-manifold M^{2n} and $f : M \to G/PL$, we have*

$$\begin{aligned} c(M, f) &= w(M) f^*(\kappa)[M] = w(M) f^*(k)[M] && (n \text{ odd}) \\ c(M, f) &= 0 = w(M) f^*\Big(\sum_{i<n} k_{2i}\Big)[M] && (n \text{ even}) \end{aligned}$$

Thus if n is even $f^*(k_{2n})[M] = 0$. So $k_{4i} = 0$ inductively.

The first assertion gives a simpler formula for c for $\mathbf{Z}_2$-manifolds: this is Sullivan's original formula. The last assertion follows since, by the lemma,

$$\begin{aligned} 0 = w(M)f^*\Big(\sum_{i<n} k_{2i}\Big)[M] &= w(M)f^*(1+Sq^2+Sq^2Sq^2)\Big(\sum_{i<n} k_{2i}\Big)[M] \\ &= w(M)f^*\Big(\sum_{i\leqslant n} \kappa_{2i}+k_{2n}\Big)[M] \\ &= w(M)f^*(\kappa)[M]+f^*k_{2n}[M] \\ &= c(M,f)+f^*k_{2n}[M] = f^*k_{2n}[M]. \end{aligned}$$

This completes the proof of (13B.5).

Proof of 13B.6. Write

$$w^{(2)} = \sum_{i\geqslant 0} w_{2i}(M), \qquad w^{(4)} = \sum_{i\geqslant 0} w_{4i}(M).$$

The desired result is equivalent to showing that for all $\mathbf{Z}_2$-manifolds M,

$$w^{(2)}(M)f^*(\xi+Sq^2\xi+Sq^2Sq^2\xi)[M] = w^{(4)}(M)(\xi)[M].$$

According to Brown and Peterson [B34], it is sufficient to know that, modulo the ideal generated by ${w_1}^2$ (which vanishes for $\mathbf{Z}_2$-manifolds), the class

$$w^{(2)}\otimes(\xi+Sq^2\xi+Sq^2Sq^2\xi)-w^{(4)}\otimes\xi$$

lies in the kernel of

$$H^*(BO;\mathbf{Z}_2)\otimes_{\mathbf{Z}_2}H^*(X;\mathbf{Z}_2)\to H^*(BO;\mathbf{Z}_2)\otimes_{\mathscr{A}_2}H^*(X;\mathbf{Z}_2),$$

and for this it is clearly sufficient to show that

$$w^{(4)} = w^{(2)}(1+Sq^2+Sq^2Sq^2),$$

or equivalently

$$w^{(2)} = w^{(4)}(1+Sq^2+Sq^2Sq^2Sq^2)$$

modulo the ideal generated by ${w_1}^2$.

Now the right action of Sq^2 on $H^*(BO;\mathbf{Z}_2)$ is given by the following formula, in which Φ, $\overline{U}$ refer to the Thom isomorphism and the Thom class for the "normal bundle" (defined as a limit)

$$\begin{aligned} xSq^2 &= \Phi^{-1}\big(\chi(Sq^2)\Phi(x)\big) \\ &= \Phi^{-1}\big(Sq^2(x\overline{U})\big) \\ &= \Phi^{-1}(Sq^2x\cdot\overline{U}+Sq^1x\cdot\overline{w}_1\overline{U}+x\cdot\overline{w}_2\overline{U}) \\ &= Sq^2x+\overline{w}_1Sq^1x+\overline{w}_2x \\ &\equiv (Sq^2+w_1Sq^1+w_2)x \qquad \bmod {w_1}^2. \end{aligned}$$

Using the Wu relations [W49] for $Sq^i w_j$, we deduce that (modulo $w_1{}^2$)

$$\begin{aligned} w_j Sq^2 &= w_1 w_{j+1} + \binom{j-1}{2} w_{j+2} \\ (w_1 w_j) Sq^2 &= \binom{j-1}{2} w_1 w_{j+2} \end{aligned}$$

Hence

$$\begin{aligned} w_{4i} Sq^2 &= w_1 w_{4i+1} \; + w_{4i+2} \\ w_{4i} Sq^2 Sq^2 &= 0 \qquad\qquad + w_1 w_{4i+3} \\ w_{4i} Sq^2 Sq^2 Sq^2 &= w_1 w_{4i+5} \,. \end{aligned}$$

Collecting these results, we have

$$\begin{aligned} w^{(4)}(1 + Sq^2 + Sq^2 Sq^2 Sq^2) &= \sum_{i \geqslant 0} (w_{4i} + w_{4i+2} + w_1 w_{4i+1} + w_1 w_{4i+5}) \\ &= \sum_{i \geqslant 0} (w_{4i} + w_{4i+2}) \;=\; w^{(2)} \,. \end{aligned}$$

For the other terms cancel in pairs, except for $w_1{}^2 = 0$.

We have now shown how to compute the θ of (13B.3) for each of the $L_m(\pi)$ given in (13A.1) with the exception of $L_{4k+3}(\mathbf{Z}_2^+) \overset{d}{\cong} \mathbf{Z}_2$. Here we find, with $\alpha_1 \in H^1\big(K(\mathbf{Z}_2, 1); \mathbf{Z}_2\big)$ the universal element,

THEOREM 13B.7. *Let $F : L \to M$ of degree 1 represent an element of $L_{4k+3}(\mathbf{Z}_2^+)$. Let $W^{4k+2} \subset M^{4k+3}$ represent the map $M \to K(\mathbf{Z}_2, 1)$; suppose F transversal on W, and $F^{-1}(W) = V$. Then the surgery obstruction (in $\mathbf{Z}_2$) of F equals that of $F\,|\,V : V \to W$.*

Hence $$d\theta : \Omega^{PL}_{4k+3}\big(K(\mathbf{Z}_2, 1) \times G/PL\big) \to L_{4k+3}(\mathbf{Z}_2^+) \to \mathbf{Z}_2$$

is given by

$$\begin{aligned} d\theta(M; f, g) &= c\theta(W, g\,|\,W) \\ &= w(W)(g\,|\,W)^*(k + Sq^2 k + Sq^2 Sq^2 k)[W] \\ &= w(M) f^*\Big(\sum_{i > 0} \alpha_1^i\Big) g^*(k + Sq^2 k + Sq^2 Sq^2 k)[M] \,. \end{aligned}$$

Proof By the second proof of (13A.1), we had isomorphisms

$$\begin{aligned} L_{4k+2}(\mathbf{Z}_2^-) &\xrightarrow{p} L_{4k+3}(1 \to \mathbf{Z}_2^+) \,, \\ L_{4k+3}(\mathbf{Z}_2^+) &\to L_{4k+3}(1 \to \mathbf{Z}_2^+) \,. \end{aligned}$$

The first states that the surgery obstruction for $V \to W$ equals that of the D^1-bundles representing their neighbourhoods in L, M respectively; the second then shows that this equals the surgery obstruction for F.

This gives the first equation for computing $d\theta$; the second follows from (13B.5), and the third since W represents the homology class dual to $f^*(\alpha)$ in M, and

$$i^*w(M) = w(W)\big(1 + (f\,|\,W)^*\alpha_1\big).$$

Next (13B.1) shows that the obstructions in $L_{2k}(\mathbf{Z}_p)$ (computed in (13A.4)) reduce in the context of (13B.3) to the simply connected obstructions, which we have already seen how to compute. Our final list of computations is given in (13A.8). We concentrate on the case where π is free abelian; it will be clear that some of the considerations below extend to the other cases also. Note that we can take $K(\mathbf{Z}^r, 1) = T^r$, the r dimensional torus.

PROPOSITION 13B.8. *The surgery obstruction*

$$\begin{aligned}\theta : \Omega_m^{PL}(T^r \times G/PL) \to L_m(\mathbf{Z}^r) &\cong \textstyle\bigoplus_{0\leqslant i\leqslant r} \binom{r}{i} L_{m-i}(1)\\ &\cong \textstyle\bigoplus_{0\leqslant i\leqslant r} H_i\big(T^r; L_{m-i}(1)\big)\end{aligned}$$

is determined by its adjoint maps

$$\phi : \Omega_m^{PL}(T^r \times G/PL) \otimes H^i(T^r;\mathbf{Z}) \to L_{m-i}(1)$$

which are given by

$$\begin{aligned}\phi(M; f, g; x) = {}& 0 && m - i \text{ odd}\\ & \tfrac{1}{8}\ell(M)f^*(x)g^*\ell(G/PL)[M] && m \equiv i \pmod 4\\ & w(M)f^*(x)g^*k[M] && m \equiv i + 2 \pmod 4\end{aligned}$$

Proof Our determination of $L_m(\mathbf{Z}^r)$ was by induction, based on the formula $L_m(\pi \times \mathbf{Z}) = L_m(\pi) \oplus L_{m-1}(\pi)$ (valid when $Wh(\pi) = 0$). Moreover, the injection of the second summand was defined by multiplying a surgery problem by S^1. Thus the projection on it is obtained by taking a surgery problem $f : L \to M$, making the maps to $K(\mathbf{Z}, 1) = S^1$ transverse to a point, and taking preimages of that point.

But when this holds, any formula for surgery obstructions in the $L_i(\pi)$ leads to one in the $L_i(\pi \times \mathbf{Z})$. For let $f : M \to K(\pi, 1)$, $f' : M \to K(\mathbf{Z}, 1) = S^1$ and $g : (M, \partial M) \to (G/PL, *)$. Obtain $W \subset M$ by making f' transversal at a point. Then the two components of the obstruction are $\theta(M, f, g)$ and $\theta(W, f\,|\,W, g\,|\,W)$ and if this last is expressible as $a(W)(f\,|\,W)^*b(g\,|\,W)^*c[W]$ with a a characteristic class, then (the normal bundle of W in M being trivial) it is also $a(M)f^*b(f')^*xg^*c[M]$, since $(f')^*x$ is dual to W. The proposition now follows by induction on r: our statement synthesises results obtained piecemeal.

Remark. If we ignore potential difficulties with Whitehead groups (or simply work modulo the class of abelian 2-groups of finite exponent), the same induction gives a computation for $L_m(\pi \times \mathbf{Z}^r)$. The above shows that this should be interpreted as

$$L_m(\pi \times \mathbf{Z}^r) \cong \textstyle\bigoplus_{0\leqslant i\leqslant r} H_i\big(T^r; L_{m-i}(\pi)\big).$$

In the case of signatures, it was observed first by Novikov [N7] that exterior algebra appeared in surgery obstructions. See also §17H.

There has been much progress in understanding the relationship between the surgery obstructions of closed manifolds, characteristic classes, the fundamental group and the surgery product formula since the first edition.

Madsen and Milgram [M1] provided a complete account of the homotopy and bordism theoretic properties of the surgery classifying spaces $G/PL, G/TOP$ and characteristic classes for simply connected surgery obstructions.

Sullivan's combination of the surgery product formula and the bordism of manifolds with singularities was extended by Wall [W34] to obtain characteristic classes for the non-simply connected surgery obstruction of normal maps of closed manifolds, in the sense of factorising the homomorphism of Theorem 13B.3

$$\theta : \Omega_m^{PL}(K(\pi,1) \times G/PL, K(\pi,1) \times *) \to L_m(\pi)$$

through the homology of π at the prime 2 and the KO-theory of $K(\pi,1)$ at the odd primes.

The assembly map. The topological surgery obstruction map factors through the assembly map A of Quinn [Q3] and Ranicki [R9]

$$\theta^{TOP} : \Omega_m^{TOP}(K(\pi,1) \times G/TOP, K(\pi,1) \times *) \to H_m(K(\pi,1), \mathbf{L}_\bullet) \xrightarrow{A} L_m(\pi)\,.$$

The first morphism is onto, so that

$$\mathrm{Im}(\theta^{TOP}) \;=\; \mathrm{Im}(A) \subseteq L_m(\pi)$$

is the subgroup consisting of the surgery obstructions $\theta(\phi, F) \in L_m(\pi)$ of the normal maps $(\phi, F) : M \to N$ of closed m-dimensional manifolds with $\pi_1(N) = \pi$; it is necessary to determine these subgroups of the L-groups in order to apply L-theory to the surgery classification of manifolds. Taylor and Williams [T3] used A to obtain the characteristic class formulae in a purely L-theoretic context. See Hambleton, Milgram, Taylor and Williams [H8] for the solution of the 'oozing conjecture' on the image of A for finite π : only the simply connected surgery obstructions along submanifolds of codimension $\leqslant 3$ are needed to detect the surgery obstructions of normal maps of closed manifolds.

14. Applications: Free Actions on Spheres

14A. General Remarks

In making explicit applications of our theory, it is necessary to know the relevant groups $L_m(\pi)$, and also to be able to solve any homotopy-theoretic problems that arise. As examples we choose the problems where π is "the only" nontrivial part of the homotopy theory, and our manifold has universal cover a sphere or euclidean space. There is, indeed, an analogy between this classification problem and the space-form problems of [W47]. In this chapter we study manifolds L^{2n-1} with fundamental group G and universal cover homotopy equivalent, hence homeomorphic to S^{2n-1}. This is equivalent to studying free actions of G on S^{2n-1}. Since we need the groups $L_i(G)$, we will assume G cyclic, though we also study for comparison purposes free actions on S^1. A discussion of known facts for G non-abelian will be found in [W26] and [T4]; see also §17E.

The following problems are of interest. Give constructions for free actions on spheres and invariants to distinguish the actions. Find enough invariants for topological or PL classification of actions, and determine all relations between the invariants. Compute the invariants on all known examples. Which free actions of H extend to free actions of $G \supset H$? Which actions are equivalent to smooth actions? We will study all of these but the last.

In §14A we introduce the concepts which will dominate our work; in §14B we extend the result of Atiyah and Singer, which we will need to justify a key calculation. Then we begin our classifications: in §14C for $G = S^1$, in §14D for G of order 2, and finally in §14E we study the general case, though the main results will only be valid for G of odd order. The classification uses ideas from many sources, which we will try to acknowledge as appropriate.

The Lefschetz fixed point theorem shows that only $\mathbf{Z}_2$ can operate freely on even-dimensional spheres: the case of $\mathbf{Z}_2$ has been, in fact, the case most studied.

A particular feature of these actions is the *join construction*, which we now describe. Free actions of G on spheres S^{m-1} and S^{n-1} induce (via the diagonal) an action of G on the join $S^{m-1} * S^{n-1} = S^{m+n-1}$, which is again free. This construction respects PL (though not differentiable) structures. As an operation on equivalence classes of actions, it is commutative and associative; and natural on passing to subgroups of G. Since $G \subset S^1$, it acts on S^1 by translations: taking the join with this action is called *suspension*. In the case $G = \mathbf{Z}_2$, it also acts freely on S^0 giving a suspension which increases dimensions by one (see Browder and Livesay [B29]). The join of this action with itself is the antipodal (i.e. natural) action on S^1; it follows by associativity of the join that double suspension in this sense is suspension in the original sense.

In the case when G is finite we have for these, as for all odd-dimensional free G-actions, the invariant ρ of (13B.2), which we regard as a complex- valued function on $G - \{1\}$. This is natural in that if we restrict the action to $H \subset G$ we obtain the restriction of ρ. Now this is also definable for free actions of S^1, though here it is necessary to use the Atiyah-Singer form of the definition with the given free action bounding an (in general non-free) action of S^1. This is valid in the PL (and topological) cases, provided we restrict the actions as in §14B. Now a free action of S^1 on M bounds the induced action on $M \times_{S^1} D^2$, so the invariant is always defined, and can be computed by formula (7.9) of [A8]. It follows from 14B that we still have naturality of ρ for subgroups.

THEOREM 14A.1. *Given free actions of G on S^{2m-1}, S^{2n-1}, the value of ρ for the join action is the (pointwise) product of the values of ρ for the given actions.*

Proof First suppose G finite. Then for some r, r times the action on S^{2m-1} bounds a free action on some M^{2m}. By adding cones to the boundary components, we get an action on a closed manifold $\overline{M}$. Then the G-signature of this (clearly the same as that of M) was defined to be $r\rho_1$. Similarly, s times S^{2n-1} bounds a free action on N^{2n}: add cones to get $\overline{N}$. The product action on $\overline{M} \times \overline{N}$ has G-signature the product of those on $\overline{M}$ and $\overline{N}$: $rs\rho_1\rho_2$. Moreover, this action is free except at rs points, and near each of these we have $D^{2m} \times D^{2n}$ with the action of G the product of conewise extensions of the given actions. The induced action on the boundary is thus the join, and on the rest is the cone on this: thus the G-signature of $\overline{M} \times \overline{N}$ is (by definition) rs times ρ for the join action. The result follows.

Now let $G = S^1$. Consider $f : S^1 - \{1\} \to \mathbf{C}$ given by the difference between ρ for the join action and the product of the ρ's. By the above, and naturality, f is zero on each element of finite order in S^1. Since each ρ is a rational, hence continuous function, and points of finite order are dense, it follows that $f = 0$, as stated. This result is the reason for our choice of sign in the definition of ρ.

Note that the proof for S^1 depends on rationality, and hence on [A8, 7.9] which is only justified in 14B below for tame actions (orbit space a manifold). It is interesting to observe that joined actions are always tame. For if their orbit spaces are $Q_1^{2n_1-2}$, $Q_2^{2n_2-2}$ then a neighbourhood in the joined orbit space Q of $x_1 \in Q_1$ is homeomorphic to $U \times \mathbf{R}^{2n_2}$ (U open in Q_1) which is euclidean since $U \times S^1$ is, as open subset of S^{2n_1-1}. It follows easily that Q is locally euclidean everywhere.

The next result shows that a knowledge of actions of S^1 is likely to be very useful in describing other actions. Suppose given a free action of S^1 on S^{2n-1} with orbit space Q; let $G \subset S^1$ be the subgroup of order N, $S^{2n-1}/G = L^{2n-1}$. There is an induced fibration $S^1 \to L \xrightarrow{p} Q$. Let $X \to Q$ be the associated bundle with fibre D^2.

LEMMA 14A.2. *Every map $L^{2n-1} \to G/PL$ (or G/TOP) extends to X (and so factors, up to homotopy, through Q).*

Proof The obstructions to extension lie in

$$H^{i+1}\big(X, L; \pi_i(G/PL)\big).$$

By the Thom isomorphism theorem, this group is isomorphic to $H^{i-1}\big(Q; \pi_i(G/PL)\big)$. But $\pi_i(G/PL)$ vanishes for i odd and $Q \simeq P_{n-1}(\mathbf{C})$, so $H^{i-1}(Q; A)$ vanishes (for any A if i is even. Thus there are no obstructions to the desired extension.

A version of this lemma is due to Lee [L8]; for $N = 2$ it is due to López de Medrano [L20] (with the above proof).

We conclude our general remarks by noting a simple fact about the normal invariant of a suspension, which is important in studying the homotopy projective spaces.

Let G be a finite cyclic group or circle acting freely on S^{2n-1} with orbit space X^{n-1}; the suspended action on S^{2n+1} then has orbit space $X^n \supset X^{n-1}$. Each element of $\mathscr{S}(X^{n-1})$ gives a homotopy equivalence $Y^{n-1} \to X^{n-1}$, and Y corresponds to another free action of G on S^{2n-1} which we can suspend, and take the orbit space to get a homotopy equivalence $Y^n \to X^n$. So suspension induces $\Sigma : \mathscr{S}(X^{n-1}) \to \mathscr{S}(X^n)$.

LEMMA 14A.3. *The following diagram is commutative*

$$\begin{array}{ccc} \mathscr{S}_{PL}(X^{n-1}) & \xrightarrow{\eta} & [X^{n-1}, G/PL] \\ \downarrow{\scriptstyle \Sigma} & & \uparrow{\scriptstyle r} \\ \mathscr{S}_{PL}(X^n) & \xrightarrow{\eta} & [X^n, G/PL] \end{array}$$

where Σ denotes suspension and r restriction.

Proof Write M for a regular neighbourhood of X^{n-1} in X^n. The suspension construction gives a PL manifold homotopy equivalent to M embedded in that for X^n, with codimension zero; clearly the normal invariant for this is the restriction of that for X^n. Thus it suffices to consider the diagram

$$\begin{array}{ccc} \mathscr{S}_{PL}(X^{n-1}) & \xrightarrow{\eta} & [X^{n-1}, G/PL] \\ \downarrow & & \uparrow{\scriptstyle r} \\ \mathscr{S}_{PL}(M) & \xrightarrow{\eta} & [M, G/PL] \end{array}$$

The first vertical map here is induced by taking the total space of a disc bundle. But now the lemma becomes a special case of a general proposition about disc bundles.

Recall the definition of η: replace a homotopy equivalence $Y^{n-1} \to X^{n-1}$ by an embedding $\varepsilon : Y^{n-1} \to X^{n-1} \times \mathbf{R}^k$; take its normal bundle, and the fibre homotopy trivialisation induced by projection on $\mathbf{R}^k$. But ε induces an embedding of the corresponding disc bundles: the normal bundle and fibre homotopy trivialisation of this are effectively the same as those for ε itself. The result follows.

Note. Both result and proof are also valid for simple suspension, when $G = \mathbf{Z}_2$.

See Madsen, Thomas and Wall [M2] and Wall [W35] for the classification of finite groups π with free actions on high-dimensional spheres.

14B. An Extension of the Atiyah-Singer G-signature Theorem

Probably the most useful application to differential topology of the Atiyah-Singer index theorem is their G-signature theorem [A8, (6.12)], which is stated as follows:

THEOREM 14B.1. *"Let X be a compact oriented (smooth) manifold of dimension $2l$, and let the compact Lie group G act (differentiably) on X preserving the orientation. Then G acts on $H^l(X;R)$ preserving the bilinear form. Let* $\operatorname{sign}(G,X)$ *be the character of G defined from this action by* (6.7) *for l even and by* (6.9) *for l odd. Let* $\operatorname{sign}(g,X)$ *be the value of* $\operatorname{sign}(G,X)$ *on an element $g \in G$. Let X^g be the fixed-point set of g, N^g the normal bundle of X^g in X, and*

$$N^g = N^g(-1) \oplus \sum_{0<\theta<\pi} N^g(\theta)$$

the decomposition of N^g determined by the eigenvalues of g. Then $N^g(-1)$ is a real vector bundle of even dimension and $N^g(\theta)$ is a complex vector bundle. Let $2t = \dim X^g$, $2r = \dim N^g(-1)$, $s(\theta) = \dim_{\mathbf{C}} N^g(\theta)$. Finally let $\mathscr{L}$ be the stable characteristic class of the orthogonal group given by (6.5), *and $\mathscr{M}^\theta$ the stable characteristic class of the unitary group given by* (6.11). *Then we have*

$$\operatorname{sign}(g,X) = \Big\{2^{t-r} \prod_{0<\theta<\pi} (i\tan\theta/2)^{-s(\theta)} \mathscr{L}(X^g)\mathscr{L}\big(N^g(-1)\big)^{-1} e\big(N^g(-1)\big) \prod_{0<\theta<\pi} \mathscr{M}^\theta\big(N^g(\theta)\big)\Big\}[X^g]. \qquad \text{(ASGSF)}$$

Here $e\big(N^g(-1)\big)$ denotes the "twisted" Euler class of $N^g(-1)$, and $[X^g]$ is the "twisted" fundamental class of X^g, both twistings being defined by the local coefficient system of orientations of X^g. (A summation over components of X^g is implicit)."

We wish to generalise this to topological actions, using the idea in the proof of (13B.1). For the assertion of the theorem to remain meaningful, it is necessary to suppose that there is a G-vector bundle N^g over the fixed point set X^g of g in X, and an equivariant homeomorphism of its total space onto a neighbourhood of X^g in X. For the proof it is necessary to assume in addition not only that X^g is a manifold but also the corresponding results for any $g' \in G$ and, in addition, that the linear structures of the $N^{g'}$ are compatible in some sense with each other.

To avoid such complicated assertions we confine ourselves to the only case which will be needed for our applications: that of *semi-free* actions – i.e. actions where there is a fixed point set $F = X^g$ for all $g \in G$, so the action on $X - F$ is free. A semi-free action will be called *tame* if F is a manifold, there is a G-vector bundle N over F equivariantly homeomorphic to a neighbourhood of F in X and if, in addition, $(X - F)/G$ is a manifold. Then our main extension of (14B.1) is

THEOREM 14B.2. *The identity* $(ASGSF)$ *holds for tame, semi-free actions on topological manifolds,* X^{2l}.

Proof Denote by $C_{2l}^{\mathrm{Top}}(G)$ the cobordism group of tame semi-free actions of G on oriented topological manifolds (the cobounding actions are of course also assumed tame and semi-free), $C_{2l}^{\mathrm{Diff}}(G)$ the corresponding group formed from differentiable actions. There is a natural forgetful homomorphism

$$\phi : C_{2l}^{\mathrm{Diff}} \to C_{2l}^{\mathrm{Top}}(G)\,.$$

Since a cobordism of (G, X) carries with it ones of F, N and since the Hirzebruch class $\ell(F)$ is well-defined for topological manifolds, each side of the formula (ASGSF) defines a homomorphism $C_{2l}^{\mathrm{Top}}(G) \to \mathbf{C}$. Now (14B.1) states that these agree on the image of ϕ. Our assertion is thus an immediate corollary of

LEMMA 14B.3. *The map* $\phi : C_{2l}^{\mathrm{Diff}}(G) \to C_{2l}^{\mathrm{Top}}(G)$ *has finite kernel and cokernel.*

Proof Let ρ run through equivalence classes of orthogonal representation of G on $\mathbf{R}^d$ $(d = \deg \rho)$ which are free on $\mathbf{R}^d - \{0\}$. For each such ρ, let $C(\rho)$ denote the centraliser of $\rho(G)$ in O_d. Then by a result of Conner [C15] there is an exact sequence

$$\cdots \to \Omega_n^{\mathrm{Diff}}(G) \to C_n^{\mathrm{Diff}}(G) \to \sum_{k+\deg \rho = n} \Omega_k^{\mathrm{Diff}}(C(\rho)) \to \Omega_{n-1}^{\mathrm{Diff}}(G) \to \ldots\,,$$

where $\Omega_n^{\mathrm{Diff}}(G)$ is the corresponding cobordism group of free oriented actions. The proof may be found in [C18] or [W13, VA, Chapter 7]: the idea is that to compute the relative cobordism group of free modulo semi-free actions one can concentrate on a neighbourhood of F: but the class ρ is constant over each component of F, so the group N can be reduced to $C(\rho)$.

This is all just as valid in the topological case, provided the semi-free actions are tame. We now see that ϕ extends to a map of exact sequences

$$\begin{array}{ccccccccc}
\cdots \to & \Omega_n^{\mathrm{Diff}}(G) & \to & C_n^{\mathrm{Diff}}(G) & \to & \sum_{k+\deg \rho = n} \Omega_k^{\mathrm{Diff}}(C(\rho)) & \to & \Omega_{n-1}^{\mathrm{Diff}}(G) & \to \cdots \\
& \downarrow & & \downarrow & & \downarrow & & \downarrow & \\
\cdots \to & \Omega_n^{\mathrm{Top}}(G) & \to & C_n^{\mathrm{Top}}(G) & \to & \sum_{k+\deg \rho = n} \Omega_k^{\mathrm{Top}}(C(\rho)) & \to & \Omega_{n-1}^{\mathrm{Top}}(G) & \to \cdots
\end{array}$$

Since, for any compact Lie G, the bordism $\Omega_*(G)$ of free G-actions can be identified with the bordism groups of $K(G,1)$, we now see that (by the 5-lemma) it will suffice to show that for any CW complex X with finite skeletons, the kernel and cokernel of $\Omega_n^{\text{Diff}}(X) \to \Omega_n^{\text{Top}}(X)$ are finite. But for $n \neq 4$, this is the map of homotopy groups of spectra

$$\pi_n(X \wedge \mathbf{MSO}) \to \pi_n(X \wedge \mathbf{MSTOP})$$

and the result follows by a simple spectral sequence argument, since $BSO \to BSTOP$ is known [K9] [K6] to be a homotopy equivalence modulo the class of finite groups.

If $n = 4$, we cannot use transversality to show that the natural map

$$\Omega_4^{\text{Top}}(X) \to \pi_4(X \wedge \mathbf{MSTOP})$$

is an isomorphism, but transversality does show it injective, and the image is at least as large as that of

$$\Omega_4^{\text{Diff}}(X) \cong \pi_4(X \wedge \mathbf{MSO}),$$

which has finite index (in fact index 2, if X is connected).

It follows, as on [A8, p. 590], that if we have a tame semi-free action of a finite group G on X^{2l} which is free on the boundary ∂X, we can use it to compute $\rho(\partial X/G)$. Note that ρ is minus the invariant σ of [A8]. And here there is no longer any need to restrict G to be finite in order to define ρ: in particular, for free actions on S^1, ρ is always defined and can be computed by [A8, 7.9]. The restriction of ρ to a subgroup then gives the value of ρ for the induced action. For this method of calculating, I am indebted to Ted Petrie [P3].

See Dovermann and Schultz [D4] and Weinberger [W41, §13] for an account of the progress in equivariant surgery theory since the first edition of this book.

14C. Free Actions of S^1

If Q is the orbit space of a free (tame) action of S^1 on S^{2n-1}, and

$$f \; : \; Q \to P_\infty(\mathbf{C}) \; = \; B(S^1)$$

the map inducing the given principal bundle, then since $\dim Q = 2n - 2$ we may assume f cellular, and so $f : Q \to P_{n-1}(\mathbf{C})$. We have an induced map of bundles

$$\begin{array}{ccc} S^{2n-1} & \xrightarrow{\widetilde{f}} & S^{2n-1} \\ \downarrow & & \downarrow \\ Q & \xrightarrow{f} & P_{n-1}(\mathbf{C}) \; : \end{array}$$

$$\begin{aligned} \text{now } H_{2n-1}(\widetilde{f}) &\cong \pi_{2n-1}(\widetilde{f}) && \text{(relative Hurewicz theorem)} \\ &\cong \pi_{2n-1}(f) && \text{(exact sequence of fibration)} \\ &\cong H_{2n-1}(f) && \text{(Hurewicz again)} \\ &= 0 && (\dim Q = 2n - 2) \end{aligned}$$

so $\widetilde{f}$ has degree 1: the homotopy properties of fibrations now show f a homotopy equivalence. Thus $(Q, f) \in \mathscr{S}\big(P_{n-1}(\mathbf{C})\big)$: we refer to Q as a fake complex projective space. Conversely, of course, given a homotopy equivalence $f : Q \to P_{n-1}(\mathbf{C})$, the induced principal bundle has total space S^{2n-1}. Thus we can identify $\mathscr{S}\big(P_{n-1}(\mathbf{C})\big)$ with the set of free (tame) actions of S^1 on S^{2n-1}.

We next compute $[P_{n-1}(\mathbf{C}), G/PL]$. We have the exact sequence

$$\pi_{2k}(G/PL) \to [P_k(\mathbf{C}), G/PL] \to [P_{k-1}(\mathbf{C}), G/PL] \to \pi_{2k-1}(G/PL) = 0\,;$$

moreover, $\theta : [P_k(\mathbf{C}), G/PL] \to L_{2k}(1) \cong \pi_{2k}(G/PL)$ splits the first map in the sequence if $k \neq 2$: for $k = 2$, a special argument shows that the sequence is isomorphic to

$$\mathbf{Z} \xrightarrow{\times 2} \mathbf{Z} \to \mathbf{Z}_2 \to 0\,.$$

We deduce inductively the following, due to Sullivan [S22].

LEMMA 14C.1. *For $f : P_n(\mathbf{C}) \to G/PL$, define $\mathrm{s}_{2r}(f) = \theta\big(f \,|\, P_r(\mathbf{C})\big)$: thus $\mathrm{s}_{4k}(f) \in \mathbf{Z}$ and $\mathrm{s}_{4k+2}(f) \in \mathbf{Z}_2$. Then $\mathrm{s}_2(f)$ is the mod 2 reduction of $\mathrm{s}_4(f)$, and the s_{2i} for $2 \leqslant i \leqslant n$ give a bijection of $[P_n(\mathbf{C}), G/PL]$.*

If G/PL is replaced by G/TOP the result is the same except that s_2 ceases to play an exceptional rôle. The formulae of §13B give the splitting invariants

$\mathbf{s}_{2r}$ in terms of characteristic numbers:

$$\mathbf{s}_{4k+2}(f) = \sum_{0\leqslant i\leqslant k} w_{4(k-i)}\big(P_{2k+1}(\mathbf{C})\big) f^*(k_{4i+2})[P_{2k+1}(\mathbf{C})]$$
$$= \sum_{0\leqslant i\leqslant k} \binom{k+1}{i+1} [k_{4i+2}](f),$$

where

$$[k_{4i+2}](f) = f^* k_{4i+2} \cdot \alpha_2^{n-2i-1}[P_n(\mathbf{C})],$$

with $\alpha_2 \in H^2\big(P_n(\mathbf{C}); \mathbf{Z}\big)$ the standard generator;

$$8\,\mathbf{s}_{4k}(f) = \sum_{0\leqslant i\leqslant k} \ell_{(k-i)}\big(P_{2k}(\mathbf{C})\big) f^* \ell_i [P_{2k}(\mathbf{C})]$$
$$= \sum_{0\leqslant i\leqslant k} c_{k,i}[\ell_i](f), \quad \text{say},$$

where

$$[\ell_i](f) = f^* \ell_i \cdot \alpha_2^{n-2i}[P_n(\mathbf{C})]$$

and the $c_{k,i}$ are certain rational numbers with odd denominators which are easily computed in principle (not in practice), but at least we have $c_{k,k} = 1$.

It is now easy to classify the actions.

THEOREM 14C.2. *The invariants* $\mathbf{s}_{2i} \in \mathbf{Z}_2$ *(n odd),* $\in \mathbf{Z}$ *(i even) for* $2 \leqslant i < n$ *define a bijection of* $\mathscr{S}^{PL}\big(P_n(\mathbf{C})\big)$.

It suffices to refer to the exact sequence

$$0 = L_{2n+1}(1) \to \mathscr{S}^{PL}\big(P_n(\mathbf{C})\big) \to [P_n(\mathbf{C}), G/PL] \xrightarrow{\theta} L_{2n}(1),$$

to the statement of the lemma, and to the fact that (by definition) $\theta(f) = \mathbf{s}_{2n}(f)$.

We obtain an analogous result for tame topological actions. I conjecture that even in the wild case the calculations can be interpreted to give a correct result. I will not discuss the smooth case: apart from [S22] and early work in [H21] the main reference seems to be [B40]

The next result follows easily.

PROPOSITION 14C.3. *The suspension* $\Sigma : \mathscr{S}^{PL}\big(P_{n-1}(\mathbf{C})\big) \to \mathscr{S}^{PL}\big(P_n(\mathbf{C})\big)$ *is injective; its image is the kernel of* $\mathbf{s}_{2(n-1)}$.

The main point is to note that for $i \leqslant n-1$ we have $\mathbf{s}_{2i}(\Sigma Q) = \mathbf{s}_{2i}(Q)$: this follows from (14A.3) or indeed from the definition. The invariant $\mathbf{s}_{2(n-1)}$ is sometimes known as the desuspension obstruction.

The next calculation is crucial for our account of free actions of cyclic groups: using it has enabled me to simplify several of my earlier arguments.

THEOREM 14C.4. *Let $h : Q \to P_{n-1}(\mathbf{C})$ be a homotopy equivalence with splitting invariants $\mathbf{s}_{2i}$. Then the corresponding free (tame) action of S^1 on S^{2n-1} has*

$$\rho(t) = f^n + \sum_{1 \leqslant r \leqslant [n/2]-1} 8\,\mathbf{s}_{4r}(f^{n-2r} - f^{n-2r-2}),$$

where $f(t) = (1+t)/(1-t)$.

Proof By (14B.2), the Atiyah-Singer formula is applicable: now (7.9) on p. 594 on [A8] gives

$$\rho(t) = \delta - 2^{n-1}\left(\frac{te^x+1}{te^x-1}\right)\mathscr{L}(Q)[Q],$$

where $\delta = 0$ (n odd) or 1 (n even) and $x \in H^2(Q;\mathbf{Z})$ is the generator. In particular, ρ is linear in the coefficients of $\mathscr{L}(Q)$, and hence in its splitting invariants, and depends only on the $\mathbf{s}_{4r}$: say

$$\rho = a_n + b_n^1\mathbf{s}_4 + b_n^2\mathbf{s}_8 + \cdots + b_n^r\mathbf{s}_{4r}, \qquad r = [n/2]-1.$$

Now for the action of S^1 on itself (Q = point), we have

$$\rho = -\frac{t+1}{t-1} = f.$$

Since ρ is multiplicative for joins, we multiply by f on suspending. Now the $\mathbf{s}_{4r}$ are unaltered by suspension. Hence

$$\begin{aligned} a_n &= f^n && \text{(all } n)\\ b_n^r &= f^{n-2r-2}b_{2r+2}^r && \text{(all } r,\ n \geqslant 2r+2). \end{aligned}$$

It remains to show that

$$b_{2r+2}^r = 8(f^2-1) = 32t/(1-t)^2.$$

Thus let $n = 2r+2$: we seek the coefficient of $\mathbf{s}_{4r}$ in $\mathscr{L}(Q)$; let it be $\alpha_r x^{2r}$. If $W^{4r} \subset Q$ is dual to x, its signature is

$$\sigma(W) = 2^{2r}\mathscr{L}(W)[W] = 2^{2r}i^*\mathscr{L}(Q)\cdot\left(\mathscr{L}(\nu)\right)^{-1}[W].$$

The constant term in $\mathscr{L}(\nu)$ is 1: thus the coefficient of $\mathbf{s}_{4r}$ in this is $2^{2r}\alpha_r$. By definition, $\sigma(W) = 1 + 8\mathbf{s}_{4r}$, thus $\alpha_r = 2^{3-2r}$. Substituting in the original formula we see that

$$b_{2r+2}^r = 2^{3-2r}\cdot -2^{2r+1}\cdot \text{coefficient of } x \text{ in } \frac{te^x+1}{te^x-1};$$

this reduces to give the formula sought.

We can use this theorem to compute $\mathbf{s}_{4r}$ for joins.

COROLLARY. *Let given free actions of S^1 on spheres have splitting invariants $\mathbf{a}_{2r}$, $\mathbf{b}_{2r}$. Then for the join we have*

$$\mathbf{s}_{4r} = \mathbf{a}_{4r} + \mathbf{b}_{4r} + 8 \sum_{i+j=r} \mathbf{a}_{4i}\mathbf{b}_{4j} - 8 \sum_{i+j=r-1} \mathbf{a}_{4i}\mathbf{b}_{4j} .$$

Proof Multiply out for ρ, and equate coefficients of powers of f.

PROBLEM *How can one compute $\mathbf{s}_{4r+2}$ for the join of the given actions?*

It seems clear that $\mathbf{s}_2 = \mathbf{a}_2 + \mathbf{b}_2$, but whether or not $\mathbf{a}_4\mathbf{b}_2$ appears as a term in $\mathbf{s}_6$, or indeed whether the formula is bilinear at all is uncertain. I *guess* that $\mathbf{s}_{4r+2} = \mathbf{a}_{4r+2} + \mathbf{b}_{4r+2}$ for all $r \geqslant 0$.

See Madsen and Milgram [M1, Chapter 8C] for a more recent account of the construction of exotic complex projective spaces.

14D. Fake Projective Spaces (Real)

The orbit space of any free action of $\mathbf{Z}_2$ on S^n is homotopy equivalent to $P_n(\mathbf{R})$, as we will see in (14E). We will now determine $\mathscr{S}^{PL}(P_n(\mathbf{R}))$ for $n \geqslant 5$, using the methods of §10. The first step is to compute $[P_n(\mathbf{R}), G/PL]$. Now $P_n(\mathbf{R})$ has only 2-torsion, and Sullivan [S22] shows that, if groups of odd (finite) order are neglected, the only nonzero k-invariant of G/PL is the first, which is δSq^2, so that if

$$Y \;=\; K(\mathbf{Z}_2, 2) \times_{\delta Sq^2} K(\mathbf{Z}, 4)\,,$$

we have

$$G/PL \simeq Y \times \prod_{i \geqslant 2} \big(K(\mathbf{Z}_2, 4i-2) \times K(\mathbf{Z}, 4i)\big)$$

modulo groups of odd order.

Now $[P_n(\mathbf{R}), Y] = [P_5(\mathbf{R}, Y]$ for $n \geqslant 5$, and by (14A.2) the composite map

$$\mathbf{s}_4 : \mathbf{Z} \cong [P_2(\mathbf{C}), Y] \to [P_5(\mathbf{R}), Y]$$

is surjective. Thus the latter group is cyclic of order 4: let y denote the isomorphism with $\mathbf{Z}_4$ induced by $\mathbf{s}_4$ reduced mod 4.

Projections on the other factors are induced by the fundamental classes of the Eilenberg-MacLane spaces: Sullivan shows that these can be taken as k_{4i-2} $(i \geqslant 2)$ (this is the same k as in (13B.5)) and $\frac{1}{8}\ell_i$: say λ_i.

LEMMA 14D.1. *Let $i \geqslant 0$. Then we have bijections*

$$[P_{2i+5}(\mathbf{R}), G/PL] \overset{r}{\cong} [P_{2i+4}(\mathbf{R}), G/PL] \overset{X}{\cong} \mathbf{Z}_4 {\textstyle\bigoplus_{j=1}^{i}} \mathbf{Z}_2\,,$$

where the components of X are y, $[k_{2j+4}]$ (j odd) and $[\lambda_{\frac{1}{2}j+1}]$ (j even), with

$$[k_{2j+4}](f) \;=\; f^* k_{2j+4} \cdot \alpha_1^{2i-2j}[P_{2i+4}(\mathbf{R})]$$

(similarly for λ), with $\alpha_1 \in H^1(P_n(\mathbf{R}); \mathbf{Z}_2)$ the generator.

This follows at once from the remarks above. We must now compute the surgery obstruction θ. There are 4 cases to consider, depending on the value of n mod 4. By (13A.1), the relevant obstruction group is

n (mod 4)	0	1	2	3
orientability	$-$	$+$	$-$	$+$
$L_n(\mathbf{Z}_2)$	$\mathbf{Z}_2$	0	$\mathbf{Z}_2$	$\mathbf{Z}_2$
invariant	c	$-$	c	d

Thus for $n = 4r+1$, θ maps to a zero group. For $n = 4r+2$ we have, by (13B.6),

$$c(P_{4r+2}(\mathbf{R}), f) = (1+\alpha_1)^{4r+3}(1+Sq^2+Sq^2Sq^2)f^*(\Sigma k_{4i+2})[P_{4r+2}(\mathbf{R})].$$

Note here first, that we can ignore terms of odd dimension; second, that Sq^2Sq^2 is always zero in $H^*(P_n(\mathbf{R}); \mathbf{Z}_2)$; and third, that on terms of dimension $4i+2$, Sq^2 acts by multiplying by ${\alpha_1}^2$. Thus

$$\begin{aligned} c(P_{4r+2}(\mathbf{R}), f) &= (1+\alpha_1)^{4r+2}(1+{\alpha_1}^2)f^*(\Sigma k_{4i+2})[P_{4r+2}(\mathbf{R})] \\ &= (1+{\alpha_1}^4)^{r+1}f^*(\Sigma k_{4i+2}[P_{4r+2}(\mathbf{R})] \\ &= \sum_{i=0}^{r}\binom{r+1}{i+1}[k_{4i+2}](f). \end{aligned}$$

For $n = 4r+3$ we have, by (13B.7),

$$\begin{aligned} d(P_{4r+3}(\mathbf{R}), f) &= c(P_{4r+2}(\mathbf{R}), f \mid P_{4r+2}(\mathbf{R})) \\ &= \sum_{i=0}^{r}\binom{r+1}{i+1}[k_{4i+2}](f). \end{aligned}$$

Finally for $n = 4r+4$ we use (13B.6) again, and the same remarks to simplify.

$$\begin{aligned} c(P_{4r+4}(\mathbf{R}), f) &= (1+\alpha_1)^{4r+5}(1+Sq^2+Sq^2Sq^2)f^*(\Sigma k_{4i+2})[P_{4r+4}(\mathbf{R})] \\ &= (1+\alpha_1)^{4r+4}(1+{\alpha_1}^2)f^*(\Sigma k_{4i+2})[P_{4r+4}(\mathbf{R})] \\ &= (1+\alpha_1)^{4r+4}{\alpha_1}^2 f^*(\Sigma k_{4i+2})[P_{4r+4}(\mathbf{R})] \end{aligned}$$

(the other terms have the wrong dimensions), and now we see that this coincides (again) with $c(P_{4r+2}(\mathbf{R}), f \mid P_{4r+2}(\mathbf{R}))$.

This computes the map θ in all cases: it is a homomorphism with respect to the obvious group structures, nonzero except when $n \equiv 1 \pmod 4$. We have obtained Lemma 3 of [W23] by direct calculation: this is a new proof of it. Since the leading term of the formula is $[k_{4r+2}(f)]$, a new notation is suggested in terms of splitting invariants: we proceed as follows.

By (14A.2), the Hopf map $\pi : P_{2k+1}(\mathbf{R}) \to P_k(\mathbf{C})$ induces a surjection $\pi^* : [P_k(\mathbf{C}), G/PL] \to [P_{2k+1}(\mathbf{R}), G/PL]$. We express this in the notation of (14D.1). For $f : P_k(\mathbf{C}) \to G/PL$, we have $y(f \circ \pi) = \mathbf{s}_4(f) \pmod 4$ by definition and, for $i \geqslant 0$,

$$\begin{aligned} [k_{4i+2}](f\circ\pi) &= \pi^* f^*(k_{4i+2}){\alpha_1}^{2k-4i-1}[P_{2k+1}(\mathbf{R})] \\ &= \pi^* f^*(k_{4i+2})[P_{4i+2}(\mathbf{R})] \\ &= f^*(k_{4i+2})[\pi_* P_{4i+2}(\mathbf{R})] \\ &= f^*(k_{4i+2})[P_{2i+1}(\mathbf{C})] \qquad \pmod 2 \\ &= f^* k_{4i+2}\cdot {\alpha_2}^{k-2i-1}[P_k(\mathbf{C})] \qquad \pmod 2 \\ &= [k_{4i+2}](f) \qquad \pmod 2, \end{aligned}$$

and a precisely similar calculation shows that $[\lambda_i](f\circ\pi)$ is the mod 2 reduction of $\frac{1}{8}[\ell_i](f)$. Using the formulae for the splitting invariants in the complex case, we now define ones in the real case by setting, for $f : P_n(\mathbf{R}) \to G/PL$,

$$\begin{aligned}
\mathbf{r}_{4k+2}(f) &= \sum_{0\leqslant i\leqslant k} \binom{k+1}{i+1}[k_{4i+2}](f)\\
\mathbf{r}_{4k}(f) &= \sum_{0\leqslant i\leqslant k} c_{k,i}[\lambda_i](f)\,.
\end{aligned}$$

Since $c_{k,k} = 1$, we are transforming by a unitriangular matrix, so the new invariants are as good as the old: (14D.1) leads to a new bijection if X is replaced by R with components y and $\mathbf{r}_{2j+4}$ $(1 \leqslant j \leqslant i)$. But the new invariants have two advantages. The first (from the definition) determines $\pi^*[P_k(\mathbf{C}), G/PL] \to [P_{2k+1}(\mathbf{R}), G/PL]$ by

$$\begin{aligned}
y(\pi^* f) &= \mathbf{s}_4(f) && \pmod 4\\
\mathbf{r}_{4r}(\pi^* f) &= \mathbf{s}_{4r}(f) && \pmod 2 \quad 1\leqslant r\leqslant k/2\\
\mathbf{r}_{4r+2}(\pi^* f) &= \mathbf{s}_{4r+2}(f) && \qquad\qquad 0\leqslant r< k/2\,.
\end{aligned}$$

Secondly, the obstruction map $\theta : [P_n(\mathbf{R}), G/PL] \to \mathbf{Z}_2$ is now given if $n = 4r+2$, $4r+3$ or $4r+4$ by the splitting invariant $\mathbf{r}_{4r+2}$.

We now complete the evaluation of $\mathscr{S}^{PL}(P_n(\mathbf{R}))$.

THEOREM 14D.2. *The map* $\eta : \mathscr{S}^{PL}(P_n(\mathbf{R})) \to [P_n(\mathbf{R}), G/PL]$ $(n \geqslant 5)$ *is surjective for* $n \equiv 1$ *(mod 4); otherwise its image is the kernel of* $\mathbf{r}_{4r+2}$, *where* $r = [(n-2)/4]$. *The map* η *is injective except when* $n \equiv -1$ *(mod 4), when* $\frac{1}{8}\rho$ *maps each fibre bijectively to* $\mathbf{Z}$.

Note that ρ is a function on $G - \{1\}$: when G only has two elements, we can identify ρ with the (unique) value which it takes.

Proof The first assertion follows, since by (10.3) the image of η is the kernel of θ, from the above determination of θ. Next, by (10.5), two elements of $\mathscr{S}^{PL}(P_n)$ have the same image under θ if and only if one can be obtained from the other by the operation of $L_{n+1}(\mathbf{Z}_2)$. Now the image of $L_{n+1}(1)$ always operates trivially; this holds for any manifold X^n, and follows from the commutative diagram

$$\begin{array}{ccc}
[S^{n+1}, G/PL] & \xrightarrow{\ \sim\ } & L_{n+1}(1)\\
\downarrow & & \downarrow\\
[\Sigma X, G/PL] & \longrightarrow & L_{n+1}(\pi X)
\end{array}$$

and the fact that (by exactness, (10.8)) the image of $[\Sigma X, G/PL]$ always operates trivially.

According to (13A.1), the group $L_{n+1}(\mathbf{Z}_2)$ is given by

$n \pmod 4$	0	1	2	3
orientability	$-$	$+$	$-$	$+$
$L_{n+1}(\mathbf{Z}_2)$	0	$\mathbf{Z}_2$	0	$\mathbf{Z} \oplus \mathbf{Z}$
invariant	$-$	c	$-$	$(\sigma/8, \widetilde{\sigma}/8)$

Since $L_{n+1}(\mathbf{Z}_2)$ vanishes in two cases, and $L_{n+1}(1)$ maps onto it in a third, we conclude that η is injective for $n \not\equiv -1 \pmod 4$.

In the case $n \equiv -1 \pmod 4$ we must use our invariant ρ. First observe that for a compact oriented $4k$-manifold W with fundamental group $\mathbf{Z}_2$, the $\mathbf{Z}_2$-signature of $\widetilde{W}$ is $2\sigma(W) - \sigma(\widetilde{W})$, as follows by decomposing the corresponding forms over $\mathbf{R}$. Thus if W is a cobordism of Q_1 to Q_2, we have

$$\rho(Q_2) - \rho(Q_1) \;=\; 2\sigma(W) - \sigma(\widetilde{W})\,.$$

Suppose in particular W a normal cobordism between fake projective spaces. If these have the same value of ρ, then $2\sigma = \widetilde{\sigma}$. But this equation characterises the image of $L_{4k}(1) \to L_{4k}(\mathbf{Z}_2^+)$. Hence Q_1 and Q_2 are PL homeomorphic.

It follows that fibres of η are mapped injectively by ρ, and that if $\eta(Q_1) = \eta(Q_2)$, then $\rho(Q_1) - \rho(Q_2)$ is an integer divisible by 8. To conclude, we must show that for any fake projective space Q, $\rho(Q)$ is divisible by 8; it will suffice to show that each normal cobordism class contains a Q with $\rho(Q) = 0$. The following direct proof is essentially due to López de Medrano [L20], [L21].

By (14D.1), restriction gives a bijection

$$r \;:\; [P_{4r+3}(\mathbf{R}), G/PL] \to [P_{4r+2}(\mathbf{R}), G/PL]\,.$$

By the calculations above, $\theta\big(r(x)\big) = \theta(x)$. Now recall the commutative diagram of (14A.3):

$$\begin{array}{ccc}
\mathscr{S}^{PL}\big(P_{4r+2}(\mathbf{R})\big) & \xrightarrow{\;\eta\;} & [P_{4r+2}(\mathbf{R}), G/PL] \\
\Big\downarrow \Sigma & & \Big\uparrow r \\
\mathscr{S}^{PL}\big(P_{4r+3}(\mathbf{R})\big) & \xrightarrow{\;\eta\;} & [P_{4r+3}(\mathbf{R}), G/PL]
\end{array}$$

and we see that each element of $\mathscr{S}^{PL}\big(P_{4r+3}(\mathbf{R})\big)$ is normally cobordant to a suspension $\big(\Sigma\eta^{-1}r\eta(Q)$: the η^{-1} can be taken since the surgery obstruction vanishes$\big)$. Now given a suspended action, interchanging the two suspension points gives an orientation-reversing homeomorphism of it onto itself. But reversing orientation changes the sign of ρ. Hence $\rho = 0$ for a suspension. This concludes the proof.

It follows that ρ, together with splitting invariants $\mathbf{r}_{2i}$ (and y), gives a complete set of invariants for oriented PL homeomorphism classification; and if we change orientation (e.g. by harmonic inversion) only the sign of ρ is altered. Note that if the same is carried through in the topological case, the only difference is the absence of the low dimensional anomaly. Thus (in dimensions $\geqslant 5$) a topological homotopy $P_n(\mathbf{R})$ admits a PL structure if and only if $\mathbf{r}_2 = \mathbf{r}_4$ (for homotopy $P_n(\mathbf{C})$: if and only if $\mathbf{s}_2$ is the mod 2 reduction of $\mathbf{s}_4$), and the PL structure is determined by choosing $y \in \mathbf{Z}_4$ reducing to $\mathbf{r}_2$ mod 2 (for $P_n(\mathbf{C})$, it is unique). This was first noted by Siebenmann.

We now use our results to study the relation between homotopy real and complex projective spaces.

PROPOSITION 14D.3. *The map* $\pi^\flat : \mathscr{S}^{PL}(P_{n-1}(\mathbf{C})) \to \mathscr{S}^{PL}(P_{2n-1}(\mathbf{R}))$ *(restricting the action on* S^{2n-1} *from* S^1 *to* $\mathbf{Z}_2$*) is given by*

$$\begin{aligned} y(\pi^\flat Q) &= \mathbf{s}_4(Q) \qquad (\mathrm{mod}\ 4) \\ \mathbf{r}_{4r}(\pi^\flat Q) &= \mathbf{s}_{4r}(Q) \qquad (\mathrm{mod}\ 2) \\ \mathbf{r}_{4r+2}(\pi^\flat Q) &= \mathbf{s}_{4r+2}(Q) \end{aligned}$$

and, if $n = 2k$ *is even,*

$$\frac{1}{8}\rho(\pi^\flat Q) = -\mathbf{s}_{4k-4}(Q)\,.$$

Consequently, the image of $\pi^\flat$ *is characterised by*

$$\begin{aligned} &(n=3)\, y=0 && (n=2k+1>3)\, \mathbf{r}_{4k}=0 \\ &(n=4)\, y=-\tfrac{1}{8}\rho \quad (\mathrm{mod}\ 4) && (n=2k+2>4)\, \mathbf{r}_{4k}=\tfrac{1}{8}\rho \quad (\mathrm{mod}\ 2)\,. \end{aligned}$$

Remark. The last relations characterise those homotopy $P_{2n+1}(\mathbf{R})$ which fibre over a homotopy $P_n(\mathbf{C})$, or equivalently, those free actions of $\mathbf{Z}_2$ on S^{2n+1} which extend to free circle actions.

Proof The first three formulae have already been obtained by computing π^*. As to ρ, we know that it is natural when we restrict actions to subgroups. We can thus apply (14C.4), and take $t = -1$, hence $f = 0$. We obtain 0 (n odd), $-8\mathbf{s}_{4(k-1)}$ ($n = 2k$).

The characterisation of the image of $\pi^\flat$ follows by inspection. Note that although π^* is surjective, if $n = 2k+1$ we have the surgery obstruction $\mathbf{s}_{4k}$ leading to a fibration obstruction; if $n = 2k+2$, the surgery obstructions $\mathbf{s}_{4k+2}$ and $\mathbf{r}_{4k+2}$ are essentially the same, but we now have ρ as well as the normal invariant, and for the fibred case they are related by

$$\tfrac{1}{8}\rho = -\mathbf{s}_{4k} = \mathbf{r}_{4k} \quad (\mathrm{mod}\ 2)\,.$$

Our other main explicit construction for fake projective spaces was the join – particularly the suspension. Our main result here is the following.

THEOREM 14D.4. *The commutativity* (14A.3) *determines all the invariants of a suspension except for*

$$\begin{aligned} &n = 4r && \text{nothing} && n = 4r+1 && \mathbf{r}_{4r+2}(\Sigma Q) = 0 \\ &n = 4r+2 && \rho(\Sigma Q) = 0 && n = 4r+3 && \mathbf{r}_{4r+4}(\Sigma Q) = \mathbf{r}_{4r}(Q) + \tfrac{1}{8}\rho(Q)\,. \end{aligned}$$

Hence Σ *is injective except for* $n = 4r+3$, *where two elements have the same suspension iff they have the same normal invariant, and values of* ρ *of the same parity. If* n *is odd,* Σ *is surjective. The image is given for* $n = 4r+2$ *by* $\rho = 0$ *and for* $n = 4r$ *by* $\mathbf{r}_{4r-2} = 0$.

Proof Our computation of $\mathscr{S}^{PL}(P_n(\mathbf{R}))$ shows that the above are indeed the only invariants not determined by $r\eta$; it remains to show that the above are the correct values. The statements about kernel and image of Σ then follow by inspection.

For $n = 4r$, there is nothing to show; for $n = 4r+1$ resp. $4r+2$ we have already seen that $\mathbf{r}_{4r+2}$ resp. ρ vanishes on suspended elements. The main assertion of the theorem is thus the calculation of $\mathbf{r}_{4r+4}(\Sigma Q)$.

First assume that $\mathbf{r}_{4r}(Q) + \frac{1}{8}\rho(Q) = 0 \pmod 2$. Then by (14D.3) Q is in $\pi^\flat(\mathscr{S}^{PL}(P_{2r+1}(\mathbf{C})))$: say $Q = \pi^\flat(X)$.

Now
$$\begin{aligned} \mathbf{r}_{4r+4}(\Sigma Q) &= \mathbf{r}_{4r+4}(\Sigma^2 Q) \\ &= \mathbf{r}_{4r+4}(\Sigma^2 \pi^\flat X) \\ &= \mathbf{r}_{4r+4}(\pi^\flat \Sigma X) \qquad \text{(naturality of joins)} \\ &= \mathbf{s}_{4r+4}(\Sigma X) \qquad \pmod 2 \\ &= 0\,, \end{aligned}$$
so the formula holds in this case.

Now suppose $\mathbf{r}_{4r}(Q) + \frac{1}{8}\rho(Q) = 1 \pmod 2$. Choose a normal cobordism of Q to Q' so that $\rho(Q') - \rho(Q) = 8$. By the above, $\mathbf{r}_{4r+4}(\Sigma Q') = 0$. It is sufficient to show $\Sigma Q \not\cong \Sigma Q'$, since only the invariant $\mathbf{r}_{4r+4}$ can differ, so $\mathbf{r}_{4r+4}(\Sigma Q) = 1$ follows. So the result will follow once we prove that

$$\Sigma Q \cong \Sigma Q' \Rightarrow \frac{1}{8}\big(\rho(Q) - \rho(Q')\big) \text{ is even}\,.$$

Given a PL homeomorphism $h : \Sigma Q \to \Sigma Q'$, we use transversality to construct a cobordism V^{4r+4} of $Q \times 0$ to $h^{-1}(Q') \times 1$ in $\Sigma Q \times I$. Attempting to do surgery to get an s-cobordism embedded in $\Sigma Q \times I$ meets an obstruction θ in

$$LN_{4r+4}(1 \to \mathbf{Z}_2^-) \cong L_{4r+4}(1) \cong \mathbf{Z}$$

whose image (by r_0) in $L_{4r+4}(\mathbf{Z}_2^+)$ is the obstruction to doing surgery, forgetting the embedding. Define $\gamma : L_{4r+4}(\mathbf{Z}_2^+) \to \mathbf{Z}$ as the signature of the quadratic form on the $\mathbf{Z}_2$-invariant part minus that on the $\mathbf{Z}_2$-skew part, i.e. the $\mathbf{Z}_2$-signature. Comparing with the definition of ρ, we have

$$\gamma r_0(\theta) = \rho(Q') - \rho(Q)\,.$$

The result thus follows from the fact that the composite

$$\mathbf{Z} \cong LN_{4r+4}(1 \to \mathbf{Z}_2^-) \xrightarrow{r_0} L_{4r+4}(\mathbf{Z}_2^+) \xrightarrow{\gamma} \mathbf{Z}$$

is multiplication by 16, which follows since by (12.9.2) this is the composite

$$\mathbf{Z} \cong L_{4r+4}(1) \xrightarrow{j} L_{4r+4}(\mathbf{Z}_2^+) \xrightarrow{\tau} L_{4r+4}(\mathbf{Z}_2^+) \xrightarrow{\gamma} \mathbf{Z}$$

and $\gamma\tau$ gives the sum of the signatures on the invariant and skew parts, i.e. $\widetilde{\sigma}$. For this argument cf. [L20, IV, 4.2], also [B11].

It is interesting to observe that it follows that if we keep suspending, all new invariants $\mathbf{s}_{4k+2}$ are 0 and all new ones $\mathbf{r}_{4k}$ are equal to each other: if they are 0, all the (odd dimensional) suspensions fibre over homotopy complex projective spaces; if they are 1, none do. One can ask more generally for invariants of joins: it is not hard to deduce from (14D.3) and (14C.4, Corollary) that $\mathbf{r}_{4k}$ is additive (apart from the effect of ρ), but we cannot yet deal with $\mathbf{r}_{4k+2}$. Of course ρ itself is multiplicative.

To conclude, I must acknowledge indebtedness to many authors for some of the details above. Another approach which has become more traditional started from the paper [B29] by Browder and Livesay in which as well as defining suspension, they investigated obstructions to existence and uniqueness of desuspension: from our point of view, the groups $LN(1 \to \mathbf{Z}_2)$: and obtained the corresponding special case of (12.9). The desuspension obstruction is then an invariant, β say. The identity $\beta(Q^{4r+3}) = \frac{1}{8}\rho(Q^{4r+3})$ (an easy consequence of the above) was obtained in unpublished work by Sullivan; a modified version appears in [L20], and a direct and natural proof in [H20] (with terminology appropriate to the smooth case, but valid in general). For the case when Q is fibred, β equals (14D.3) the desuspension obstruction for the fake complex projective space: this was shown directly by Montgomery and Yang [M19] (again referring unnecessarily to the smooth case). Our treatment (14D.4) is complete and independent of these. It is interesting to note (as we did in [W23], which has however sometimes been misquoted) that the obstruction of [B29] to uniqueness of desuspension of Q^{4r+2} is, in fact, bogus.

Much work has also been done on the smooth case, where matters are much more complicated. The main difference lies, of course, in the homotopy theory rather than the surgery: the application of our methods to this problem is adequately discussed in [L20]. The most interesting examples are those of Brieskorn (see [H19]), whose homotopy theory is clearly expounded in Giffen [G2].

See also López de Medrano [L21, Chaps. IV, V] for the combinatorial and smooth classifications of involutions on homotopy spheres.

14E. Fake Lens Spaces

By 'fake lens space' we mean a manifold with cyclic fundamental group and universal cover a sphere. These give the best examples for application of our techniques, but the problem is of substantial complexity: this section has repeatedly held up completion of the book. At present, results are still only partial, for we sometimes need to assume the fundamental group of odd order, though a substantial number of results can be obtained without this restriction. Compared with §14D we find ρ plays a more important part, and that we also must watch Reidemeister torsion. We begin by investigating this.

Let G be a cyclic group of order N with preferred generator T: let χ be the faithful representation $\chi : G \to S^1$ with $\chi(T) = \exp(2\pi i/N)$. The standard action of $S^1 \subset \mathbf{C}$ on $S^{2n-1} \subset \mathbf{C}^n$ is given by multiplying each coordinate. Restrict the action (via χ) to G: the orbit space is the *standard lens space* $L_0^{2n-1}(N)$, or L_0 when no confusion is to be feared. This has CW structure with one cell in each dimension: the odd skeletons are the images of the spheres $S^{2i-1} \subset \mathbf{C}$ (last $(n-i)$ coordinates zero) and the $2i$-cell the image of the subset of S^{2i+1} with last coordinate real and $\geqslant 0$.

For any (triangulated) fake lens space, Milnor [M14] defines a Reidemeister torsion as follows. Choose orientations for the cells of L, and liftings of them to cells in the universal cover $\widetilde{L}$: the result gives a free $\mathbf{Z}G$-base for $C_*(\widetilde{L})$. Let $Z \in \mathbf{Z}G$ be the sum of all group elements, R_G be the quotient of $\mathbf{Z}G$ by the ideal generated by Z (which is just the set of integer multiples of Z), $\mathbf{Q}R_G = \mathbf{Q} \otimes R_G = \mathbf{Q}G/\langle Z \rangle$. Then

$$C_*(\widetilde{L}) \otimes_{\mathbf{Z}G} \mathbf{Q}R_G$$

is acyclic, with a preferred base. Thus it has Reidemeister torsion $\Delta(L)$, which is a unit of $\mathbf{Q}R_G$, and is determined uniquely up to sign, and up to multiplication by elements of G.

To obtain further information, we utilise our knowledge about the homotopy type of L, and argue following the idea at the beginning of (14C). The isomorphism $\pi_1(L) \to G$ induces a homotopy class of maps $L \to K(G,1)$. We may suppose L mapped into the $(2n-1)$-skeleton of $K(G,1)$: but this is just $L_0^{2n-1}(N)$, so

$$f : L \to L_0 .$$

Since $\widetilde{L}$ is $(2n-2)$-connected, f is $(2n-1)$-connected. It follows by elementary obstruction theory that on the $(2n-2)$-skeleton we can construct

$$g : L_0^{2n-2}(N) \to L ,$$

unique up to homotopy, with fg homotopic to the inclusion. Then g is $(2n-2)$-connected, so all homology and cohomology groups of g, with respect to any coefficient bundle, vanish in dimensions other than $(2n-1)$. By (2.3 (a)), $H_{2n-1}(g;\mathbf{Z}G)$ is a projective $\mathbf{Z}G$-module. Since both complexes are finite, by (2.3 (c)) it is stably free. Now by a result of Swan [S25] it is a free module; it clearly has rank 1. Thus (see [W14]) there exist a map $\phi : S^{2n-2} \to L_0^{2n-2}(N)$ and an extension of g to a homotopy equivalence

$$g' : L_0^{2n-2}(N) \cup_\phi e^{2n-1} \to L\,.$$

Now (see [B7, (7.3), p. 623]) the natural map from the units of $\mathbf{Z}G$ to $K_1(\mathbf{Z}G)$ is an isomorphism. Thus by re-choosing the generator of the above free $\mathbf{Z}G$-module $\pi_{2n-1}(g) = H_{2n-1}(g;\mathbf{Z}G)$ we can arrange that g' is a simple homotopy equivalence. This gives a normal form for L which we can use to calculate torsions.

For $L_0^{2n-1}(N)$, with the cells above, we have the chain complex

$$\partial_{2i+1}e_{2i+1} \;=\; e_{2i}(T-1) \qquad \partial_{2i}e_{2i} \;=\; e_{2i-1}Z\,.$$

In $\mathbf{Q}R_G$, Z becomes zero, so

$$\Delta\bigl(L_0^{2n-1}(N)\bigr) \;=\; (T-1)^n\,.$$

Now for L, normalised as above, only the top cell is attached differently. And even here, both boundary maps ∂_{2n-1} have the same image: viz. Ker ∂_{2n-2}, since $H_{2n-2}(L;\mathbf{Z}G) = H_{2n-2}\bigl(L_0^{2n-1}(N);\mathbf{Z}G\bigr) = 0$. This image is isomorphic to R_G. So one value of $\partial_{2n-1}e_{2n-1}$ differs from the other by multiplying by a unit u of R_G. hence $\Delta(L) = (T-1)^n u$.

Conversely, since
$$\text{Ker }\partial_{2n-2} \cong H_{2n-2}\bigl(\widetilde{L}_0^{2n-2}(N)\bigr) \cong \pi_{2n-2}\bigl(L_0^{2n-2}(N)\bigr)\,,$$
to any unit u of R_G we have the spherical homology class $e_{2n}(T-1)u$, and can attach a $(2n-1)$-cell to obtain a complex L with fundamental group G, universal cover homotopy equivalent to S^{2n-1} and torsion $(T-1)^n u$. Note moreover that if cells of $L_0^{2n-1}(N)$ are oriented, then all cells of L are oriented except the top one; but an orientation of this gives one of $\widetilde{L}$. Thus if we prescribe an orientation of $\widetilde{L}$, we lose the ambiguity in sign of $\Delta(L)$: suppose this done from now on.

Now we have the commutative diagram of ring epimorphisms, which is in fact a pullback diagram:

$$\begin{array}{ccc} \mathbf{Z}G & \xrightarrow{\;\eta\;} & R_G \\ {\scriptstyle\varepsilon}\downarrow & & \downarrow{\scriptstyle\varepsilon'} \\ \mathbf{Z} & \xrightarrow{\;\eta'\;} & \mathbf{Z}_N \end{array}$$

with ε the augmentation map. The kernel of each horizontal map is infinite cyclic (generated by Z and $\varepsilon(Z) = N$ respectively); the kernel I_G of ε is generated by $T-1$ and is mapped isomorphically by η onto Ker ε'. We deduce first, an exact sequence of groups of units

$$0 \to (\mathbf{Z}G)^\times \to R_G^\times \oplus \mathbf{Z}^\times \to \mathbf{Z}_N^\times ,$$

so units in $\mathbf{Z}G$ mapped by ε to 1 are sent isomorphically by η onto $\mathrm{Ker}(R_G^\times \to \mathbf{Z}_N^\times)$. Also $R_G^\times$ maps onto $\mathbf{Z}_N^\times$: for if, say, $dr = 1 + kN$ then

$$(1 + T + T^2 + \cdots + T^{d-1})(1 + T^d + T^{2d} + \cdots + T^{(r-1)d}) \;=\; 1 + kZ$$

in $\mathbf{Z}G$, so $(1 + T + \cdots + T^{d-1})$ is a unit in R_G mapping to the unit d in $\mathbf{Z}_N$.

I now claim that the homotopy type of L above is determined by the unit $\varepsilon'(u)$ of $\mathbf{Z}_N$. For if $\eta(x) = u$, we have a map of L to L_0 which is the identity on the $(2n-2)$-skeleton, and maps the chain e_{2n-1} to xe_{2n-1}. The induced map of universal covers then has degree $\varepsilon(x)$. It is now well known that the homotopy type is determined by (and determines) the reduction

$$\eta'\varepsilon(x) \;=\; \varepsilon'\eta(x) \;=\; \varepsilon'(u) \;:$$

if we have L, L' corresponding to u, u' they have the same homotopy type if and only if $\varepsilon'(u) = \varepsilon'(u')$. There is then a (unique) unit of $\mathbf{Z}G$ mapping by η to u'/u: this clearly represents the Whitehead torsion of a homotopy equivalence $L' \to L$ (compute the chain map as above).

We can make the homotopy statement more explicit by determining the first k-invariant of L (which, in this case, suffices for homotopy classification). This lies in $H^{2n}(G; \mathbf{Z})$. Now there are natural isomorphisms

$$I_G/I_G^2 \cong G \cong H_1(G; \mathbf{Z}) \cong \widehat{H}^{-2}(G; \mathbf{Z}) .$$

Also I_G is an invertible ideal in $\mathbf{Q}R_G$, since $1 - T$ has inverse

$$-N^{-1}(1 + 2T + 3T^2 + \cdots + NT^{N-1}) .$$

We can thus identify

$$I_G^n/I_G^{n+1} \cong \widehat{H}^{-2n}(G; \mathbf{Z}) \qquad \text{for all } n \in \mathbf{Z} .$$

LEMMA 14E.1. $\Delta(L) \in I_G^n$: *its class mod* I_G^{n+1} *corresponds to the inverse of the first* k*-invariant of* L.

Proof First consider L_0^{2n-1}. Here, $\Delta = (T-1)^n$. This comes from raising to the nth power $(T-1) \in I_G$, which corresponds under the above isomorphisms to $T \in G$ and its image in $H_1(G; \mathbf{Z})$. Now the natural multiplicative pairing of $\widehat{H}^{-2}(G; \mathbf{Z})$ and $\widehat{H}^2(G; \mathbf{Z})$ to

$$\widehat{H}^0(G; \mathbf{Z}) \cong \mathbf{Z}_N$$

can be identified via the Bockstein isomorphism

$$\beta \ : \ H^1(G; \mathbf{Z}_N) \to H^2(G; \mathbf{Z})$$

with the Kronecker product

$$\langle \ , \ \rangle \ : \ H_1(G; \mathbf{Z}) \otimes H^1(G; \mathbf{Z}_N) \to \mathbf{Z}_N \, .$$

For the case $n = 1$, $L \simeq S^1$ but is presented as base of a fibration $G \to S^1 \to L$ whose "k-invariant" is $\beta(\iota)$, where $\iota \in H^1(G; \mathbf{Z}_N) \cong \mathrm{Hom}(G, \mathbf{Z}_N)$ satisfies $\iota(T) = 1$. Hence the Kronecker product $\langle T, \iota \rangle = \iota(T) = 1$, so the class T is indeed dual to $\beta(\iota)$, as asserted. For $n > 1$, the k-invariant of ${L_0}^{2n-1}(N)$ is the nth power of the above, which also agrees with our claim.

For general L, we have $\Delta(L) = (T-1)^n u$: the class in $\widehat{H}^{-2n}(G; \mathbf{Z})$ is thus $\varepsilon'(u)$ times that for L_0. We also have a map $L \to L_0$ such that the induced map of universal covers has degree d, with $\eta'(d) = \varepsilon(u)$. It follows that the first k-invariant of L_0 is d times that of L. This completes the proof.

It remains to discuss duality. Each of the complexes L discussed above is a Poincaré complex: this is trivial, since duality certainly holds in $\widetilde{L}$. Indeed, each is homotopy equivalent to a classical lens space: if $\eta'(d) = \varepsilon'(u)$, one can choose

$$L^{2n-1}(N; d, 1, \ldots, 1) \, .$$

I assert moreover that each L is a simple Poincaré complex. Use an asterisk to denote the involutions of $\mathbf{Z}G$, R_G etc. induced by $T \mapsto T^{-1}$. Now in case $\varepsilon'(u) = 1$, so we have a homotopy equivalence $f : L \to L_0$, and $u = \eta(x)$ for a unique unit x which represents the Whitehead torsion of f, consider the diagram

$$\begin{array}{ccc} C^*(\widetilde{L}) & \xleftarrow{\ f^*\ } & C^*(\widetilde{L}_0) \\ {\scriptstyle [L] \cap -}\downarrow & & \downarrow{\scriptstyle [L_0] \cap -} \\ C_*(\widetilde{L}) & \xrightarrow{\ f_*\ } & C_*(\widetilde{L}_0) \, . \end{array}$$

Since L_0 is a closed smooth manifold, and so a simple Poincaré complex, $[L_0] \cap$ is a simple equivalence. But f_* has torsion x^{-1}. By duality (the dimension being odd), the torsion of f^* is x^*. Thus if u is as above, $[L] \cap$ has torsion x/x^*. We obtain a similar result in the other cases on replacing L_0 by a suitable lens space. The assertion now follows from

LEMMA 14E.2. *Any unit of $\mathbf{Z}G$ is of the form $T^i v$, where $v^* = v$.*

Proof Let x be the given unit. For each complex nth root of 1, ζ say, $T \mapsto \zeta$ defines a homomorphism $\phi : \mathbf{Z}G \to \mathbf{Z}[\zeta]$, and $\phi(x)$ is a unit of $\mathbf{Z}[\zeta]$. The involution $T \mapsto \overline{T} = T^{-1}$ induces an involution $\zeta \mapsto \zeta^{-1}$; the field $\mathbf{Q}[\zeta]$ is totally complex, and the fixed field under the involution is totally real. Thus $\phi(x^*)$ and $\phi(x)$ have the same value at all infinite primes; hence $\phi(x^*/x)$ is of finite order (see proof of the Dirichlet unit theorem).

The set of all ϕ as above defines a monomorphism of $\mathbf{Z}G$ (in fact $\mathbf{Q}G$ is isomorphic to a direct sum of $\mathbf{Q}[\zeta]$'s), so x^*/x itself has finite order. By a theorem of Higman [H14], $x^* = \pm x T^j$ for some j.

Write $x = \sum_0^{N-1} a_i T^i$. Then the above gives

$$\sum_0^{N-1} a_i T^{-i} = \pm \sum_0^{N-1} a_i T^{i+j}.$$

If we took the minus sign, it would follow that $\sum a_i = 0$, i.e. $\varepsilon(x) = 0$, contradicting the hypothesis that x is a unit. If we can solve $j = 2i \pmod N$, then writing $v = T^{-i}x$ gives the desired conclusion: this is always possible if N is odd. If, finally, N is even and j odd we find, equating coefficients, that $a_r = a_s$ for $r + s = -j$. Since we cannot have $r = s$ here, all a_i are equal in pairs, so $\varepsilon(x) = \sum a_i$ is even, again contradicting the fact that x is a unit.

As well as proving the desired result, this lemma suggests normalising powers of T by imposing some restriction such as $v^* = v$. For N odd, we normalise Δ by requiring $\Delta(L)$ to satisfy $\delta^* = \delta$ (we call δ real) when n is even, and $\delta^* = -\delta$ (when δ is termed 'imaginary') when n is odd: there is then no indeterminacy left. For N even we can fix $\delta^* = \delta$ or $-T\delta$, but an indeterminacy of $T^{N/2}$ still remains. (Check the possibility of this normalisation for $n = 1$, $\Delta = T^r - 1$, r prime to N: products then give the classical lens spaces, and multiplication by a real unit of $\mathbf{Z}G$ the general case).

We can summarise our results so far as follows.

THEOREM 14E.3. *Let L be a finite CW complex with a generator T (of order N) of $\pi_1(L) = G$ and a homotopy equivalence $e : \widetilde{L} \to S^{2n-1}$: we refer to (T, e) as a polarisation of L. There exist ϕ and a simple homotopy equivalence*

$$f : L_0^{2n-2}(N) \cup_\phi e^{2n-1} \to L$$

preserving the polarisation; these are unique up to homotopy and the action of G. The chain map is $\partial_{2n-1} e_{2n-1} = e_{2n-2}(T-1)U$, where $U \in \mathbf{Z}G$ maps to a unit u of R_G unique up to powers of T; there is a bijection between classes of u and simple homotopy types L. L is a simple Poincaré complex, with torsion $\Delta(L) = (T-1)^n$ which can be normalised as described above. The homotopy type of L is determined by $\varepsilon'(u)$, as made precise in (14E.1).

We now consider our key problem of classifying fake lens spaces. This can be done in the following stages: first, the simple homotopy classification (we have already done this), next, the normal invariant, then the surgery obstruction for existence and finally that for uniqueness. For the normal invariant we have also several techniques: use of Sullivan's determination [S22] of the homotopy type of G/PL, direct application of his 'characteristic variety theorem', comparison (using (14A.2)) with fake $P_{n-1}(\mathbf{C})$, and *ad hoc* arguments involving ρ. We aim to show how these all fit together.

For the homotopy type of G/PL at odd primes, we have the same as BO. The spectral sequence for $KO^0(L^{2n-1})$ gives $[(n-1)/2]$ factors each cyclic of order N (modulo the class of 2-groups), but the extensions are not trivial: if $N = 2r+1$ is prime, for example, there are only r cyclic summands, with orders N^x, $x = [(n+2i-3)/(N-1)]$, $1 \leqslant i \leqslant r$. At the prime 2, we have a product as described in §14D, and thus a string of splitting obstructions $\mathbf{t}_{2i} \in \mathbf{Z}_2$ (i odd, N even) $\in \mathbf{Z}_N$ (i even), $i < n$, except that $\mathbf{t}_2$ and $\mathbf{t}_4$ combine (in the PL case) to give an invariant in $\mathbf{Z}_{2N}$.

The above tells us little more than the order of the group $[L_0, G/PL]$. We next describe more explicit invariants, with which we ought to be able to compute, though our results will not in fact be expressed in these terms. Note that for any lens space $L = L_0^{2n-2}(N) \cup_\phi e^{2n-1}$ as above, and any map $f : L_0^{2n-1}(N) \to G/PL$, then $f \mid L_0^{2n-2}(N)$ extends to a map $L \to G/PL$, unique up to homotopy; thus we have a natural bijection $[L_0, G/PL] \to [L, G/PL]$, and can concentrate on L_0.

Write $N = 2^e M$, with M odd. We define invariants for the 2-primary part of $[L_0, G/PL]$, following §14D, as follows. Let $f : L_0^{2n-1} \to G/PL$ factorise as $g \circ \pi$, $g : P_{n-1}(\mathbf{C}) \to G/PL$.

Define
$$\begin{aligned} \mathbf{t}_{4r}(f) &= \mathbf{s}_{4r}(g) && \pmod{2^e} && 1 \leqslant r \leqslant (n-1)/2 \\ \mathbf{t}_{4r+2}(f) &= \mathbf{s}_{4r+2}(g) \in \mathbf{Z}_2 && (e \geqslant 1) && 0 \leqslant r \leqslant (n/2)-1 \\ \mathbf{T}(f) &= \mathbf{s}_4(g) && \pmod{2^{e+1}} \end{aligned}$$

(the latter is needed in the PL, though not the topological case, and then $\mathbf{t}_2(f) = \mathbf{T}(f) \pmod 2$). As before, these invariants (with the specified relations in low dimensions) give a bijection of the 2-primary part of $[L_0, G/PL]$.

For the odd part, we refer to Sullivan's notion of characteristic variety [S22]. Recall that, regarding $\mathbf{Z}$ as Ω_*-module via the signature, Sullivan constructs a natural isomorphism

$$\Omega_*(X) \otimes_{\Omega_*} \mathbf{Z}[\tfrac{1}{2}] \to KO_*(X) \otimes \mathbf{Z}[\tfrac{1}{2}]$$

of $\mathbf{Z}_4$-graded functors. One then seeks a basis of the odd torsion part of $KO_{-1}(X)$, representative manifolds $V_\alpha^{4k-1} \to X$ giving bordism classes of finite orders r_α, and bordisms $W_\alpha^{4k} \to X$, $\partial W_\alpha = r_\alpha V_\alpha$. Then given $f : X \to G/PL$ one can construct a normal map $e : X' \to X$, make it transversal to W_α to induce $e_\alpha : W'_\alpha \to W_\alpha$, and define the splitting obstruction $\mathbf{s}_\alpha(f)$ as $\frac{1}{8}\big(\sigma(W'_\alpha) - \sigma(W_\alpha)\big) \pmod{r_\alpha}$, or rather (for our purposes) as

$$\big(\sigma(W'_\alpha) - \sigma(W_\alpha)\big)/8r_\alpha \pmod 1.$$

In our case, while there are complications in obtaining an explicit basis with elements of given orders, it is easy to specify generators: since the classes of the sub-lens-spaces L_0^{2k-1} generate the $H_{2k-1}(L_0^{2n-1}; \mathbf{Z})$, they generate $\Omega_*(L_0^{2n-1})$ as Ω_*-module. Thus we can take the V_* to be the L_0^{4k-1}, $1 \leqslant$

$k \leqslant n/2$. The bordism spectral sequence shows that these do have finite order (again, it is somewhat complicated to determine the orders exactly), so appropriate W_α can be constructed.

As all our manifolds are mapped into L_0^{2n-1}, and thence into $K(G,1)$ we can, instead of just taking the signature, lift to the universal cover and take the G-signature: $\sigma(\widetilde{W}^{4k}, G)$. This is a function on G. Since $\partial\widetilde{W}^{4k} = r_k\widetilde{V}^{4k-1}$, the restriction to $G - \{1\}$ of σ is $r_k\rho(V^{4k-1})$. Also, $\sigma(W^{4k})$ is the coefficient of the trivial representation, which is the mean value of $\sigma(\widetilde{W}^{4k}, G)$ on all the elements of G; so the term in the splitting obstruction for f, $\sigma(W^{4k})/8r_k$, is (one eighth of) the mean value of $\rho(V^{4k-1})$ plus $\sigma(\widetilde{W}^{4k})/Nr_k$. As we are studying odd torsion, the factor 8 is innocuous. If we can ignore this last term, we then get the expression $\{(8N)^{-1}\times$ sum of values of $\rho({V'}^{4k-1}) - \rho(V^{4k-1})\}$ (mod 1) for the splitting invariant. In fact, though we can show that $\sigma(\widetilde{W}^{4k})/r_k$ is an integer, it need not be divisible by N: nor is $\rho({V'}^{4k-1})$ well-determined by transversality, though modulo representations it appears to be obtained from $\rho(L)$ by multiplying by the appropriate power of f^{-1}.

Hence although this strongly suggests a connection between ρ and the normal invariant, we will use a different method to obtain it.

We next study the surgery obstructions. Our discussion of the obstruction to existence is based on an inductive argument due to Browder [B23] (see also [B30]): for the purpose of starting the induction we have to go back to homotopy type rather than simple homotopy type. Some discussion of this modified surgery problem will be found in §17D; we also show that for an odd dimensional surgery problem with fundamental group cyclic of odd order, if surgery is possible to get a homotopy equivalence, then one can get a simple homotopy equivalence. I conjecture that this is also true for N even.

If we are only studying homotopy types, we do know that in each one there is a classical lens space L^{2n-1}. Thus it suffices to compute $\theta(f)$ for $f : L^{2n-1} \to G/PL$ (or G/TOP).

THEOREM 14E.4. *Let L^{2n-1} be a lens space with fundamental group of order N, $f : L^{2n-1} \to G/PL$ a map. Then the obstruction to surgery on a corresponding normal map to get a homotopy equivalence vanishes unless n and N are even, and in this case it equals $\mathbf{t}_{2n-2}(f) \in \mathbf{Z}_2$.*

Proof First suppose n, N even. Then the image of the surgery obstruction under

$$L_{2n-1}(\mathbf{Z}_N) \to L_{2n-1}(\mathbf{Z}_2) \xrightarrow{d} \mathbf{Z}_2$$

can be calculated by (13B.7): we find

$$d\theta(f) = w(L^{2n-1}) \sum_{i>0} {\alpha_1}^i f^*(k + Sq^2k + Sq^2Sq^2k)[L^{2n-1}],$$

where $\alpha_i \in H^i(L^{2n-1}; \mathbf{Z}_2)$ is the non-zero class: we have $\alpha_2\alpha_i = \alpha_{i+2}$ if $i+2 \leqslant 2n-1$, and ${\alpha_1}^2 = \alpha_2$ $\big(N \equiv 2 \pmod 4\big)$ or ${\alpha_1}^2 = 0$ $\big(N \equiv 0 \pmod 4\big)$. Thus

$Sq^2\alpha_{4k-2} = \alpha_{4k}$, $Sq^2\alpha_{4k} = 0$. Also, $w(L^{2n-1}) = (1+\alpha_2)^n$. Thus

$$d\theta(f) = \sum_{i>0} \alpha_1{}^i (1+\alpha_2)^{n+1} f^*(k)[L^{2n-1}]$$

and the first term is α_1 if $N \equiv 0 \pmod 4$; otherwise, its odd dimensional component is $\alpha_1(1+\alpha_2)^{-1}$. Since n is even, say $n = 2r$, we can ignore terms of the wrong dimension, and find

$$\begin{aligned} d\theta(f) &= \alpha_1(1+\alpha_4)^r f^*(k)[L^{2n-1}] \\ &= \sum \binom{r}{i+1}[k_{4i+2}](f) = \mathbf{t}_{4r-2}(f) = \mathbf{t}_{2n-2}(f). \end{aligned}$$

Thus for n, N even, $\mathbf{t}_{2n-2}(f)$ certainly is an obstruction to surgery. Now let $n = 2$. Then $[L^3, G/PL]$ is trivial for N odd: for N even, $\mathbf{t}_2$ defines an isomorphism of it with $\mathbf{Z}_2$. Thus the theorem is true in this case.

Now suppose the theorem proved for n: consider a sub-lens-space $L^{2n-1} \subset L^{2n+1} \xrightarrow{f} G/PL$. Let M be a tubular neighbourhood of L^{2n-1} in L^{2n+1}. We have the surgery obstruction $\theta(f \mid L^{2n-1})$ for the sub-lens-space; that for M is $p\big(\theta(f \mid L^{2n-1})\big)$, where

$$p : L_{2n-1}(\mathbf{Z}_N) \to L_{2n+1}(\mathbf{Z} \to \mathbf{Z}_N)$$

is the map defined in (11.6). We will see below that $p\big(\theta(f \mid L^{2n-1})\big) = 0$. Thus one can do surgery to get a homotopy equivalence on M. Now by exactness of

$$L_{2n+1}(\mathbf{Z}) \xrightarrow{\alpha} L_{2n+1}(\mathbf{Z}_N) \to L_{2n+1}(\mathbf{Z} \to \mathbf{Z}_N)$$

it follows that $\theta(f) \in \operatorname{Im} \alpha$. Since this holds inductively, to justify the assertion above, it will suffice to prove

(14E.5 (a)) $L_{2n-1}(\mathbf{Z}) \xrightarrow{\alpha} L_{2n-1}(\mathbf{Z}_N) \xrightarrow{p} L_{2n+1}(\mathbf{Z} \to \mathbf{Z}_N)$ is zero.

It then remains to investigate α. We will show

(14E.5 (b)) The map $\alpha : L_{2n-1}(\mathbf{Z}) \to L_{2n-1}(\mathbf{Z}_N)$ is zero unless n, N are even.

Since, in this case, the composite

$$L_{2n-1}(\mathbf{Z}) \to L_{2n-1}(\mathbf{Z}_N) \to L_{2n-1}(\mathbf{Z}_2) \cong \mathbf{Z}_2$$

is an isomorphism (by (13A.9)), it follows that the only obstruction to surgery is as described in the theorem. It thus remains only to prove (14E.5).

LEMMA 14E.5.

(a) $L_{2n-1}(\mathbf{Z}) \xrightarrow{\alpha} L_{2n-1}(\mathbf{Z}_N) \xrightarrow{p} L_{2n+1}(\mathbf{Z} \to \mathbf{Z}_N)$ *is zero*.

(b) $L_{2n-1}(\mathbf{Z}) \xrightarrow{\alpha} L_{2n-1}(\mathbf{Z}_N)$ *is* 0 *except when* n, N *are even*.

Proof of (b). This is equivalent to showing

$$L_{2n}(\mathbf{Z} \to \mathbf{Z}_N) \xrightarrow{\partial} L_{2n-1}(\mathbf{Z})$$

surjective; it suffices to prove the composite

$$L_{2n-2}(\mathbf{Z}_N) \xrightarrow{p} L_{2n}(\mathbf{Z} \to \mathbf{Z}_N) \xrightarrow{\partial} L_{2n-1}(\mathbf{Z}) \cong L_{2n-2}(1)$$

surjective. Now p is defined by taking the universal D^2-bundle: for ∂ we take its boundary S^1-bundle, and then split along a submanifold generating H^1, which can be chosen to be the N-fold cover of what we started with. Thus the composite is the transfer. Under this, Arf invariants are multiplied by N: this proves the result when n is even. For n odd, it is stated in (13A.4) that $\tau : L_0(\mathbf{Z}_N) \to L_0(1)$ is surjective.

Proof of (a). Because of (b), we can suppose n even so that $L_{2n-1}(\mathbf{Z}) \cong \mathbf{Z}_2$. In view of the above geometrical interpretation of (a), it will suffice to give one example of lens spaces and maps $L^{2n+1} \xrightarrow{f} G/PL$ for which $\theta(f \,|\, L^{2n-1}) \neq 0$, but $\theta(f)$ – and thus *a fortiori* $p\theta(f \,|\, L^{2n-1})$ – vanishes. We choose the standard lens space fibred over $P_{n-1}(\mathbf{C}) \subset P_n(\mathbf{C})$, and choose a map $g : P_n(\mathbf{C}) \to G/PL$ with $\mathbf{s}_{2n}(g) = 0$, $\mathbf{s}_{2n-2}(g) \neq 0$. Then one can do surgery on g, hence on $f = g \circ \pi$, but

$$d\theta(f \,|\, L^{2n-1}) \;=\; \mathbf{s}_{2n-2}(f) \;=\; \mathbf{s}_{2n-2}(g) \neq 0\,.$$

The result follows.

Note that the one reason we cannot work with simple homotopy type throughout is the difficulty of starting the induction.

We will now assume N odd, and outline another proof of (14E.4): the new proof is somewhat more constructive, and also shows that the normal invariant of a fake lens space with N odd is determined by the invariant ρ, but it does not apply when N is even.

We have seen that there are $\phi(N)$ (the order of the multiplicative group of $\mathbf{Z}_N$) different homotopy types of lens spaces L^{2n-1} with fundamental group of order N, and for each of these, $[L^{2n-1}, G/PL]$ has order N^a, with $a = [(n-1)/2]$. Now given two lens spaces L_1, L_2 with the same homotopy type and the same normal invariant, there is a normal cobordism W^{2n} between them. Then (cf. (13B.2)), $\rho(L_1) - \rho(L_2)$ equals the G-signature of $\widetilde{W}$; in particular, it is the restriction of an actual representation of G. We will prove

PROPOSITION 14E.6. *There exist (at least) $\phi(N)N^a$ ($a = [(n-1)/2]$) fake lens spaces L_i^{2n-1} such that if $i \neq j$, $\rho(L_i) - \rho(L_j)$ is not the restriction representation of G.*

It follows that no two L_i are in the same normal cobordism class; since there are so many L_i, every normal cobordism class contains one of them. Thus surgery is possible in each normal cobordism class, which re-proves (14E.4) for this case. It also follows that for any two fake lens spaces M, M' in different normal cobordism classes we can find i, j $(i \neq j)$ so that L_i and M are in the same class, as are L_j and M'. Then

$$\rho M - \rho M' = (\rho M - \rho L_i) + (\rho L_i - \rho L_j) + (\rho L_j - \rho M') :$$

the first and third terms here are restrictions of representations; the middle is not. Hence nor is $\rho M - \rho M'$. Thus $\rho(M)$ determines the normal cobordism class of M.

COROLLARY. *For any fake lens space M, $\rho(M)$ determines the odd torsion part of the normal cobordism class of M.*

For if H is the 2-complement in the fundamental group G of M, and L the corresponding covering space, $\rho(M)$ determines $\rho(L)$ by restriction; this determines the normal cobordism class of L as above, and $[M, G/PL] \to [L, G/PL]$ is an isomorphism modulo the class of finite 2-groups.

Proof of Proposition 14E.6. Recalling the definition of ρ, we can rephrase it as follows. Let $\widehat{G} = \mathrm{Hom}_{\mathbf{Z}}(G, \mathbf{Q}/\mathbf{Z})$ be the Pontrjagin dual of G: it is cyclic, generated by χ. The representation ring of G is $\mathbf{Z}\widehat{G}$. Restricting to $G - \{1\}$ means factoring out the regular representation – i.e. defining $R_{\widehat{G}}$ by $1 + \chi + \cdots + \chi^{N-1} = 0$. Finally, ρ lies in $\mathbf{Q}R_{\widehat{G}}$. For the natural action of G on S^1, by (14C.4), $\rho = (1+\chi)/(1-\chi)$. By (14A.1), ρ is multiplied by this factor on taking suspensions. Since N is odd, $1+\chi$ is invertible in $R_{\widehat{G}}$ (its inverse is $1 + \chi^2 + \chi^4 + \cdots + \chi^{N-1}$). Hence if L_i, L_j are fake lens spaces with suspensions ΣL_i, ΣL_j such that $\rho(\Sigma L_i) - \rho(\Sigma L_j) = \lambda \in R_{\widehat{G}}$, then $\rho(L_i) - \rho(L_j) = \lambda(1-\chi)/(1+\chi)$ belongs to $R_{\widehat{G}}$ too.

We will prove the proposition inductively. If it is true in dimension $4k+1$ $(n = 2k+1)$, then the ΣL_i provide the right number of examples in dimension $4k+3$ and, by the remark above, have the desired property. In dimension 3, we can take the standard lens spaces: there are just enough of them. Finally we must show that from each L_i in dimension $4k-1$ we can construct N examples in dimension $4k+1$. Subject L_i to normal cobordisms given by the $x_j \in L_0(G)$ with signatures

$$\sigma(x_j) = 4j(\chi + \chi^{-1}) \qquad (0 \leqslant j \leqslant N) :$$

(this is possible by (13A.4)).

We get fake lens spaces L_{ij} with $\rho(L_{ij}) = \rho(L_i) + 4j(\chi + \chi^{-1})$, and thus

$$\rho(\Sigma L_{ij}) = \rho(\Sigma L_i) + \frac{4j(\chi + \chi^{-1})(1+\chi)}{(1-\chi)} .$$

Since

$$\frac{4(\chi + \chi^{-1})(1+\chi)}{(1-\chi)} = \left\{ 8 \sum_{1}^{N-1} \left(1 - \frac{2t}{N}\right) \chi^t \right\} - 4(\chi + \chi^{-1}),$$

no two of these are congruent mod $R_{\widehat{G}}$, as claimed.

This fails in the lowest dimension $k = 1$, since we are unable to construct the normal cobordism. There are several ways to get round this technical difficulty; none are very neat. Here is one. By (14E.4), we can find fake lens spaces in the normal cobordism classes, so all we need do is compute ρ for them. (One can also argue avoiding (14E.4), but this is not necessary here). For classes homotopy equivalent to $L_0{}^5(N)$, we have the examples of manifolds fibred over fake complex projective spaces: for these, ρ is computed by (14C.4). For the rest (as in (14E.9) below) take the join of L^5 with the free action of G on S^1 given by χ^4, where d is chosen that the result is homotopy equivalent to $L_0{}^7(N)$, and hence normally cobordant to a fake lens space which fibres over $P_3(\mathbf{C})$. Since $[L_0{}^7(N), G/PL] \to [L^5, G/PL]$ is bijective, we can represent all normal cobordism classes this way.
Thus

$$\frac{\chi^d+1}{\chi^d-1}\rho(L) \equiv \left(\frac{\chi+1}{\chi-1}\right)^4 + \frac{32\chi}{(\chi-1)^2}\mathbf{s}_4 \qquad (\text{mod } R_{\widehat{G}})\,.$$

where $\mathbf{s}_4$ can take any integral value, as L varies. The desired result follows, by an easy algebraic exercise.

We have now a complete set of invariants which determine the normal invariant of fake lens spaces: in fact, combining the results above we find that for $N = 2^e M$ and L^{2n-1} with fundamental group G of order N, invariants are:

$$\mathbf{t}_{4r} \in \mathbf{Z}_{2^e} \quad (1 \leqslant r \leqslant (n-1)/2)\,, \quad \mathbf{t}_{4r+2} \in \mathbf{Z}_2 \quad (e \geqslant 1,\, 0 \leqslant r \leqslant (n/2)-1)\,;$$

in the PL case $\mathbf{T} \in \mathbf{Z}_{2^{e+1}}$ with reductions $\mathbf{t}_4$ mod 2^e and $\mathbf{t}_2$ mod 2; and $\rho \in \mathbf{Q}[\chi \mid 1+\chi+\chi^2+\cdots+\chi^{N-1} = 0] = \mathbf{Q}R_{\widehat{G}}$: ρ mod $\mathbf{Z}[\chi]$ is part of the normal invariant. We also have the torsion

$$\Delta \in \mathbf{Q}[T \mid 1 + T + \cdots + T^{N-1} = 0] = \mathbf{Q}R_G$$

which determines the simple homotopy type of L. Given two fake lens spaces L, L' for which all these invariants agree, there is a normal cobordism W^{2n} between them. If we wish to do surgery on W so as to obtain an s-cobordism between L and L', we meet a surgery obstruction in $L_{2n}(G) \cong L_{2n}(\mathbf{Z}_N)$. We need to know this group before we can proceed further. At the time of writing, the answer is known for N odd but not for N even; although I conjecture that the answer takes the same form for N even, all the other details are more complicated then, so I now restrict to the case when N is odd.

By (13A.4), if we write

$$L_{2n}(G) \;=\; L_{2n}(1) \oplus \widetilde{L}_{2n}(G)\,,$$

the multisignature maps the second summand isomorphically to the subgroup of $R_{\widehat{G}}$ consisting of real elements (n even) or imaginary elements (n odd). Now as observed in the proof of (14D.2), the image of $L_{2n}(1)$ acts trivially on the set of fake lens spaces. Also, if W^{2n} is, as above, a normal cobordism from L to

L', then $\rho(L') - \rho(L) = \sigma(\widetilde{W})$, the G-signature of $\widetilde{W}$. Thus if $\rho(L) = \rho(L')$, the class of W in $L_{2n}(G)$ lies in the summand $L_{2n}(1)$, hence L and L' are (PL)-homeomorphic. This proves the first part (for which see also [B30]) of

THEOREM 14E.7. *Let L^{2n-1} and L'^{2n-1} be oriented fake lens spaces, with fundamental group G cyclic of odd order N. Then there is an orientation preserving homeomorphism $L \to L'$ inducing the identity on G if and only if $\Delta(L) = \Delta(L')$ and $\rho(L) = \rho(L')$.*

Given $\Delta \in R_G$, $\rho \in \mathbf{Q}R_{\widehat{G}}$, there exists a corresponding fake lens L^{2n-1} if and only if

(i) *Δ and ρ are both real (n even) or imaginary (n odd).*

(ii) *Δ generates I_G^n, $\rho \in I_{\widehat{G}}^{-n}$.*

(iii) *The classes of ρ mod $I_{\widehat{G}}^{-n+1}$, $(-2)^n\Delta$ mod I_G^{n+1} correspond under*

$$I_{\widehat{G}}^{-n}/I_{\widehat{G}}^{-n+1} \cong \widehat{H}^{2n}(\widehat{G};\mathbf{Z}) \cong \widehat{H}^{-2n}(G;\mathbf{Z}) \cong I_G^n/I_G^{n+1}.$$

(iv) $\rho \equiv - \sum_{\phi \in \widehat{G},\, \phi \neq 1} \operatorname{sign}\left(i^n \phi(\Delta)\right)\phi \pmod 4$.

Note. The isomorphism in (iii) comes since $\widehat{H}^{2n}(G;\mathbf{Z})$ is dually paired with both $\widehat{H}^{-2n}(G;\mathbf{Z})$ and $\widehat{H}^{2n}(\widehat{G};\mathbf{Z})$: under it, $(\chi-1)^{-n}$ corresponds to $(T-1)^n$. The congruence in (iv) can be interpreted in $R_{\widehat{G}}$ localised at $\langle \chi - 1\rangle$, or equivalently at $\langle N\rangle$: it can also be written

$$\rho \equiv - \sum_{r=1}^{N-1} \operatorname{sign}\left(i^n \chi^r(\Delta)\right)\chi^r \pmod 4.$$

Proof By (14E.3), simple homotopy types of polarised CW complexes L correspond bijectively with generators Δ of I_G^n which are real (n even) or imaginary (n odd). Each such L is a simple Poincaré complex, with homotopy type as described in (14E.1).

For each Δ there exist normal invariants and (since by (14E.4) the surgery obstructions vanish) also fake lens spaces which, by the above, are classified by ρ. From the definition of ρ, it is real for n even and imaginary for n odd. By (13A.4), with normal cobordisms from $L_{2n}(G)$ we can add to ρ an arbitrary real resp. imaginary element of $4R_{\widehat{G}}$. It remains to determine, in each normal cobordism class (with Δ given) the class of ρ mod $4R_{\widehat{G}}$.

To determine ρ we need a construction (direct or indirect) for all the fake lens spaces, which must be fairly explicit, and a calculation of the effect on ρ of our constructions. We will give these in reverse order, and then return to the proof of the theorem.

PROPOSITION 14E.8.

(a) *The invariants* $\Delta(L) \in R_G$, $\rho(L) \in \mathbf{Q}R_{\widehat{G}}$ *are invariant under change of generator of* G. *Both change sign under a change of orientation.*

(b) *Let* W^{2n} *be a normal cobordism from* L *to* L' *whose class in* $L^h_{2n}(G)$ *has invariants* D, σ. *Then*

$$\Delta(L') = D\Delta(L), \qquad \rho(L') = \sigma + \rho(L).$$

(c) *If* L^{2n-1} *is fibred over* $Q \simeq P_{n-1}(\mathbf{C})$, *with invariants* $\mathbf{s}_{2r}$, *then*

$$\Delta(L) = T^{n(N-1)/2}(T-1)^n,$$

$$\rho(L) = f^n + \sum_{1 \leqslant r \leqslant [n/2]-1} 8\,\mathbf{s}_{4r}(f^{n-2r} - f^{n-2r-2}), \qquad f = \frac{1+\chi}{1-\chi}.$$

(d) *If* L *is constructed by the join of actions defining* L_1, L_2 *then*

$$\Delta(L) = \Delta(L_1)\Delta(L_2) \qquad \rho(L) = \rho(L_1)\rho(L_2).$$

Proof The statements (a) are immediate from the definitions. In (b), the formula for ρ was obtained in the proof of (13B.2). The formula for Δ amounts to saying that D is the (Whitehead) torsion of the homotopy equivalence $L \to L'$. Now W is obtained from L by attaching n-cells and from L' by attaching the dual cells. But D is (up to sign) the determinant of the change of basis from the cells to their duals (this is the algebraic interpretation of the intersection numbers, cf. [W13, IV]); the result follows, for the sign is + since $\varepsilon(D) = 1$ and we have the same homotopy type. As to (c), the second assertion is (14C.4). For the first, we note that $Q \to P_{n-1}(\mathbf{C})$ is a simple homotopy equivalence, hence so is $L \to L_0^{2n-1}(N)$: the power of T is inserted to make $\Delta(L)$ real. The value of ρ in (d) is given by (14A.1). It remains to determine Δ for (d).

Let L_1, L_2 come from (free) actions of G_1, G_2 on S_1, S_2 respectively: consider G-triangulations. The corresponding chain complex for $S_1 * S_2$ has a subcomplex coming from $S_1 \cup S_2$, and the quotient can be identified with the tensor product of chain complexes for S_1, S_2. But each of S_1, S_2 has zero Euler characteristic. By the standard multiplicative property of torsion [K17], this quotient has zero torsion as based complex of $(G_1 \times G_2)$-modules, hence also as G-modules ($G \cong G_1 \cong G_2$ diagonally embedded). Hence the torsion of $S_1 * S_2$ equals that of $S_1 \cup S_2$, the product of the torsions of S_1 and S_2.

Note that the proof above makes no reference to the parity of N. We now need a construction to make (14E.4) more explicit: as is shown even more clearly in (14E.6), the idea there is that L is obtained from S^1 by repeatedly suspending and taking normal cobordisms. Here, partly for variety and partly to avoid the messy induction with special arguments for $n = 5$, we proceed differently.

LEMMA 14E.9. (i) *Let L^{2n-1} be a fake lens space. We can choose d uniquely such that if L^* is the join of L with the free action on S^1 given by χ^d, $L^* \simeq L_0^{2n+1}(N)$.*

(ii) *Let $L^{2n-1} \simeq L_0$. If n is even, L is normally cobordant to a fake lens space which fibres over a fake complex projective space. Thus if n is odd, the same holds for ΣL.*

Proof (i) $\Delta(L^*) = T^a(T^{d'} - 1)\Delta(L)$ by (14E.8), where $dd' \equiv 1 \pmod N$. Now $T^{d'} - 1 \equiv d'(T-1) \pmod{I_G^2}$. Thus if $\Delta(L) \equiv d(T-1)^n \pmod{I_G^{n+1}}$, $\Delta(L^*) \equiv (T-1)^{n+1} \pmod{I_G^{n+2}}$, so by (14E.1) $L^* \simeq L_0^{2n+1}(N)$. Since both $\Delta(L)$ and $(T-1)^n$ generate I_G^n, and I_G^n/I_G^{n+1} is cyclic, this must hold for a unique (mod N) d prime to N.

(ii) We know that the map $L_0 \to G/PL$ corresponding to the homotopy equivalence factors through $P_{n-1}(\mathbf{C})$. The obstruction to surgery to produce a corresponding homotopy equivalence $Q^{2n-2} \to P_{n-1}(\mathbf{C})$ is $\mathbf{s}_{2n-2}$. Now suppose n is even. If N is even, $\mathbf{s}_{2n-2} = 0$ by (14E.4), since there is no surgery obstruction to the existence of L^{2n-1}. If N is odd, alter the map $P_{n-1}(\mathbf{C}) \to G/PL$ on the top cell to change $\mathbf{s}_{2n-2}$ to 0: this does not alter the class of the composite map $L_0^{2n-1} \to G/PL$. Hence in either case we may suppose $\mathbf{s}_{2n-2} = 0$, and that Q exists. The total space of the induced bundle over Q is then normally cobordant to L^{2n-1}, since both correspond to homotopic maps $L_0^{2n-1}(N) \to G/PL$.

Proof of 14E.7. Recall that N is odd, $1+\chi$ is invertible in $R_{\widehat{G}}$. We first show that for L^{2n-1}, $(1-\chi)^n\rho \in R_{\widehat{G}}$, i.e. that $\rho \in I_{\widehat{G}}^{-n}$, and moreover that (14E.7 (iii)) holds.

First let L fibre over $Q^{2n-2} \simeq P_{n-1}(\mathbf{C})$. By (14E.8 (c)), $\rho(L) \in I_{\widehat{G}}^{-n}$ and is congruent mod $I_{\widehat{G}}^{-n+2}$ to $(1+\chi)^n/(1-\chi)^n$, and hence mod $I_{\widehat{G}}^{-n+1}$ to $(-2)^n(\chi-1)^{-n}$. Now the class of $\Delta(L)$ mod I_G^{n+1} is that of $(T-1)^n$, and the classes of $(T-1)$ and of $(\chi-1)^{-1}$ correspond under our isomorphism, so the result holds for this case. If L' is normally cobordant to such an L, then $\rho(L') - \rho(L) \in R_G$, and $\Delta(L')/\Delta(L) = \eta(D)$, where D is a unit in $\mathbf{Z}G$ with $\varepsilon(D) = 1$, and hence $\varepsilon'(\eta(D)) = \eta'(\varepsilon(D)) = 1$, so $\eta(D) - 1 \in I_G$, and $\Delta(L') - \Delta(L) \in I_G^{n+1}$. Hence the formula holds for L'. By (14E.9), it remains only to show that if the result holds for L^*, it holds for L (this includes desuspension as a special case). Combining results of (14E.8), and writing $dd' = 1 \pmod N$, we have

$$\begin{aligned} \rho(L^*) &= \frac{1+\chi^d}{1-\chi^d}\rho(L) \\ \Delta(L^*) &= T^{\frac{1}{2}d'(N-1)}(T^{d'}-1)\Delta(L). \end{aligned}$$

Now $\Delta(L^*) = (T-1)^{n+1} \bmod I_G^{n+2}$, and $T^{d'} - 1 \equiv d'(T-1) \bmod I_G^2$; it follows that $\Delta(L) \equiv d(T-1)^n \bmod I_G^{n+1}$, as in the proof of (14E.9). Similarly,

$\rho(L^*) \equiv (-2)^{n+1}(\chi-1)^{-(n+1)} \bmod I_{\widehat{G}}^{-n}$ and

$$\frac{1-\chi^d}{1+\chi^d} \equiv -\frac{d}{2}(\chi-1) \bmod I_G^2,$$

whence

$$\rho(L) \equiv (-2)^n d(\chi-1)^{-n} \bmod I_{\widehat{G}}^{-n+1},$$

establishing the assertion.

It follows in particular that ρ has odd order modulo $R_{\widehat{G}}$, so we can discuss this class independently of the condition at the prime 2. Now of the N^n classes in $I_{\widehat{G}}^{-n}/R_{\widehat{G}}$, just $N^{[n/2]}$ are real and $N^{[(n+1)/2]}$ imaginary, so if $n=2k$ there are N^k real classes and if $n=2k-1$, N^k imaginary ones. The above condition cuts this down by a factor of N, so in each homotopy class there are N^{k-1} values of ρ mod $R_{\widehat{G}}$, which is just the right number for the normal invariant. Hence there are no further congruence conditions mod $R_{\widehat{G}}$. It thus remains only to prove (14E.7 (iv)), where the congruence may be interpreted in the localised ring $(R_{\widehat{G}})^{\langle 1-\chi\rangle}$, or equivalently, in $R_{\widehat{G}}[N^{-1}]$.

Part of the assertion is that $\rho - 1$ is divisible by 2, say ρ is odd. Note that if α, β are both odd, then $(\alpha-1)(\beta-1) \equiv 0 \pmod 4$; also that α^{-1} is odd. Let L_1 and L_2 have join L_3: write $\dim L_i = 2n_i - 1$, $\rho(L_i) = \rho_i$ and $\Delta(L_i) = \Delta_i$ for $i = 1, 2, 3$ so that (14E.8) gives $\Delta_1\Delta_2 = \Delta_3$ and $\rho_1\rho_2 = \rho_3$. If two of the ρ_i are odd so is the third, and $\rho_3 + 1 \equiv \rho_1 + \rho_2 \pmod 4$. As to the right hand sides of the equations,

$$\begin{aligned}
&\sum_1^{N-1} \{\operatorname{sign}\left(i^{n_1}\chi^r(\Delta_1)\right) + \operatorname{sign}\left(i^{n_2}\chi^r(\Delta_2)\right)\}\chi^r \\
&\quad = \sum_1^{N-1} \{\operatorname{sign}\left(i^{n_1+n_2}\chi^r(\Delta_1\Delta_2)\right) + 1\}\chi^r \pmod 4 \\
&\quad = \sum_1^{N-1} \operatorname{sign}\left(i^{n_3}\chi^r(\Delta_3)\right)\chi^r - 1,
\end{aligned}$$

so if the formula holds for two of the L_i, it holds for the third.

Next, for the standard action on S^1,

$$\rho = \frac{1+\chi}{1-\chi} = \sum_1^{N-1}\left(1-\frac{2r}{N}\right)\chi^r \equiv \sum_1^{N-1}(-1)^r\chi^r \pmod 4.$$

$$\begin{aligned}
i\chi^r(\Delta) &= i\chi^r(T^{(N+1)/2} - T^{(N-1)/2}) \\
&= i\left\{\exp\frac{r(N+1)i\pi}{N} - \exp\frac{r(N-1)i\pi}{N}\right\} = (-1)^{r+1}\, 2\sin r\pi/N,
\end{aligned}$$

so the formula holds here too. Taking repeated joins, we see that it holds for the standard action defining $L_0^{2n-1}(N)$. By (14E.8 (c)), it holds for all fibred fake lens spaces. We now show that if L and L' are normally cobordant and the formula holds for L, then it holds for L': (14E.9) then implies that it holds for all fake lens spaces (since the action by χ^d on S^1 is obtained from the standard action by changing the preferred generator, and our formula is natural). Let the normal cobordism have invariants (D, σ). By (13A.5), we have

$$\sigma = \sum_1^{N-1} 2\alpha_r \chi^r, \qquad (-1)^{\alpha_r}\chi^r(D) > 0.$$

By (14E.8 (b)),

$$\begin{aligned} \rho(L') &= \rho(L) + \sigma \\ &\equiv -\sum_1^{N-1} \{\operatorname{sign}\left(i^n\chi^r(\Delta)\right) + 2\alpha_r\}\chi^r \qquad (\mathrm{mod}\ 4), \\ \operatorname{sign}\left(i^n\chi^r(\Delta')\right) &\equiv \operatorname{sign}\left(i^n\chi^r(\Delta)\right) + \operatorname{sign}\left(\chi^r(D)\right) - 1 \quad (\mathrm{mod}\ 4) \\ &\equiv \operatorname{sign}\left(i^n\chi^r(\Delta)\right) + 2\alpha_r \qquad (\mathrm{mod}\ 4), \end{aligned}$$

which proves the result.

COROLLARY. Σ *gives a bijection of the set of homeomorphism classes of fake lens spaces of dimension* $2n-1$ *and* $2n+1$ $(n \geqslant 3)$ $(N$ *odd*$)$.

We refer to homeomorphism rather than PL homeomorphism since for N odd, the obstructions of [K6] to existence and uniqueness of PL structures lie in zero groups. Injectivity follows from the first assertion of the theorem and the fact that, by (14E.8d), ρ and Δ for ΣL determine those for L. As to surjectivity, it is shown in the proof of (14E.7) that if $[(1-\chi)/(1+\chi)]\rho$ and $T^{\frac{1}{2}(N-1)}(T-1)\Delta$ satisfy the conditions, then so do ρ and Δ.

Observe that $\Delta(L) \in R_G$, $\rho(L) \in \mathbf{Q}R_{\widehat{G}}$ are invariants of the polarised fake lens space L. To see if there exists any homeomorphism between L and L', by (14E.8 (a)) check if there is an automorphism of G taking $\Delta(L)$ to $\varepsilon\Delta(L')$ and $\rho(L)$ to $\varepsilon\rho(L')$ $(\varepsilon = \pm 1)$. For example, L has an orientation reversing homeomorphism if and only if there exists d (prime to N, and with even order (multiplicatively) modulo N) such that if

$$\Delta(L) = \sum_1^{N-1} a_i T^i, \qquad \rho(L) = \sum_1^{N-1} b_i \chi^i,$$

then $a_{di} = -a_i$ and $b_{di} = -b_i$ for all i, $1 \leqslant i < N$. There cannot be one inducing the identity on G.

Now we consider which fake lens spaces fibre over fake complex projective spaces – or equivalently, which free actions of G on spheres extend to free actions of S^1. (14E.8 (c)) gives a necessary condition: that

$$\Delta(L) = \{T^{\frac{1}{2}(N-1)}(T-1)\}^n :$$

we assume this satisfied for the rest of our discussion. If the action does extend, ρ for G is the restriction of ρ for S^1. Now the latter is given by (14E.8 (c)), and so depends additively on at most $[n/2]-1$ parameters whereas ρ for G can vary by at least the image of $L_{2n}(G)$, so depends on $\frac{1}{2}(N-1)$ additive parameters. Thus for $n \leqslant N$, there are 'not enough' fake complex projective spaces: the fake space with invariant ρ can be written in the form (14E.8 (c)) and (as is easily seen, cf. also below) the coefficients $\mathbf{s}_{4k}$ are then uniquely determined.

THEOREM 14E.10. *Let $n > N$. Then the fake lens space L^{2n-1} fibres over a fake complex projective space if and only if its suspension does. The possible values of $\rho - f^n$ lie in a certain free abelian group: L fibres if and only if it lies in a certain subgroup of finite index such that the invariant factors of the quotient are 2×4^r, $1 \leqslant r \leqslant \frac{1}{2}(N-1)$, and $\Delta(L)$ is as above.*

Proof If L is fibred then, by (14E.8 (c)),

$$P(L) \equiv \frac{f^{4-n}\big(\rho(L) - f^n\big)}{8(f^2-1)} = \sum_{0 \leqslant r \leqslant [n/2]-2} \mathbf{s}_{4(r+1)}(f^{-2})^r\,,$$

and (14E.7) implies that if $P(L)$ has this form (with Δ as given earlier) then L is fibred, provided the $\mathbf{s}_{4r} \in \mathbf{Z}$.

Now by (14A.1), $P(\Sigma L) = P(L)$; by (14E.7 (i)), $P(L)$ is real. If $P(L)$ can be written as a polynomial (of any degree) in f^{-2} with integer coefficients, then since f^{-2} satisfies a monic equation over $\mathbf{Z}$ of degree $\frac{1}{2}(N-1)$, we can write $P(L)$ also as a polynomial of degree $< \frac{1}{2}(N-1)$: but since $n > N$, this is $\leqslant [n/2]-2$. The first assertion of the theorem follows.

The equation satisfied by f^{-2} is obtained on noting that

$$(f+1)^N - (f-1)^N = \left(\frac{2}{1-\chi}\right)^N - \left(\frac{2\chi}{1-\chi}\right)^N = \left(\frac{2}{1-\chi}\right)^N (1-\chi^N) = 0\,,$$

so $\sum_{r=0}^{\frac{1}{2}(N-1)} \binom{N}{2r} f^{2r} = 0$ and so, dividing by f^{N-1} and setting $2(r+s) = N-1$,

$$\sum_{s=0}^{\frac{1}{2}(N-1)} \binom{N}{2s+1} (f^{-2})^s = 0$$

which is of the required form. It is easily seen (cf. below) to be the minimal polynomial of f^{-2}.

The possible values of $\rho(L)$ $\big($subject to $\Delta(L)$ being as above$\big)$ are given by (14E.7): namely

(i) ρ is real (n even) or imaginary (n odd)

(ii) $\rho \in I_{\widehat{G}}^{-n}$

(iii) $(\rho - f^n) \in I_{\widehat{G}}^{-n+1}$

(iv) $(\rho - f^n) \equiv 0 \pmod 4$,

the latter two since we know f^n satisfies the same congruences as ρ. Clearly (iii) implies (ii). Moreover, it follows from (i) that (iii) implies the apparently stronger

(iii)′ $(\rho - f^n) \in I_{\widehat{G}}^{-n+2}$.

Turning to $P(L)$, we note that $8(f^2 - 1) = 32\chi/(1-\chi)^2$, and recall again that $(1+\chi)$ is a unit of $R_{\widehat{G}}$. Thus since

$$P(L) = \frac{(1-\chi)^{n-2}}{(1+\chi)^{n-4} \cdot 32\chi} \big(\rho(L) - f^n\big)$$

we see that as $\big(\rho(L) - f^n\big)$ runs through the real (imaginary) part of $4I_{\widehat{G}}^{-n+2}$ (which is, by the above, its range), $P(L)$ runs through precisely the real part of $\frac{1}{8}R_{\widehat{G}}$.

It remains to determine the subgroup of the real part of $R_{\widehat{G}}$ spanned by powers of f^{-2}. A basis for this real part is given by the $(\chi^r + \chi^{-r})$, $1 \leqslant r \leqslant \frac{1}{2}(N-1)$. Changing base by a unitriangular matrix, we obtain $(\chi + 2 + \chi^{-1})^r$, $1 \leqslant r \leqslant \frac{1}{2}(N-3)$ with -1, or, reordering, $(\chi + 2 + \chi^{-1})^r$, $0 \leqslant r \leqslant \frac{1}{2}(N-3)$. Now $\chi + 2 + \chi^{-1} = (1+\chi)^2/\chi$ is a unit – since $(1+\chi)$ is – so multiplying these by the same real unit $[\chi/(1+\chi)^2]^{\frac{1}{2}(N-3)}$ we find that a basis for the real part of $R_{\widehat{G}}$ is given by the $[\chi/(1+\chi)^2]^r$, $0 \leqslant r \leqslant \frac{1}{2}(N-3)$. On the other hand we have the $(f^{-2})^r$, $0 \leqslant r \leqslant \frac{1}{2}(N-3)$, and changing base again by a unitriangular matrix we find they span the same subgroup as the $(1+f^{-2})^r$, $0 \leqslant r \leqslant \frac{1}{2}(N-3)$. But $1 + f^{-2} = 4\chi/(1+\chi)^2$. We can now read off the desired invariant factors, since we have obtained stacked bases for the group and the subgroup (don't forget the extra factor of 8) – and also incidentally shown that the $(f^{-2})^r$, $0 \leqslant r \leqslant \frac{1}{2}(N-3)$, are linearly independent, as mentioned earlier.

15. Applications: Free Uniform Actions on Euclidean Space

We now consider free actions of groups G on $\mathbf{R}^m$, with compact orbit space M^m (by 'uniform' I mean that the orbit space is compact). Thus M is a compact topological manifold, and is an Eilenberg-MacLane space $K(G, 1)$. We first study the case G free abelian, where $L_i(G)$ is known from (12.6) and (13A.8); then a somewhat more general case, where we can use the problem to compute $L_i(G)$.

The case G free abelian is important in view of the spectacular application in the work of Kirby and Siebenmann [K9] [K10] [K6] [K7] on topological manifolds. My results were announced in [W25]; an independent discovery was announced by Hsiang and Shaneson [H23] [H24]. See also [H26].

15A. Fake Tori

Let $X = T^n$ be a product of n copies of S^1. The results of §10 give us an exact sequence

$$[\Sigma X, G/PL] \xrightarrow{\theta_1} L_{n+1}(n\mathbf{Z}^+) \xrightarrow{\partial} \mathscr{S}^{PL}(X) \xrightarrow{\eta} [X, G/PL] \xrightarrow{\theta} L_n(n\mathbf{Z}^+)\,.$$

Here, it is easy to compute the homotopy groups, and (13A.8) computes the L-groups and (13B.8) the maps θ. We first use these computations to show

LEMMA 15A.1. *The maps θ and θ_1 are injective. The cokernel of θ_1 is naturally isomorphic to $H^3(X; \mathbf{Z}_2)$.*

Proof Since the attaching maps of the cells in the natural cell decomposition (with 2^n cells) of T^n have trivial suspensions and since G/PL is a loop space [B12], say of Y, we can write

$$[T^n, G/PL] \;=\; [T^n, \Omega Y] \;=\; [\Sigma T^n, Y]\,.$$

But ΣT^n has the homotopy type of a wedge of spheres: $\binom{n}{i}$ of dimension $i+1$. Hence

$$[\Sigma T^n, Y] \cong [\vee S^{i+1}, Y] \cong \bigoplus \pi_{i+1}(Y) \cong \bigoplus \pi_i(G/PL)\,,$$

with $\binom{n}{i}$ summands π_i for each i. But we can write $L_n(n\mathbf{Z})$ as a corresponding sum $\bigoplus L_i(1)$, and moreover $\pi_i(G/PL) \cong L_i(1) \cong \mathbf{Z}$, 0, $\mathbf{Z}_2$ or 0, depending on the value of i modulo 4. More functorially, our first argument shows that we have an isomorphism

$$[T^n, G/PL] \cong \bigoplus_{0 \leqslant i \leqslant n} H^i\big(T^n; \pi_i(G/PL)\big)\,.$$

We can make this explicit as follows. For i odd, our group is zero. For $i \equiv 2 \pmod 4$, the class $k_i \in H^i(G/PL; \mathbf{Z}_2)$ induces a nonzero class in S^i via the generator $S^i \to G/PL$; hence for $f : T^n \to G/PL$, we can take $f^*(k_i)$ as our ith component. For $i = 4j$, the class $\lambda_j = \frac{1}{8} l_j \in H^{4j}(G/PL; \mathbf{Q})$ induces the orientation class in S^i via the generator $S^i \to G/PL$, if $j > 1$; and twice this class if $j = 1$. For this, see e.g. [S22]. Thus for $f : T^n \to G/PL$, the class $\frac{1}{8} f^*(l_j)$ is an integral class, and we take it as our $(4j)$th component, where we identify $\pi_{4j}(G/PL)$ with $\mathbf{Z}$ for $j > 1$, but with the subgroup E of even integers if $j = 1$.

Now by (13B.8), $\theta(f)$ determines the element of

$$L_n(n\mathbf{Z}) \cong \bigoplus_{0 \leqslant i \leqslant n} \operatorname{Hom}\big(H^i(T^n), L_{n-i}(1)\big)$$

given as follows. Let $x \in H^i(T^n;\mathbf{Z})$. Then

$$\begin{aligned}\theta(f)(x) &= 0 && n-i \text{ odd}\\ &= \tfrac{1}{8}\ell(T^n)\cdot f^*\ell(G/PL)\cdot x \;=\; \tfrac{1}{8}f^*\ell_j\cdot x && n-i=4j\\ &= w(T^n)f^*k(G/PL)\cdot x \;=\; f^*k_{n-i}\cdot x && n-i\equiv 2 \pmod 4\,.\end{aligned}$$

But Poincaré duality shows that cup product induces isomorphisms

$$\begin{aligned}H^{n-i}(T^n;\mathbf{Z}_2) &\to \operatorname{Hom}\big(H^i(T^n;\mathbf{Z}),\mathbf{Z}_2\big)\\ H^{n-i}(T^n;\mathbf{Z}) &\to \operatorname{Hom}\big(H^i(T^n;\mathbf{Z}),\mathbf{Z}\big)\,.\end{aligned}$$

Hence our map θ is an isomorphism on all components except that involving l_1, where it is injective.

Repeating the argument for $T^n \times I$, we find essentially the same situation, except for a dimension shift. It follows that θ_1 is injective; the cokernel comes from the component involving l_1, where θ_1 reduces to the natural map

$$H^3(T^n;E) \subset H^3(T^n;\mathbf{Z}) \to \operatorname{Hom}\big(H^{n-3}(T^n;\mathbf{Z}),\mathbf{Z}\big)\,.$$

The cokernel can be identified with $H^3(T^n;\mathbf{Z}_2)$. The lemma is thus established.

It follows immediately from the lemma, and from the exact sequence which we quoted preceding it, that ∂ induces a bijection $H^3(T^n;\mathbf{Z}) \cong \mathscr{S}^{PL}(T^n)$, for $n \geqslant 5$. We can describe directly how this arises as follows. The lemma shows that any two fake tori have the same normal invariant, so define the same element of the degree 1 bordism set of T^n. Let $W^{n+1} \xrightarrow{g} T^n$ be an appropriate bordism. Make g transverse to a factor T^3 of T^n; let $V^4 = g^{-1}(T^3)$. If we can suppose each component of ∂V a torus, then the signature of V depends only on W and the homology class x of T^3; $\frac{1}{8}\sigma(V)$ is an integer, and its class modulo 2 is $\langle x, a\rangle$, where $a \in H^3(T^n;\mathbf{Z}_2)$ is the invariant discriminating the two ends of W.

Unfortunately, PL manifolds V as above with $\frac{1}{8}\sigma(V)$ odd do not exist, and in order to obtain a correct version of the above, one must adopt some trick such as the following: due, for this problem, to Browder. Multiply W by $P_2(\mathbf{C})$. We can then arrange that each end of the corresponding V^8 is (up to homotopy) $T^3 \times P_2(\mathbf{C})$, and the rest of the description is now unchanged.

We next present an extension of this result, which is also used by Kirby and Siebenmann. In fact, we compute $\mathscr{S}(T^n \times D^k, T^n \times \partial D^k)$.

Here, we have the exact sequence

$$[\Sigma^{k+1}(T^{n+}), G/PL] \xrightarrow{\theta_{k+1}} L_{n+k+1}(n\mathbf{Z}^+) \xrightarrow{\partial} \mathscr{S}^{PL}(T^n \times D^k, T^n \times \partial D^k)$$

$$\xrightarrow{\eta} [\Sigma^k(T^{n+}), G/PL] \xrightarrow{\theta_k} L_{n+k}(n\mathbf{Z}^+)\,.$$

Exactly as before, it is seen that each θ_k is injective; and we have

$$\begin{aligned}
\mathscr{S}^{PL}(T^n) &\cong \text{Coker } \theta_1 \simeq H^3(T^n;\mathbf{Z}_2)\\
\mathscr{S}^{PL}(T^n\times I, T^n\times\partial I) &\cong \text{Coker } \theta_2 \cong H^2(T^n;\mathbf{Z}_2)\\
\mathscr{S}^{PL}(T^n\times D^2, T^n\times S^1) &\cong \text{Coker } \theta_3 \cong H^1(T^n;\mathbf{Z}_2)\\
\mathscr{S}^{PL}(T^n\times D^3, T^n\times S^2) &\cong \text{Coker } \theta_4 \cong H^0(T^n;\mathbf{Z}_2)\,, \quad \text{and}\\
\mathscr{S}^{PL}(T^n\times D^k, T^n\times S^{k-1}) &\cong \text{Coker } \theta_{k+1} = 0 \quad \text{for} \quad k \geqslant 4\,.
\end{aligned}$$

This contains most of our result, viz.

THEOREM 15A.2. *We have bijections, natural for finite coverings,*

$$\mathscr{S}^{PL}(T^n\times D^k, T^n\times S^{k-1}) \;\cong\; H^{3-k}(T^n;\mathbf{Z}_2) \quad \textit{for} \quad n+k\geqslant 5\,.$$

Proof It remains to show the naturality. Let $\pi : T^n \to T^n$ be a finite covering, τ the transfer induced by the corresponding monomorphism of fundamental groups (see §17). We have the commutative diagram with exact rows

$$\begin{array}{ccccccccc}
0 \to & [\Sigma^{k+1}(T^{n+}), G/PL] & \xrightarrow{\theta_{k+1}} & L_{n+k+1}(n\mathbf{Z}^+) & \xrightarrow{\partial} & \mathcal{S}^{PL}(T^n\times D^k, T^n\times\partial D^k) & \to 0\\
& \downarrow{\scriptstyle \pi^*} & & \downarrow{\scriptstyle \tau} & & \downarrow{\scriptstyle \pi^\flat} & \\
0 \to & [\Sigma^{k+1}(T^{n+}), G/PL] & \xrightarrow{\theta_{k+1}} & L_{n+k+1}(n\mathbf{Z}^+) & \xrightarrow{\partial} & \mathcal{S}^{PL}(T^n\times D^k, T^n\times\partial D^k) & \to 0
\end{array}$$

Now (13A.8) gives isomorphisms

$$L_{n+k+1}(n\mathbf{Z}^+) \cong \bigoplus_{0\leqslant i\leqslant n} \text{Hom}\big(H^i(T^n), L_{n+k+1-i}(1)\big)$$

which are made more explicit in (13B.8). We claim that if this is written (setting $j = n - i$), using duality, as

$$\bigoplus_{0\leqslant j\leqslant n} H^j\big(T^n; L_{j+k+1}(1)\big)\,,$$

then τ is the cohomology map induced by π. The theorem will then follow. In fact, as we are only interested in the $L_4(1)$ component, it suffices to consider only the components $L_r(1) \cong \mathbf{Z}$.

By (13B.8), if $(M; f, g)$ defines an element of $\Omega_m^{PL}(G/PL \times T^n)$, then for $x \in H^i(T^n)$ we have, for $m \equiv i \pmod 4$,

$$\theta(M; f, g)(x) \;=\; \tfrac{1}{8}\ell(M) f^*(x) g^*\ell(G/PL)[M]\,.$$

Note that since (by the first part of the theorem) the cokernel of θ is finite for suitable (bounded) M, and τ is a homomorphism, it is enough to evaluate τ on the surgery obstructions $\theta(M; f, g)$. Now if $\pi : T^n \to T^n$ induces $M \overset{\pi_M}{\leftarrow} \overline{M} \overset{\overline{f}}{\to} T^n$ we have

$$\begin{aligned}\tau\big(\theta(M; f, g)(\pi^*(x))\big) &= \theta(\overline{M}; \overline{f}, g \circ \pi_M)\big(\pi^*(x)\big) \\ &= \tfrac{1}{8}\ell(\overline{M})\overline{f}^*\big(\pi^*(x)\big) \cdot \pi_M^* g^* \ell(G/PL)[\overline{M}] \\ &= \tfrac{1}{8}\pi_M^*\ell(M) \cdot \pi_M^* f^*(x) \cdot \pi_M^* g^* \ell(G/PL)[\overline{M}] \\ &= \tfrac{1}{8}\ell(M) f^*(x) g^* \ell(G/PL)[\pi_{M*}[\overline{M}]] \\ &= d\,\theta(M; f, g)(x),\end{aligned}$$

where d is the degree of the covering, so that $\pi_{M*}[\overline{M}] = d[M]$. Let us write $\theta'(M; f, g)$ for the cohomology class dual to $\theta(M; f, g)$, i.e. defined by the identity

$$\theta'(M; f, g) \cdot x[T^n] = \theta(M; f, g)(x).$$

Then

$$\begin{aligned}\tau\big(\theta'(M; f, g)\big) \cdot \pi^*(x)[T^n] &= d\,\theta'(M; f, g) \cdot x[T^n] \\ &= \theta'(M; f, g) \cdot x[\pi_*(T^n)] \\ &= \pi^*\theta'(M; f, g) \cdot \pi^*(x)[T^n].\end{aligned}$$

Since π is a finite covering, as x runs through $H^*(T^n)$, $\pi^*(x)$ runs through a subgroup of finite index. Hence, modulo torsion,

$$\tau\big(\theta'(M; f, g)\big) = \pi^*\big(\theta'(M; f, g)\big),$$

and as observed above, this suffices to complete the proof of the theorem.

We now concentrate on describing the 'fake' tori. One further naturality result should be noted, which is important for geometrical arguments on fake tori.

LEMMA 15A.3. *Let $(U^n, U^{n-1}) \xrightarrow{f} (S^1 \times T^{n-1}, T^{n-1})$ be a homotopy equivalence of pairs; let f have invariant $\theta(f) \in H^3(T^n; \mathbf{Z}_2)$. Then $f|U^{n-1}$ has invariant $i^*\theta(f) \in H^3(T^{n-1}; \mathbf{Z}_2)$.*

For since the invariant $\theta(f)$ is defined as a quotient of the invariant in $L_{n+1}(n\mathbf{Z})$ defined by a normal cobordism, it suffices to consider the latter: the result is now immediate from the inductive determination of $L_i(n\mathbf{Z})$ (based on this same geometric situation) (13B.8).

Since $Wh(n\mathbf{Z}) = 0$, our results classify all closed PL manifolds homotopy equivalent to T^n. For a PL homeomorphism classification, one must study the action of the group of self-homotopy equivalences of T^n on $\mathscr{S}^{PL}(T^n)$. These induce automorphisms of the exterior algebra $H^*(T^n; \mathbf{Z}_2)$ via the natural action of the general linear group $GL_n(\mathbf{Z}_2) \cong GL\big(H^1(T^n; \mathbf{Z}_2)\big)$. There appears to be no known canonical shape for 3-forms analogous to

$$x_1 \wedge x_2 + x_3 \wedge x_4 + \cdots + x_{2r-1} \wedge x_{2r}$$

for 2-forms, but using duality the orbits of the action are easily found for $n \leqslant 5$; apart from 0 there is one for $n = 3$, 4 and two for $n = 5$. For $n = 6$, there are 6 orbits. I give in each case a representative, and the size of the orbit:

0	(1)	$x_1 \wedge x_2 \wedge x_3$	$(1{,}395)$
$x_1 \wedge (x_2 \wedge x_3 + y_2 \wedge y_3)$	$(54{,}684)$	$x_1 \wedge x_2 \wedge x_3 + y_1 \wedge y_2 \wedge y_3$	$(357{,}120)$
$\Sigma\,(x_1 \wedge x_2 \wedge y_3)$	$(468{,}720)$	$\Sigma\,(x_1 \wedge x_2 \wedge y_3 + y_1 \wedge y_2 \wedge x_3)$	$(166{,}656)$

where Σ denotes symmetrising with respect to the suffices.

Now if $n \leqslant 2$, homotopy equivalence is well known to imply (PL or smooth) homeomorphism, for closed manifolds. For $n = 3$, the Stallings fibration theorem [S15] shows that any fake T^3 is the connected sum of a standard one with a fake 3-sphere. For $n = 4$, we have shown that s-cobordism classes of fake T^4's correspond injectively to $H^3(T^4; \mathbf{Z}_2)$. Since $GL_4(\mathbf{Z}_2)$ has only 2 orbits here, either all elements of the group are realised by fake T^4's or only the zero element is.

The above discussion concentrates on the PL classification. For topological manifolds, it is known [H26] that a closed manifold homotopy equivalent to T^n ($n \geqslant 5$) is homeomorphic to it: see also the next section. Similarly, the PL automorphisms of T^n which are not PL concordant to the identity are topologically so, and Siebenmann has given an elegant proof of this. Another easy argument, which has been observed by Siebenmann and by Hsiang, using these PL automorphisms in low dimensions, shows that the s-cobordism theorem fails for PL (or smooth) s-cobordisms of T^3 or of T^4.

Since our fake tori all had the same PL normal invariant, all are parallelisable, and hence smoothable. The set of smoothings of each corresponds bijectively to $[T^n, PL/O]$. As above for G/PL, since PL/O is a loop space, this group is isomorphic to $\bigoplus_{i \leqslant n} H^i(T^n; \Gamma_i)$, where $\Gamma_i = \pi_i(PL/O)$ as usual.

This has large (but finite) order, and does not seem very interesting. Note, however, that since these invariants are again natural for coverings, every closed smooth manifold homotopy equivalent to T^n ($n \geqslant 5$) has a finite covering diffeomorphic to it.

The assembly maps

$$A \; : \; H_*(T^n; \mathbf{L}_\bullet) \to L_*(n\mathbf{Z}^+)$$

are isomorphisms, and the topological structure sets are given by

$$\begin{aligned} \mathscr{S}^{TOP}(T^n) \;&\cong\; 0\,, \\ \mathscr{S}^{TOP}(T^n \times D^k, T^n \times S^{k-1}) \;&\cong\; 0 \text{ for } k \geqslant 1\,. \end{aligned}$$

The surgery classification of fake PL tori was an essential step in the key result

$$TOP/PL \simeq K(\mathbf{Z}_2, 3)$$

of Kirby and Siebenmann [K11].

15B. Polycyclic Groups

If $\mathscr{A}$ is a class of groups, a group G is said to be a poly-$\mathscr{A}$ group if there is a series of subgroups

$$1 = G_n < G_{n-1} < \cdots < G_1 < G_0 = G,$$

each normal in the next, with $(G_{i-1}/G_i) \in \mathscr{A}$ for each i. Thus 'solvable' is synonymous with 'polyabelian', the term polycyclic is defined; and, as we usually denote infinite cyclic groups by $\mathbf{Z}$, so is the class of poly-$\mathbf{Z}$ groups. Several characterisations of polycyclic groups are summarised (with references) in a paper by Wolf [W48]. See also [S2, Chapter 12] for some standard group theory for these groups. Our main result refers only to poly-$\mathbf{Z}$ groups.

Let G be a poly-$\mathbf{Z}$ group with a series as above, G_{i-1}/G_i infinite cyclic for each i, $1 \leqslant i \leqslant n$. Then G is said to have rank n; it is also noetherian (this holds for any polycyclic group), hence regular. Now Farrell and Hsiang [H22], [F3] have generalised the work of Bass, Heller and Swan [B9] to non-commutative polynomial rings and show, by induction on the rank, that if G is a poly-$\mathbf{Z}$ group then $\widetilde{K}_0(G)$ and $Wh(G)$ vanish (this has also been since proved by Waldhausen [W3]).

For the argument below, it is essential to work with topological surgery: this is justified by the work of Kirby and Siebenmann [K6] (which depends on §15A, so we do not obtain an alternative proof of the results there).

THEOREM 15B.1. *Let G be a poly-$\mathbf{Z}$ group of rank n.*

(a) *There exists a closed n-manifold M_G with fundamental group G and universal cover homeomorphic to $\mathbf{R}^n$.*

(b) *The surgery obstruction map induces isomorphisms for all $i \geqslant 0$*

$$\theta : [\Sigma^i M_G^+, G/TOP] \to L_{n+i}(G).$$

(c) *For $n \neq 3$, 4 any homotopy equivalence $h : X \to M_G$ with X a closed manifold is homotopic to a homeomorphism.*

For $n = 3$, this holds if and only if X is irreducible.

For $n = 4$, there is an s-cobordism W of X to M_G and a retraction on M_G whose restriction to X is h.

Remarks. This result was announced in [W26]. To avoid confusion, the classifying space G/TOP should be regarded as a single symbol. The homotopy classes of maps in (b) do respect base points; the superscript $^+$ denotes a base point added to M_G. The result in (c) for $n = 3$ implies M_G irreducible; for $n = 4$ it is enough to obtain an h-cobordism of X to M_G since $Wh(G) = 0$.

We have suppressed mention of w: it is the unique homomorphism such that $H_n(M_G; \mathbf{Z}^t) \cong \mathbf{Z}$ for the corresponding twisted coefficients $\mathbf{Z}^t$.

Proof The proof uses induction on the rank of G. We show that (a) $\Rightarrow$ (b) $\Rightarrow$ (c), and that (c) for G_1 (defined above) implies (a) for G.

(c) $\Rightarrow$ (a). Choose $g \in G$ which generates G mod G_1; let conjugation by g induce the automorphism α of G_1. Then α induces a homotopy self-equivalence of M_{G_1} which by (c) is homotopic to a homeomorphism α_* (for $n = 4$, note that M_{G_1} is irreducible). We obtain M_G from $M_{G_1} \times I$ by glueing the two ends together by α_*: this clearly has fundamental group G. In case $n = 5$, we glue the two ends of the h-cobordism W together instead.

(a) $\Rightarrow$ (b). We have, up to homotopy, a fibration

$$M_{G_1} \to M_G \to S^1$$

and hence a coexact sequence of spaces

$$M_{G_1}^+ \to M_G^+ \to \Sigma M_{G_1}^+ \to \Sigma M_{G_1}^+ \to \ldots$$

in which the last map is easily seen to be $1 \pm \Sigma\alpha_*$, where $\alpha_* : M_{G_1} \to M_{G_1}$ is the characteristic map of the above fibration.

We also have the exact sequence of L-groups corresponding to the inclusion $G_1 \to G$, and since $Wh(G_1) = 0$, (12.6) gives isomorphisms

$$L_m(G_1) \;\cong\; L_{m+1}(G_1 \cup G_1 \to G_1) \;\cong\; L_{m+1}(G_1 \to G)\,.$$

Hence we have a diagram

$$\begin{array}{ccccccccc}
[M_{G_1}^+, G/TOP] & \leftarrow & [M_G^+, G/TOP] & \leftarrow & [\Sigma M_{G_1}^+, G/TOP] & \leftarrow & [\Sigma M_{G_1}^+, G/TOP] & \leftarrow & \cdots \\
\downarrow \theta & & \downarrow \theta & & \downarrow \theta & & \downarrow \theta & & \\
L_{n-1}(G_1) & \longleftarrow & L_n(G) & \longleftarrow & L_n(G_1) & \longleftarrow & L_n(G_1) & \longleftarrow & \cdots
\end{array}$$

The vertical maps involving G_1 are isomorphisms, by the induction hypothesis. If we can show that the diagram is commutative, it follows by the Five Lemma that the other vertical maps are also isomorphisms, which is the desired conclusion. Strictly, this does not give surjectivity of $\theta : [M_G^+, G/TOP] \to L_n(G)$, but this follows since we can extend the sequence a term using ad hoc arguments (G/TOP is a homotopy commutative H-space) or, more simply, using periodicity of $L_n(G)$ and of G/TOP.

For the first square note that in (12.6) the map $L_n(G) \to L_{n-1}(G_1)$ is to be interpreted as: take the map to $K(G, 1) \simeq M_G \to S^1$ given by the fundamental group; make transverse on a point, and take the surgery obstruction of the preimage. Since $M_{G_1} \subset M_G$ with trivial normal bundle (see construction), this preimage is just the induced problem for G_1: commutativity follows.

The commutativity of the second square follows similarly from its geometrical interpretation: from a surgery problem for $M_{G_1} \times I$ (with ends fixed), glue the ends together to obtain one for M_G.

I feel that geometry ought to be used for the third square too, but at least it is easy algebraically: the upper map was computed above, and the lower is, by (12.6), $1 - w(g)\alpha_*$ where $g \in G$, α are as above. On inspection, the sign $-w(g)$ is seen to be that of the $\pm$ in the $1 \pm \Sigma\alpha_*$ above.

Commutativity of the other squares follows from the same arguments with M replaced by $M \times D^r$, relative to $M \times \partial D^r$.

(b) $\Rightarrow$ (c). The result for $n \leqslant 2$ is trivial, and for $n = 3$ follows from Stallings [S15] and Neuwirth [N3] (note that although Neuwirth only asserts existence of a homeomorphism, the proof gives one inducing a prescribed isomorphism of fundamental groups). For $n \geqslant 4$ we see, combining (b) with the surgery results of §10, that there is an s-cobordism of h to the identity map of M_G. This proves our assertion for $n = 4$; for $n \geqslant 5$, it follows from the s-cobordism theorem.

We must show inductively that M_G has universal cover euclidean space. Now M_G is obtained from $M_{G_1} \times I$ (or, if $n = 5$, from W) by glueing the ends together. Thus M_G has an infinite cyclic cover $M_{G_1} \times \mathbf{R}$, and the assertion for G follows from that for G_1. In case $n = 5$ the result follows from the 'open h-cobordism theorem' which implies, in particular, that an infinite composite of h-cobordisms is a product with $\mathbf{R}$ [S16].

It is interesting to study which structures the above manifolds, one for each G, admit: one expects rich structure. However, I have not obtained any results other than those which follow trivially from the existence of certain homogeneous spaces of Lie groups (below): I cannot even show that every M_G is triangulable, though we may note that for $n = 4$ they are (the Neuwirth-Stallings theorem gives PL homeomorphisms).

For Lie group examples, poly-**Z** groups are not the most natural class to study. Begin with the larger class of poly- (finite *or* cyclic) groups. For free action on $\mathbf{R}^n$, a necessary condition is that the group be torsion-free. But this does *not* imply it is poly-**Z**, for any finite group occurs as quotient group of the fundamental group of a flat manifold (see e.g. Wolf [W47, p. 110]) so the group need not even be soluble. And even torsion free polycyclic groups need not be poly-**Z** [B14].

At the other end of the scale, the class of fundamental groups of compact solvmanifolds (homogeneous spaces of connected solvable Lie groups) has been determined by H. C. Wang [W39]: these are the groups G with normal nilpotent subgroup N finitely generated and torsion free, and G/N free abelian of finite rank. Every poly- (finite or cyclic) group has a subgroup of finite index of this form. But not all poly-**Z** groups are: one which is not is

$$G = \{w, x, y, z \,|\, x^w = xz^2,\ y^w = y,\ z^w = z,\ y^x = y,\ z^x = zy^{-1},\ y^z = y^{-1}\}$$

with $G_3 = \{y\}$, $G_2 = \{y, z\}$, $G_1 = \{x, y, z\}$. For if G had this form, N would contain the commutator subgroup $\{y, z^2\}$; since G/N is torsion-free, we would have $z \in N$. But $\{y, z\}$ is not nilpotent. Thus M_G is not a solvmanifold. But for every G in the theorem, some finite covering of M_G is a solvmanifold.

Better results can be obtained if G is a poly- (finite or cyclic) group which has a nilpotent subgroup H of finite index. We can suppose H torsion free: it is then automatically a poly-$\mathbf{Z}$ group, and there is a unique natural embedding (Mal'cev) of H in a 1-connected nilpotent Lie group L, homeomorphic to $\mathbf{R}^n$, where n is the rank of G. One can then define a 'pushout'

$$\begin{array}{ccc} H & \longrightarrow & G \\ \downarrow & & \downarrow \\ L & \longrightarrow & L' , \end{array}$$

and write L' as a semi-direct product $L \cdot F$ with F finite. Then L' acts on L with L acting by translations and F by automorphisms; the isotropy subgroups are finite. We obtain an induced action of G on $L \cong \mathbf{R}^n$, which is free if G is torsion-free, and has compact orbit space. For all this, see [A9], also [Q1, A6]. In particular if G is poly-$\mathbf{Z}$, M_G has the structure of a homogeneous space of L', and is a (so-called) infranilmanifold, with a preferred smooth structure.

The G above has a nilpotent subgroup of finite index, viz. $\{w, x, y, z^2\}$. It is easy to find an example (direct product of above G with $\{a, b, c \,|\, c^b = c, b^a = b^2c^3, c^a = bc^2\}$) which is neither the fundamental group of a solvmanifold nor of an infranilmanifold, though one can well conjecture that all torsion-free poly- (cyclic or finite) groups should be fundamental groups of "infrasolvmanifolds".

It is easy to use (15B.1) to construct examples of L-groups having torsion: for example, if

$$Gc = \{x, y, z \,|\, x^y = x,\ x^z = x^{n+1}y^n,\ y^z = xy\} ,$$

then $L_1(Gc)$ has a cyclic subgroup of order n.

Polycyclic groups G are Poincaré duality groups, i.e. discrete groups such that the Eilenberg-MacLane space $K(G,1)$ is a Poincaré complex. Such groups are necessarily infinite (or else trivial) and torsion-free. See Davis [D3] for a survey of Poincaré duality groups. The main result of this chapter verifies that the assembly maps $A : H_*(K(G,1); \mathbf{L}_\bullet) \to L_*(\mathbf{Z}[G])$ are isomorphisms for polycyclic groups G – see Farrell and Jones [F6], [F7] for generalisations (e.g. for the fundamental groups of compact infrasolvmanifolds and nonpositively curved manifolds) proved using a combination of algebra, differential geometry and controlled topology. See Stark [S18] for a survey of the computations of the L-groups of infinite groups. See the notes at the end of §17H for the connection with the Borel and Novikov conjectures.

16. Applications to 4-manifolds

Although the techniques of this book do not apply directly to 4-manifolds, we can use suitable slight modifications of our ideas to obtain useful results. In this chapter, we present two techniques, the first proving that any h-cobordism of $S^1 \times S^3$ to itself is a product. The corresponding result with $S^1 \times S^3$ replaced by a 1-connected 4-manifold is due to D. Barden [B4]: our proof follows the same plan. The result was announced by me in [W19]: it has also been published by J. Shaneson [S5]. It follows, as was shown in [W12], that any submanifold S of S^5, diffeomorphic to S^3, such that $S^5 - S$ is a homotopy circle, is unknotted.

The other part of the chapter is devoted to a substantial reformulation of the underlying ideas of [W10], [W11], leading on to a partial extension of my earlier results to the non-simply connected case.

THEOREM 16.1. *Any h-cobordism of $S^3 \times S^1$ to itself is diffeomorphic to $S^3 \times S^1 \times I$.*

We give two proofs, the first following the outline already suggested (which is part of a general technique), the second using geometrical properties of $S^3 \times S^1$. The first proof begins with a lemma.

LEMMA 16.2. *Any self-homotopy equivalence of $S^1 \times S^3$ is homotopic to a diffeomorphism.*

Proof $[S^1 \times S^3, S^1 \times S^3] = [S^1 \times S^3, S^1] \times [S^1 \times S^3, S^3]$.

The first factor is isomorphic to $H^1(S^1 \times S^3) \approx \mathbf{Z}$. In the second, as S^3 is 2-connected, we can shrink $S^1 \times 1$ to a point. But $S^1 \times S^3 / S^1 \times 1 \simeq S^3 \vee S^4$, and

$$[S^3 \vee S^4, S^3] \;=\; [S^3, S^3] \times [S^4, S^3] \approx \mathbf{Z} \times \mathbf{Z}_2 .$$

(Note that $[S^1 \times S^3, S^1 \times S^3]$ is a group, since $S^1 \times S^3$ is). Now a homotopy equivalence must at least be a homology equivalence, so the $\mathbf{Z}$-components must be ± 1. Thus there are 8 classes of self-homotopy equivalences. The group of these is clearly generated by the following 3 diffeomorphisms,

$$(x, y) \mapsto (R_1 x, y), \qquad R_1 \text{ a reflection of } S^1 .$$

$$(x, y) \mapsto (x, R_2 y), \qquad R_2 \text{ a reflection of } S^3 .$$

$$(x, y) \mapsto \big(x, T(x) \cdot y\big), \quad T : S^1 \to SO_3 \text{ an essential map} .$$

This proves the lemma. (The proof works for $S^1 \times S^n$ for any $n \geqslant 2$).

Now let W be an h-cobordism; h_0, $h_1 : S^1 \times S^3 \to W$ diffeomorphisms onto the two ends. Then h_0 and h_1 are homotopy equivalences, hence have homotopy inverses, and $h_1^{-1}h_0$ is a homotopy equivalence. By the lemma, it is homotopic to a diffeomorphism f. Replacing h_1 by $h_1 f$, we can suppose $h_0 \simeq h_1$.

Define M from $W \cup S^1 \times S^3 \times I$ by identifying (for $x \in S^1$, $y \in S^3$) $(x, y, 0)$ with $h_0(x, y)$ and $(x, y, 1)$ with $h_1(x, y)$. A homotopy of h_0 to h_1 in W now defines part of a map $S^1 \times S^3 \times S^1 \xrightarrow{H} M$ which is clearly a homotopy equivalence. Now $S^1 \times S^3$ (as a group) is parallelisable. We choose a framing of the stable normal bundle of $S^1 \times S^3 \times 0$. As h_0 and the inclusion in $S^1 \times S^3 \times I$ are homotopy equivalences, it extends uniquely to framings of W and $S^1 \times S^3 \times I$. The induced framings on $S^1 \times S^3 \times I$ are homotopic (indeed, to the original framing), so we may suppose they agree, and define a framing of M. This framing, together with the map

$$r : M \xrightarrow{H^{-1}} S^1 \times S^3 \times S^1 \xrightarrow{p_1 \times p_2} S^1 \times S^3 \,,$$

defines an element α of the framed bordism group $F^5(S^1 \times S^3)$.

The element α need not be zero. In fact, by Thom theory, we have an isomorphism

$$F^m(X) \cong \pi_{m+N}(\Sigma^N X^+)$$

(for any X) for N large, where X^+ is the disjoint union of X and a point. The suspension $\Sigma^N(S^1 \times S^3)^+$ splits (up to homotopy) as a wedge $S^N \vee S^{N+1} \vee S^{N+3} \vee S^{N+4}$, so

$$\begin{aligned} F^5(S^1 \times S^3) &\cong \pi_{N+5}(S^N) \oplus \pi_{N+5}(S^{N+1}) \oplus \pi_{N+5}(S^{N+3}) \oplus \pi_{N+5}(S^{N+4}) \\ &\cong 0 \oplus 0 \oplus \mathbf{Z}_2 \oplus \mathbf{Z}_2 \,. \end{aligned}$$

We can be more explicit. The last two summands can be identified with $F^5(S^3) \approx F^2(pt)$ and $F^5(S^4) \approx F^1(pt)$, and the maps as induced by $S^1 \times S^3 \xrightarrow{p_2} S^3$ and $S^1 \times S^3 \to S^1 \wedge S^3 \cong S^4$. Thus the two invariants of an element of $F^5(S^1 \times S^3)$ can be computed as follows. First, make $M^5 \xrightarrow{r} S^1 \times S^3 \xrightarrow{p_2} S^3$ transverse to $P \in S^3$; then take the class of the framed submanifold $A^2 \subset M^5$, preimage of P. Similarly, make r transverse to $(Q, P) \in S^1 \times S^3$, and take the class of the framed submanifold $B^1 \subset M^5$, preimage of (Q, P).

In our case, we can annihilate the second class be re-choosing the framing. Indeed, we used a homotopy of framings of $S^1 \times S^3$. Multiply this homotopy by a homotopy $S^1 \times S^3 \times I \xrightarrow{p_3} I \xrightarrow{l} O_n$, where l is a non-trivial loop in O_n. Then the component of B^1 meeting $S^1 \times S^3 \times I$ acquires the opposite class of framings, so the class of B^1 changes. Hence we can suppose this class zero.

Similarly, we can now make the first class zero, if it is not already so, by re-choosing the homotopy. For choose a disc $D^5 \subset S^1 \times S^3 \times I$, disjoint from A^2 and B^1, and also from the preimage of $Q \times S^3$. We alter the homotopy by changing it on D^5. Keep the component map $D^5 \to S^1$ the same (thus B^1, and the second obstruction class, are unaltered), but alter the map $D^5 \to S^3$ so that the difference map represents the nontrivial class in $\pi_5(S^3)$. Then the preimage A^2 of P acquires a new component inside D^4, with nonzero Kervaire-Arf invariant, and so the class of A^2 is changed as required. This completes the proof of

LEMMA 16.3. *Let W be an h-cobordism of $S^1 \times S^3$ to itself. Then we can attach $S^1 \times S^3 \times I$ to W along the boundary to obtain a closed framed manifold M, bounding a framed manifold N which has $S^1 \times S^3 \times 0$ as a retract.*

For the assertions about N are the geometric formulation of the proposition that $\alpha = 0$ in $F^5(S^1 \times S^3)$.

We thus have a commutative diagram

$$\begin{array}{ccc} M & \xrightarrow{H^{-1}} & S^1 \times S^3 \times S^1 \\ \big\downarrow & & \big\downarrow p_1 \times p_2 \\ N & \longrightarrow & S^1 \times S^3 . \end{array}$$

Extend $p_3 : M \to S^1$ to a map of N to D^2. This combines with the above to give a map of connected bounded manifolds,

$$\phi : (N, M) \to S^1 \times S^3 \times (D^2, S^1) .$$

Since $\phi \,|\, M$ has degree 1, so has ϕ. Since both manifolds are framed, we can take ν to be any trivial bundle, and define F using the framings. Now $\phi \,|\, M = H^{-1}$ is a homotopy equivalence and, since $\pi_1(S^1 \times S^3 \times S^1) \cong \mathbf{Z} \times \mathbf{Z}$ has vanishing Whitehead group, as follows from [B9], a simple homotopy equivalence: this also follows from its explicit construction.

By our main theorem (3.2), we can now do framed surgery on ϕ, leaving M fixed, to obtain a simple homotopy equivalence, if and only if a certain obstruction vanishes, $\theta(\phi) \in L_6(\mathbf{Z})$, since $\mathbf{Z} \cong \pi_1(S^1 \times S^3 \times D^2)$. By (13A.8) we have $L_6(1) \cong L_6\mathbf{Z}) \cong \mathbf{Z}_2$, with the isomorphism defined by the Arf invariant. But in dimension 6, we can ignore this, for if we take a map of degree 1, $\phi' : S^3 \times S^3 \to S^6$, provided with the natural framing, this has zero Arf invariant. Change the framing on each sphere by a generator of $\pi_3(SO_n)$, however, and the Arf invariant becomes 1 (cf. [M10]). Replacing ϕ by its connected sum with ϕ' does not change the boundary, but gets rid of the obstruction. Thus, after doing this if necessary, we may suppose ϕ a simple homotopy equivalence of pairs.

Now introduce corners in M at $S^1 \times S^3 \times i$, for $i = 0, \frac{1}{3}, \frac{2}{3}, 1$. Then we can regard N as an s-cobordism of W to $S^1 \times S^3 \times [\frac{1}{3}, \frac{2}{3}]$, which is a product on the boundary. By the s-cobordism theorem N is diffeomorphic to a product, and W to $S^1 \times S^3 \times [\frac{1}{3}, \frac{2}{3}]$.

Second proof of 16.1. Glue copies of $D^4 \times S^1$ to the two ends of the h-cobordism M. The resulting manifold, $\overline{M}$ say, is then homotopy equivalent to $S^4 \times S^1$. We next seek an embedding $S^4 \to \overline{M}$ determining the same Poincaré embedding as the inclusion $S^4 \times I \subset S^4 \times S^1$: assume this possible. Cutting $\overline{M}$ along S^4 we obtain an h-cobordism of S^4: if D^5 is glued to each end, we have a homotopy

5-sphere, which must be S^5; hence this h-cobordism is diffeomorphic (removing the two 5-discs again) to $S^4 \times I$. But, according to Smale [S10], $\Gamma_5 = \Theta_5 = 0$, so any diffeomorphism of S^4 preserving orientation is concordant to the identity. Thus $\overline{M}$ is diffeomorphic to $S^4 \times S^1$. But by general position, any two homotopic embeddings $S^1 \cup S^1 \to S^4 \times S^1$ are isotopic; thus we can suppose the $D^4 \times S^1$'s attached to M at the beginning of the proof in standard position. Hence M is diffeomorphic to $S^3 \times S^1 \times I$, as stated.

Now the "obstruction" to constructing the desired embedding of S^4 lies in

$$LS_6\begin{pmatrix} 1 \cup 1 & \longrightarrow & I \\ \downarrow & & \downarrow \\ 1 & \longrightarrow & \mathbf{Z} \end{pmatrix},$$

which vanishes by (12.5.1) (using $Wh(\mathbf{Z}) = 0$) and periodicity. Unfortunately, the theory of §§ 11, 12 does not apply without modification: we now indicate the necessary modifications. The basic construction of embeddings always follows the pattern of (11.3), and the only argument here which encounters difficulties on account of the low dimension is the first, as follows. Make a map $\overline{M} \simeq S^4 \times S^1 \xrightarrow{p_2} S^1$ transverse to a point: its inverse image is then a submanifold $N^4 \subset \overline{M}$. We wish to have a cobordism A of N^4 and S^4 so that we can glue $A \times I$ to $\overline{M} \times I$ along $N^4 \times I \subset \overline{M} \times 1$, and then proceed as in §§ 11, 12: further, it is easily seen (as in the first proof of (16.1)) that $\overline{M}$ can be framed, and we need a framed cobordism which retracts on S^4. First do ordinary framed surgery to make N^4 1-connected. Now N^4 must have zero signature (either since the surgery obstruction vanishes or, more directly, applying [W20, Lemma 3] in the universal cover of $\overline{M}$) so, by [W11] (see also below) it is h-cobordant to a connected sum of copies of $S^2 \times S^2$. The rest follows.

As we have already said, it was shown in [W12] that (16.1) had the following corollary.

THEOREM 16.4. *A submanifold S of S^5, diffeomorphic to S^3, is unknotted if and only if $S^5 - S$ is a homotopy circle.*

This completes the arguments of [W19] (where the result was announced).

The first proof of (16.1) can be reformulated to clarify the nature of the problem of showing that an s-cobordism of any 4-manifold V^4 to itself is a product. First, our techniques only apply with the strengthened hypothesis that W^5 is an s-cobordism and h_0, h_1 are diffeomorphisms of V^4 on its boundary components, which are homotopic as maps from V to W. The homotopy $V \times I \to W$ is a simple homotopy equivalence; an inverse defines an element of $\mathscr{S}(V \times I, V \times \partial I)$. Now by §10 we have the exact sequence

$$L_6\big(\pi(V)\big) \to \mathscr{S}(V \times I, V \times \partial I) \to [\Sigma V, G/O] \to L_5\big(\pi(V)\big)$$

so an analysis of the problem from this viewpoint depends on computation of $L_6\big(\pi(V)\big)$ (which is difficult) and of $[\Sigma V, G/O] \cong H^1(V; \mathbf{Z}_2) \times H^3(V; \mathbf{Z})$. Here again, the obstructions depend on choice of homotopy and framing – the choice of framing, for example, can be varied by any element of $[V, O]$. The full result is not yet clear.

The next results are not applications of the preceding theory, but extensions of it. We begin with the simplest case, which is deducible also from [W11]: the argument below is founded on the same construction but is more explicit.

THEOREM 16.5. *Let X^4 be a closed 1-connected smooth or PL manifold. Then the sequence of* (10.3) *is exact:*

$$\mathscr{S}(X) \xrightarrow{\eta} \mathscr{T}(X) \xrightarrow{\theta} L_4(1)\,.$$

Remarks. In this dimension, the smooth and PL cases are equivalent. The result is proved for manifolds but *not* for Poincaré complexes, even if 1-connected; however, such Poincaré complexes are classified by nonsingular symmetric bilinear forms over $\mathbf{Z}$, and it follows from [M8] and [R15] that there is a corresponding manifold, at least if the form is indefinite, provided the necessary condition, that if the form is even the signature is divisible by 16, is satisfied. If $\mathscr{S}(X)$ is defined via s-cobordism classes (rather than diffeomorphism classes) of structures, partial results were obtained in §10 about extensions of the exact sequence to the left; since $L_5(1)$ is trivial, these imply that η above is injective.

Proof We first calculate $\mathscr{T}(X)$ by the method of Sullivan [S22]. Since X is 1-connected, its homology is torsion free, and we obtain a characteristic variety by choosing submanifolds V^2 representing a base of $H_2(X; \mathbf{Z}_2)$ – or of the subgroup annihilated by $w_2(X)$ if this is nonzero – together with X itself. Then an element of $\mathscr{T}(X)$, or what is the same a map $X \to G/PL$, is determined by its splitting invariants, one element of $\mathbf{Z}_2$ for each V^2, and an integer (an even integer if $w_2(X) = 0$) for X. This integer is the surgery obstruction in $L_4(1)$. Thus what we must show is that given an assignation of elements of $\mathbf{Z}_2$ to V^2's we can find a corresponding element of $\mathscr{S}(X)$. We construct, in fact, a homotopy equivalence $X \to X$ which is homologous (but not, of course, homotopic) to the identity.

Choose an embedding $D^4 \subset M$. If D^4 is shrunk to a point, the result is homeomorphic to M. Shrink instead ∂D^4 to a point to give a map $c : M \to M \vee S^4$. Now let $\eta^2 : S^4 \to S^2$ be an essential map and $x : S^2 \to M$; our map is the composite F of

$$M \xrightarrow{c} M \vee S^4 \xrightarrow{1 \vee \eta^2} M \vee S^2 \xrightarrow{(1,x)} M\,.$$

This is clearly homologous to 1_M, and so a homotopy equivalence. To compute its splitting invariant along $V^2 \subset M$, assume V disjoint from D^4. Then $F^{-1}(V) = V \cup W$, with W framed in D^4, and the splitting invariant is the Arf invariant of W. Now if $x(S^2)$ meets V transversely in n_V points, W is the union of n_V preimages of points under $\eta^2 : S^4 \to S^2$, and each of these has Arf invariant one. Thus the required F is obtained if the x is dual to a mod

2 cohomology class assigning to each V^2 the given corresponding element of $\mathbf{Z}_2$: since $\pi_2(X) \to H_2(X;\mathbf{Z}_2)$ is surjective, such an x exists. This proves the theorem.

Now most of the arguments are evidently valid if X is any closed 4-manifold – and indeed, up to a point, even for any compact 4-manifold. First, we can compute $[X/\partial X, G/PL]$. Since the only relevant homotopy groups of G/PL are $\pi_2 \cong \mathbf{Z}_2$ and $\pi_4 \cong \mathbf{Z}$, with first k-invariant δSq^2, we have an exact sequence

$$\to H^4(X,\partial X;\mathbf{Z}) \to [X/\partial X, G/PL] \to H^2(X,\partial X;\mathbf{Z}_2) \to$$

where the outer maps are easily seen to vanish. Let us assume X connected and orientable. Then the first term is $\cong \mathbf{Z}$ and maps injectively to $L_4(1)$; moreover, the characteristic class of the extension of abelian groups is, essentially, $w_2(X)$. Thus we seek to realise each element of $H^2(X,\partial X;\mathbf{Z}_2)$ or, by duality, of $H_2(X;\mathbf{Z}_2)$ – orthogonal to $w_2(X)$. Now our construction (plus the calculation above) shows that we can realise the spherical elements. How close this is to realising all elements is shown by the exact sequence (from terms of low degree of the spectral sequence of the universal covering of X)

$$\pi_2(X) \to H_2(X;\mathbf{Z}_2) \to H_2\big(\pi_1(X);\mathbf{Z}_2\big) \to 0\,.$$

The situation in the nonorientable case is much the same, except that here the term ($\cong \mathbf{Z}_2$) corresponding to the top cell cannot be dismissed. Of course in each case we have ignored the $L_4\big(\pi_1(X)\big)$ obstruction. In the nonorientable case we can use $L_4(\mathbf{Z}_2^-) \cong \mathbf{Z}_2$: the obstruction is computed by (13B.5) as $k_2w_2 + k_2^2 = k_2{w_1}^2$. This does not simplify things, however. We observe merely that if $H_2\big(\pi_1(X);\mathbf{Z}_2\big)$ vanishes, then so does $H^2\big(\pi_1(X);\mathbf{Z}_2\big)$ and hence also ${w_1}^2$, as w_1 comes from $H^1\big(\pi_1(X);\mathbf{Z}_2\big)$.

The best result to be obtained from these ideas seems to be the following

THEOREM 16.6. *Let X^4 be a compact connected oriented smooth or PL 4-manifold; suppose $H_2\big(\pi_1(X);\mathbf{Z}_2\big) = 0$. Then the sequence of* (10.3) *is exact:*

$$\mathscr{S}(X,\partial X) \to \mathscr{T}(X,\partial X) \xrightarrow{\theta} L_4\big(\pi_1(X)\big)\,.$$

The argument above proves exactness with $L_4(1)$ for $L_4\big(\pi_1(X)\big)$, but since the above has order 2, its exactness follows. This leads to the conjecture that $\operatorname{Im}\theta \subset L_4(1) \subset L_4\big(\pi_1(X)\big)$.

The hypothesis on $\pi_1(X)$ is moderately, but not unduly restrictive: it is satisfied, for example, if $\pi_1(X)$ is infinite cyclic, finite of odd order, the fundamental group of a homology sphere (or of any space X with $H_2(X;\mathbf{Z}_2) = 0$) or the free product of any such. It fails – and so does the whole argument – for $\mathbf{Z}_2$ or $4\mathbf{Z}$, so we cannot take $X = P_4(\mathbf{R})$ or T^4.

There are corresponding extensions of our surgery results for n-ads, when the lowest dimension for surgery is 4; we do not state these here, but note particularly the hypothesis in (16.6) that X is a manifold, not just a Poincaré complex.

To conclude this chapter, we remark that if the results and ideas of [W10] on embeddings of S^4 in 4-manifolds are fully exploited, the result is that one can construct normal cobordisms (with just 2- and 3-handles) of 4-manifolds corresponding to any element of $RU(\Lambda)$: the most that exist without giving useful information!

Cappell and Shaneson [C8] established stable surgery theory in dimension 4: a normal map $(\phi, F) : M^4 \to X$ *has surgery obstruction* $\theta(\phi, F) = 0 \in L_4\big(\pi_1(X)\big)$ *if and only if for some integer* $t \geqslant 0$ *the normal map*

$$(\phi', F') \;=\; (\phi, F)\#1 \;:\; M\#\#_t(S^2 \times S^2) \to X\#\#_t(S^2 \times S^2)$$

is normal bordant to a homotopy equivalence.

Freedman proved the 4-dimensional topological Poincaré conjecture in 1982. This led to the extension of surgery theory to 4-dimensional topological manifolds with 'good' fundamental group – see Freedman and Quinn [F11]. Donaldson proved in 1982 that 4-dimensional differentiable manifolds must have diagonalisable intersection forms on account of the restrictions imposed by gauge theory, subsequently showing that the h-cobordism theorem fails for 4-dimensional differentiable manifolds. Thus surgery theoretic invariants play only a small part in the classification of 4-dimensional differentiable manifolds – see Donaldson and Kronheimer [D2].

See Kirby and Taylor [K13] and Quinn [Q8] for recent accounts of 4-dimensional surgery theory.

Part 4

Postscript

17. Further Ideas and Suggestions: Recent Work

Thanks are due to the many researchers who have communicated their recent results to me before publication: the very brief descriptions below are, of course, intended to advertise these papers, and not in any way to replace them.

17A. Function Space Methods

See the notes at the end of 17B for the applications of surgery to topological manifolds which motivated the initial development of the function space methods.

Some of our results – particularly those involving exact sequences – can be more conveniently stated in a 'functional' form, using a space $\mathbf{L}(\pi)$ whose homotopy groups are the surgery obstruction groups $L_i(\pi)$. A preliminary version of this was developed by Andrew Casson (unpublished) in 1967–68; an account in the simply connected case, with applications (due to Sullivan) to the homotopy theory of G/TOP, appears in [R16]. A more satisfactory account is given in the 1969 Princeton Ph.D. thesis of Frank Quinn [Q2], [Q3]. Here is a summary of the main results.

For each n-ad K with $w^1 \in H^1(|K|;\mathbf{Z}_2)$, $\mathbf{L}_m(K)$ is the (incomplete) semi-simplicial complex (alias Δ-set) whose k-simplices are the objects over $s_0^k K$ of type $n+k$ and dimension $(m+k)$, in the sense of §9. The boundary operators are the first $(k+1)$ of those of the object. Clearly, $\pi_0\big(\mathbf{L}_m(K)\big)$ is the $L_m^1(K)$ of §9. In fact Quinn follows the argument of our §9 to show that (modulo low-dimensional difficulties – i.e. assuming $m-n+k \geqslant 5$ or 6) the homotopy type of $\mathbf{L}_m(K)$ depends only on $\pi(K)$ (cf. (9.7)). It is easy to see that $\mathbf{L}_{m+1}(K) \simeq \Omega\mathbf{L}_m(K)$ and hence $\pi_k\big(\mathbf{L}_m(K)\big) \cong L_{m+k}\big(\pi(K)\big)$. I will ignore the low-dimensional difficulties in the following, and sometimes write $\mathbf{L}_m\big(\pi(K)\big)$ for $\mathbf{L}_m(K)$.

One can now interpret (9.6) as giving homotopy fibrations

$$\mathbf{L}_m(\partial_n K) \to \mathbf{L}_m(\delta_n K) \to \mathbf{L}_m(K)\,.$$

Also the natural extension of the exact sequence of §10 leads to another homotopy fibration. Take – for simplicity – a closed manifold M of dimension $m \geqslant 5$ with S-structure ($S = TOP$, PL, or O – the last corresponding to differential structure). Let $\mathscr{S}^{G/S}(M)$ be the semi-simplicial complex whose k-simplices are simple homotopy equivalences $X \to M \times \Delta^k$, X an S-manifold $(k+2)$-ad, with the obvious boundary. Then there is a homotopy fibration

$$\mathscr{S}^{G/S}(M) \xrightarrow{\eta} (G/S)^M \xrightarrow{\theta} \mathbf{L}_m(M)\,.$$

Since, by the s-cobordism theorem, X can be identified with $M \times \Delta^k$, (at least, if one restricts to the component of $\mathscr{S}^{G/S}(M)$ containing the identity map of M as a 0-simplex), the result can be reformulated. Let $\widetilde{S}(M)$ be the semi-simplicial group, a k-simplex of which is an S-automorphism of $M \times \Delta^k$ preserving faces, but not necessarily projection on Δ^k; similarly for $\widetilde{G}(M)$ (which is only a monoid). Then there is a homotopy equivalence of the quotient space

$$\widetilde{G}(M)/\widetilde{S}(M) \to \mathscr{S}^{G/S}(M)\,.$$

These results should be compared with other structure theorems such as

$$P\widetilde{L}(M)/\widetilde{O}(M) \simeq \mathscr{S}^{PL/O}(M) \simeq (PL/O)^M\,,$$

and similarly for TOP/PL, TOP/O.

This can be interpreted geometrically. Suppose given a homotopy fibration $M^m \to E^{m+b} \to B^b$ of (closed) S-manifolds: when does E have the structure of S-block bundle over B with M as fibre? This can be thought of as reducing the structural group from $\widetilde{G}(M)$ to $\widetilde{S}(M)$. It turns out that a better problem is to seek an S-block bundle with fibre homotopy equivalent to M. Making $E \to B$ transverse to simplices and attempting to surger their preimages leads to a surgery obstruction, given by a section over B of a fibration with fibre $\mathbf{L}_m(M)$.

Observe that piecing together simplices from such a section defines an element of $L_{m+b}(E)$: clearly zero in the case above. Taking this semi-simplicially, one obtains a map

$$\Gamma_B\big(\mathbf{L}_m(M)\big) \to \mathbf{L}_{m+b}(E)$$

called by Quinn the 'assembly'. The construction of this map is technically one of the most interesting aspects of the function space technique. Very little is known about it, beyond some formal properties (compatibility with the fibrations mentioned above).

Another construction, cases of which have been observed by several other authors, notably Lopez de Medrano [L20], is a transfer-like homomorphism coming from bundles. Quinn's formulation seems particularly neat. "Suppose $\pi : E \to B$ is a block fibration over a CW n-ad B, with fibre a compact manifold k-ad M^m. If $N \to X \to B$ is a surgery map over B, then we can form the pullback fibrations

$$\begin{array}{ccccc} \pi^*N & \longrightarrow & \pi^*X & \longrightarrow & E \\ \big\downarrow & & \big\downarrow & & \big\downarrow \\ N & \longrightarrow & X & \longrightarrow & B\,. \end{array}$$

Since the fibre is a compact manifold, this is a surgery map over E, with dimension raised by m". This induces a map

$$\pi^\sharp : \mathbf{L}_j(B) \to \mathbf{L}_{j+m}(E)$$

called the *pullback* map. As with the assembly, we get commutative diagrams (for closed manifolds, say)

$$\begin{array}{ccccc}
\mathscr{S}^{G/S}(B) & \longrightarrow & (G/S)^B & \longrightarrow & \mathbf{L}_b(B) \\
\downarrow \pi^\flat & & \downarrow \pi^* & & \downarrow \pi^\sharp \\
\mathscr{S}^{G/S}(E) & \longrightarrow & (G/S)^E & \longrightarrow & \mathbf{L}_{b+m}(E) .
\end{array}$$

Commutativity of the first square holds by the argument of (14A.3); for the second, it is immediate from the definition. For the second square in a special (non-trivial) case, compare [M13] and (14D.4).

The first important special case of the pullback is simply the product of §9, including as a special case the periodicity isomorphism. This has been further studied by Williamson [W46].

Next is the case when E is a covering space of B, giving the transfer map (referred to in (13A.4 (iii) and (14E.5 (b)). This has the usual properties of a transfer [B7, Ch. III] (see also Sylvia de Christ's thesis, UNAM, Mexico 1967, and [C14]): indeed Charles Thomas [T3] has shown that for π finite, $L_m(\pi)$ is a Frobenius module (in the sense of Lam) over an appropriate Frobenius functor.

Key cases for us too are when the fibre is I or D^2: this defines the homomorphism p_0 of (11.6). In this connexion, we note a problem. In (12.9.2) we give an isomorphism of exact sequences (caution: an earlier version of this was incorrect). We conjecture that another such can be obtained using (13A.7). Indeed, there are isomorphisms between corresponding terms of two exact sequences, but I have no proof that any square is commutative* :

$$\begin{array}{ccccccc}
L_i(\pi \times \mathbf{Z}_2^-) & \xrightarrow{p_0} & L_{i+1}(\pi \to \pi \times \mathbf{Z}_2^+) & \xrightarrow{q_0} & LN_{i-1}(\pi \to \pi \times \mathbf{Z}_2^+) & \xrightarrow{r_0} & L_{i-1}(\pi \times \mathbf{Z}_2^-) \\
\downarrow (13A.7) & & \downarrow (12.9.2) & & \downarrow (12.9.1) & & \downarrow \\
L_{i+2}(\pi \times \mathbf{Z}_2^-) & \to & L_{i+2}(\pi \to \pi \times \mathbf{Z}_2^-) & \longrightarrow & L_{i+1}(\pi) & \longrightarrow & L_{i+1}(\pi \times \mathbf{Z}_2^-)
\end{array}$$

I feel that a proper understanding of this diagram would clarify some obscure points in §14D and lead to substantial generalisation of some of the results there.

There are, of course, function space analogues of the results referring to embeddings, and Quinn lays particular stress on splitting theorems of the Farrell type. One neat result which is immediate from (15B.1) is that for π a poly-$\mathbf{Z}$ group of rank n,

$$\mathbf{L}_n(\pi) \simeq (G/TOP)^{M_\pi} ,$$

* *The isomorphism of the exact sequences was subsequently obtained by Hambleton [H7] and Ranicki [R7, Proposition 7.6.4]*

where $M_\pi \simeq K(\pi,1)$. In general no attempt has been made to determine the homotopy type in particular cases, but let us note two results. First, the pullback map corresponding to product with $P_2(\mathbf{C})$ gives homotopy equivalences

$$\mathbf{L}_n(K) \to \mathbf{L}_{n+4}(K)\,,$$

so the spaces $\mathbf{L}_n(K)$ are all periodic, with period 4: combining with (13A.7) we see that some have period 2. Now Sullivan's proof that the 2-adic k-invariants of $\mathbf{L}\,(1) \simeq G/TOP$ are trivial, with this, shows that all k-invariants of $\mathbf{L}\,(\mathbf{Z}_2^-)$ vanish.

The non-simply connected surgery classifying spaces $\mathbf{L}_m(\Lambda)$ of a ring with involution Λ were defined in Ranicki [R9] as Kan Δ-sets of quadratic Poincaré n-ads over Λ (see the notes at the end of §17G), such that

$$\pi_k(\mathbf{L}_m(\Lambda)) = L_{m+k}(\Lambda)\,,\ \mathbf{L}_{m+1}(\Lambda) \simeq \Omega\mathbf{L}_m(\Lambda)\,,\ \mathbf{L}_m(\Lambda) \simeq \mathbf{L}_{m+4}(\Lambda)\,.$$

Thus $\mathbf{L}_\bullet(\Lambda) = \{\mathbf{L}_{-k}(\Lambda)\,|\,k \in \mathbf{Z}\}$ is an Ω-spectrum with homotopy groups

$$\pi_m\big(\mathbf{L}_\bullet(\Lambda)\big) = \pi_{m+k}\big(\mathbf{L}_{-k}(\Lambda)\big) = L_m(\Lambda)\ .$$

For a space K the surgery obstruction functions define homotopy equivalences

$$\mathbf{L}_m(K) \simeq \mathbf{L}_m\big(\mathbf{Z}[\pi_1(K)]\big)\,.$$

See the notes at the end of §10 for a brief account of the algebraic assembly map in the special case $M = \{1\}$ (= the 0-dimensional manifold consisting of a single point), $E = B$ a closed b-dimensional manifold

$$A\ :\ \Gamma_B\big(\mathbf{L}_m(M)\big) \simeq B_+ \wedge \mathbf{L}_{m+b}(\mathbf{Z}) \to \mathbf{L}_{m+b}(B) \simeq \mathbf{L}_{m+b}\big(\mathbf{Z}[\pi_1(B)]\big)\ .$$

The spectrum $\mathbf{L}_\bullet$ appearing in the algebraic surgery exact sequence of [R9] is the 1-connective cover of $\mathbf{L}_\bullet(\mathbf{Z})$, and the structure space $\mathscr{S}^{TOP}(B)$ is the fibre of the assembly map

$$A\ :\ B_+ \wedge \mathbf{L}_\bullet \to \mathbf{L}_\bullet\big(\mathbf{Z}[\pi_1(B)]\big)\ .$$

The homotopy types of the Ω-spectra $\mathbf{L}_\bullet(\Lambda)$ were determined by Taylor and Williams [T1].

The pullback maps $\pi^\sharp : \mathbf{L}_j(B) \to \mathbf{L}_{j+m}(E)$ of a block fibration $\pi : E \to B$ with fibre a compact m-dimensional manifold M induce the surgery transfer maps in the L-groups

$$\pi^\sharp : L_j\big(\mathbf{Z}[\pi_1(B)]\big) \to L_{j+m}\big(\mathbf{Z}[\pi_1(E)]\big)\,.$$

See Lück and Ranicki [L23] for the algebraic description of $\pi^\sharp$, which depends on the chain homotopy action of $\pi_1(B)$ on the Poincaré duality chain complex $C(\widetilde{M})$ of the cover $\widetilde{M}$ of the fibre M induced from the universal cover $\widetilde{E}$ of the total space E.

17B. Topological Manifolds

Since Parts 1 and 2 of this book were written, there has taken place the celebrated breakthrough in the theory of topological manifolds due to Kirby and Siebenmann, to which reference has already been made at several points in the text. The ultimate result of this has been that all our arguments can be justified in the topological category. Since the definitive paper [K12] by Kirby and Siebenmann is not yet completely written (let alone published), I will now give an outline of some of the existing results*.

The fundamental basic theorem seems to be a product theorem (analogous to the Cairns-Hirsch theorem in ordinary smooth theory) which states, in a very precise relative form, that for M^m a manifold, $m \geqslant 5$, any PL or smooth structure on $M \times \mathbf{R}$ can be deformed (by a small isotopy) to one induced from a PL or smooth structure on M. The proof uses induction on handles, the main diagram of [K6], [K7] and the s-cobordism theorem, but not surgery.

This result has many applications. Combining it with the stable arguments of Milnor [M11] we see (precisely as in the theory of smoothing PL manifolds) that concordance classes of smooth (resp. PL structures) on a topological manifold M correspond bijectively to reductions of the structure group of the tangent bundle, or equivalently, to homotopy classes of liftings of $M \to BTOP$ through BO (resp. BPL). Since it is also shown that $\pi_i(TOP/PL) = 0$ for $i \neq 3$ and has order 2 if $i = 3$, there is an obstruction in $H^4(M; \mathbf{Z}_2)$ to existence of a PL structure; in $H^3(M; \mathbf{Z}_2)$ to uniqueness. One has the full relative form of these results. Kirby and Siebenmann have also shown – again as a corollary of the main product theorem – that concordance of smooth (or PL) structures implies isotopy.

Next, the theorem implies that a topological manifold of dimension $\geqslant 6$ has a handle decomposition. The proof is by induction over coordinate patches, rather like the argument below for transversality.

Third – and this is important for our development – a closed topological manifold Q has in a natural way the structure of a simple Poincaré complex, and one with boundary a simple Poincaré pair. Here no dimension restriction is necessary, as we may multiply by a disc if needed to raise the dimension. For the proof, embed Q in euclidean space with a normal disc bundle $D(\nu)$. Its boundary $S(\nu)$ has a product neighbourhood; we can thus deform it to a PL submanifold, $\Sigma(\nu)$. The deformation takes $D(\nu)$ to $\Delta(\nu)$, say, bounded by $\Sigma(\nu)$; we assign Q the simple homotopy type of $\Delta(\nu)$. It is not hard to show that this is independent of the embedding. One can see this defines a simple Poincaré complex by using a handle decomposition of Q (to which the argument of §2

* *The paper [K12] was published as Essay I of the book of Kirby and Siebenmann [K11].*

applies) and noting that the simple homotopy type given by the cores of the handles must coincide with the above.

Finally, one has a transversality theorem. I outline this in a little more detail, since I have not yet seen the proof written down anywhere*. Let ξ be a bundle with fibre $(\mathbf{R}^n, 0)$ and projection $\pi : E(\xi) \to X$. A map $f : M^m \to E(\xi)$ is called transverse to the zero-section $S(X)$ if $f^{-1}(S(X)) = L^{m-n}$ is a submanifold with normal bundle the pullback of ξ. Then for $m - n \geqslant 5$, any f is close to a transverse map. The idea is to cover M by charts (hence, smoothable open sets U_α) whose images lie in trivial sub-bundles of ξ, given say by charts $X_\alpha \times \mathbf{R} \to E(\xi)$: denote by π_α the induced local projection on $\mathbf{R}^n$. If now f is already transversal on the U_β with $\beta < \alpha$, these meet U_α in an open subset V_α; $L \cap V_\alpha$ is a submanifold, and we have a local projection on it of a neighbourhood in V_α, induced from the trivial bundle over X_α. Using the product structure theorem, construct an isotopy of V_α which takes $L \cap V_\alpha$ to a smooth submanifold and the local projection to a smooth submersion. We can now (after shrinking V_α a little) deform $f \,|\, U_\alpha$ to be smooth, transverse to $X_\alpha \times 0$ in $X_\alpha \times \mathbf{R}^n$ in the smooth sense, and extending the given $f \,|\, V_\alpha$. The result follows by induction.

We also need, for §11, good properties of neighbourhoods of locally flat submanifolds of topological manifolds. Again in codimension 1 or 2 there is no essential difference from the smooth theory: this follows essentially by Brown [B37] for codimension 1 and Kirby [K8] for codimension 2. In higher codimensions, germs of 'neighbourhoods' of manifolds of codimension k are classified by a space $\widetilde{BTOP}_k$ such that we have a homotopy pullback diagram

$$\begin{array}{ccc} \widetilde{BTOP}_k & \longrightarrow & BG_k \\ \downarrow & & \downarrow \\ BTOP & \longrightarrow & BG \end{array}$$

according to a recent paper by Rourke and Sanderson [R19], so the results of §11 for the PL case carry over here also.

Finally, a word about references. The basic works by Kirby and Siebenmann at present available are [K7], [K6] and [K9]. The account above is closer to a talk by Kirby at the 1970 Bonn Arbeitstagung, which follows the paper [K12].

Casson [C11] and Sullivan [S22] used the torus trick of Novikov [N6], the invariant of Rochlin [R15] and the simply connected surgery classifying spaces G/TOP, G/PL to disprove the manifold Hauptvermutung. They showed that the classifying space for topologically trivialised PL block bundles is $TOP/PL \simeq K(\mathbf{Z}_2, 3)$, and proved that for $m \geqslant 5$ a homeomorphism of m-dimensional PL manifolds $h : N^m \to M^m$ is homotopic to a PL homeomorphism if and only if the classifying element $c(h) \in [M, TOP/PL] = H^3(M; \mathbf{Z}_2)$ is 0. See also Armstrong, Cooke and Rourke [A3], Kirby and Siebenmann [K11] and Ranicki [R11], [R12].

See Marin [M3] and Quinn [Q7].

17C. Poincaré Embeddings

First we observe that the definition of Poincaré embedding used by Levitt [L16] is simpler than the one given in §11. For any finite complex K, Poincaré complex P and map $f : K \to P$ we say that f is homotopic to an embedding if there exist Poincaré pairs (P_1, Q) and (P_2, Q) (with $P_1 \cap P_2 = Q$) and homotopy equivalences $e : K \to P_1$ and $g : P_1 \cup P_2 \to P$ with $g \circ e \simeq f$. Thus in particular if K is a Poincaré complex, Spivak's theorem shows that (modulo troubles from fundamental groups) Q is equivalent to the total space of a spherical fibration over K, so we can regard the pair (P_1, Q) as fibred over K with fibre (D^q, S^{q-1}), and thus recover the formulation of §11.

More fundamental work has been done by Lowell Jones in a rather difficult paper [J1]: again we give a summary, since the main results of this paper are not mentioned in its introduction. The idea is to replace the study of Poincaré complexes by that of the more geometrical 'patch spaces', which consist of sets of smooth manifolds with some attaching data. The first main result (Lemma 3.1) states that this replacement is possible in dimensions $\geqslant 15$; there is also a uniqueness clause. The second main result (Lemma 1.1) states that for patch spaces the problem of Poincaré embeddings in the middle dimensions ($K^k \to P^p$, $|p - 2k| \leqslant 1$) meets the same obstructions as that for smooth embeddings. This shows (and this is the result emphasised by Jones himself), that the arguments in this book are now directly applicable to Poincaré complexes: the existence in general of handlebody decompositions, and hence (following arguments similar to those in [L16]) the traditional embedding theorems for Poincaré complexes in the metastable range; though at present his results are only asserted in dimension $\geqslant 15^*$.

Finally both Jones [J1] and Levitt [L17], [L18] have considered bordism of Poincaré complexes: my impression is that Levitt's statements can be simplified by using Jones's results. The general philosophy seems to be that the obstructions to validity of transversality theorems for Poincaré complexes are the surgery obstruction groups themselves, and that this carries over to bordism. For example, if $\mathbf{M}(G, \pi)$ denotes the Thom spectrum of the universal spherical fibration over the pullback of

$$BG \times K(\pi, 1) \xrightarrow{w} K(\mathbf{Z}_2, 1),$$

I conjecture an exact sequence

$$\cdots \to L_n(\pi) \to \Omega_n^{\mathrm{Poinc}}(\pi) \to \pi_n\big(\mathbf{M}(G, \pi)\big) \to L_{n-1}(\pi) \to \ldots$$

* *The dimension restriction* $\geqslant 15$ *in the preprint version of [J1] was improved to* $\geqslant 5$ *in the published version.*

The conjectured sequence is a special case of the Levitt-Jones-Quinn exact sequence

$$\cdots \to L_n(\pi) \to \Omega_n^{\text{Poinc}}(K) \to \pi_n\big(\mathbf{M}(G,K)\big) \to L_{n-1}(\pi) \to \dots ,$$

which is defined for any space K with $\pi_1(K) = \pi$ and an orientation character $w \in H^1(K;\mathbf{Z}_2)$, and $\mathbf{M}(G,K)$ the Thom spectrum of the spherical fibration over the pullback of

$$BG \times K \xrightarrow{w} K(\mathbf{Z}_2, 1) .$$

Quinn [Q3] proposed a proof of exactness using a homotopy theoretic approach to surgery on Poincaré complexes involving the extension of cofibrations to the left, generalising the Poincaré π_1-surgery method of Browder [B25]. Hausmann and Vogel [H11] proved exactness for $n \geqslant 5$ using manifold surgery methods, such as the π-π theorem of §4. See Klein [K14] for a survey of Poincaré surgery.

17D. Homotopy and Simple Homotopy

It is now time to mention the version of our theory in which homotopy equivalences are used throughout in place of simple homotopy equivalences. This is not altogether new: it is, indeed, the theory envisaged in [W18]. It has been observed by Shaneson [S6], following suggestions of Rothenberg, that on can develop the whole theory along these lines. For example, the 'simple unitary' groups of §6 are replaced by unitary groups. I had originally intended to include a detailed comparison of theories with slightly differing assumptions. Here is an outline, probably incorrect: I hope some reader can put it right.

We contemplate three kinds of spaces:

Poincaré complexes,
Those which are finite CW complexes,
Simple Poincaré complexes.

With each of these, there are corresponding notions of Poincaré pair, cobordism, etc. We can define equivalence in five cases:

A Poincaré complex, homotopy equivalence.
B Finite Poincaré complex, homotopy equivalence.
C Finite Poincaré complex, simple homotopy equivalence.
D Simple Poincaré complex, homotopy equivalence.
E Simple Poincaré complex, simple homotopy equivalence.

In each case we can define 'objects' as in §9 (now using Poincaré complexes only, not manifolds), and define obstruction groups L – or even, following Quinn, spaces – accordingly. The proof that these depend only on the fundamental group follows the pattern in §9, once an analogue to the theorem of §4 is established. My impression is that the recent results of Levitt and Jones to which I referred in §17C make this possible: hitherto this has been the difficult point in validating these ideas.

For a Poincaré complex to be finite, there is an obstruction θ in $\widetilde{K}_0(\mathbf{Z}[\pi])$ which satisfies a symmetry condition $\theta^* = (-1)^m\theta$ coming from duality; only the class of this modulo elements $\phi+(-1)^m\phi^*$ is cobordism invariant. The result is homotopy unique, but the simple homotopy types are classified by $Wh(\pi)$. For a finite Poincaré complex to be simple, there is an obstruction ϕ in $Wh(\pi)$ with $\phi^* = (-1)^m\phi$; if only the homotopy type is given, this is only determined modulo elements $\psi + (-1)^m\psi^*$.

Such arguments lead us to expect exact sequences, which can be formulated as follows. The group of order 2 acts on each of $\widetilde{K}_0(\mathbf{Z}[\pi])$, $Wh(\pi)$ by the duality map. For any group M with an endomorphism T satisfying $T^2 = 1$, we define a complex by $C^n(M) = M$, $d^n = 1 - (-1)^n T$, and use the usual notations

$$\begin{aligned} Z^n(M) &= \operatorname{Ker}\bigl(d^n : C^n(M) \to C^{n+1}(M)\bigr)\,, \\ B^n(M) &= \operatorname{Im}\bigl(d^{n-1} : C^{n-1}(M) \to C^n(M)\bigr) \end{aligned}$$

for cocycles and coboundaries, with cohomology groups

$$\widehat{H}^n(\mathbf{Z}_2; M) = Z^n(M)/B^n(M)\,.$$

Then we expect an exact commutative diagram

$$\begin{array}{ccccccc} & & 0 & & 0 & & 0 \\ & & \downarrow & & \downarrow & & \downarrow \\ Z^{m+1}\bigl(Wh(\pi)\bigr) & \longrightarrow & L_m^E(\pi) & \longrightarrow & L_m^D(\pi) & \longrightarrow & 0 \\ \downarrow & & \downarrow & & \downarrow & & \\ Wh(\pi) & \longrightarrow & L_m^C(\pi) & \longrightarrow & L_m^B(\pi) & \longrightarrow & 0 \\ \downarrow & & \downarrow & & \downarrow & & \\ B^m\bigl(Wh(\pi)\bigr) & \longrightarrow & Z^m\bigl(Wh(\pi)\bigr) & \longrightarrow & \widehat{H}^m\bigl(\mathbf{Z}_2; Wh(\pi)\bigr) & \longrightarrow & 0\,. \end{array}$$

For a second relation, one expects further groups $L_m^F(\pi)$ coming from general Poincaré complexes but 'ordinary' surgeries (a relative finiteness condition) and sequences

$$\begin{aligned} &0 \to L_m^B(\pi) \to L_m^F(\pi) \to Z^m\bigl(\widetilde{K}_0(\mathbf{Z}[\pi])\bigr)\,, \\ &\widetilde{K}_0\bigl(\mathbf{Z}[\pi]\bigr) \to L_m^F(\pi) \to L_m^A(\pi) \to 0\,. \end{aligned}$$

In fact Shaneson gives [S6, 4.1] an exact sequence of Rothenberg

$$\cdots \to \widehat{H}^{m+1}\bigl(\mathbf{Z}_2; Wh(\pi)\bigr) \to L_m^E(\pi) \to L_m^B(\pi) \to \widehat{H}^m\bigl(\mathbf{Z}_2; Wh(\pi)\bigr) \to \ldots$$

and we would expect also

$$\cdots \to \widehat{H}^{m+1}\bigl(\mathbf{Z}_2; \widetilde{K}_0(\mathbf{Z}[\pi])\bigr) \to L_m^B(\pi) \to L_m^A(\pi) \to \widehat{H}^m\bigl(\mathbf{Z}_2; \widetilde{K}_0(\mathbf{Z}[\pi])\bigr) \to \ldots$$

Presumably the images in $\widehat{H}^m\bigl(\mathbf{Z}_2; Wh(\pi)\bigr)$, $\widehat{H}^m\bigl(\mathbf{Z}_2; \widetilde{K}_0(\mathbf{Z}[\pi])\bigr)$ depend only on m modulo 4: it would be interesting to have an algebraic characterisation. In even dimensions I expect all groups to be equivalence classes of quadratic forms:

A On projective modules, modulo hyperbolic forms $H(P)$, P projective.
B On free modules, modulo $H(F)$, F free.
C On based free modules, modulo $H(F)$, F free and based.
D As E, modulo equivalence given by change of basis.
E Simple forms on based free modules, modulo $H(F)$, F free and based.
F On projective modules, modulo hyperbolic forms $H(F)$, F free.

It is not so simple to conjecture algebraic versions in the odd-dimensional case.

Note that the groups L^E are those previously used in this book, and denoted simply by L. The groups of [W18], [S6] and here denoted by L^B. A more usual notation is to write

$$L^s = L^E, \qquad L^h = L^B.$$

Shaneson uses the two families of obstruction groups to give a neat formulation of the splitting theorem. Namely, there are natural isomorphisms

$$L^E_m(\pi \times \mathbf{Z}) \cong L^E_m(\pi) \oplus L^B_{m-1}(\pi).$$

Note how this fits with the sequence of Rothenberg on the previous page and with (12.6). This does not apply to our generalisation (12.6), since $\widehat{H}^n(\mathbf{Z}_2; Wh(\pi))$ is there replaced by $\widehat{H}^n(\mathbf{Z}_2; Wh(\pi)^\alpha)$. See also Farrell and Hsiang [F3].

We observe finally that the necessity of referring to simple equivalences has sometimes complicated our proofs, so although I prefer the theory as presented above, it is better to have both. For example, a proof of (11.3.4) ought not to need the discussion of torsion involved in our argument. Also, the theory of normal invariants does not need reference to simple equivalence. Finally, we made crucial use of the L^B theory in (14E.4).

Projective L-theory. Novikov [N8] and Ranicki [R1], [R2] defined the projective L-groups

$$L^p = L^A$$

using quadratic forms on f. g. projective modules in the even-dimensional case, and projective formations in the odd-dimensional case – see the notes at the end of §6 for a brief account of formations. The projective and free L-groups are related by a Rothenberg-type exact sequence

$$\cdots \to \widehat{H}^{m+1}(\mathbf{Z}_2; \widetilde{K}_0(\mathbf{Z}[\pi])) \to L^h_m(\pi) \to L^p_m(\pi) \to \widehat{H}^m(\mathbf{Z}_2; \widetilde{K}_0(\mathbf{Z}[\pi])) \to \dots$$

as was conjectured above, with natural isomorphisms

$$L^h_m(\pi \times \mathbf{Z}) \cong L^h_m(\pi) \oplus L^p_{m-1}(\pi)$$

analogous to the natural isomorphisms of Shaneson [S6]

$$L^s_m(\pi \times \mathbf{Z}) \cong L^s_m(\pi) \oplus L^h_{m-1}(\pi).$$

Pedersen and Ranicki [P2] gave a geometric interpretation of the projective L-groups, using normal maps from compact manifolds to finitely dominated Poincaré complexes. The lower L-groups $L^{\langle -i\rangle}_*(\pi)$ of Ranicki [R2], [R8] are defined for $i \geqslant 1$ by analogy with the lower K-groups K_{-i} of Bass [B7, Chapter XII], with splitting theorems

$$L^{\langle -i\rangle}_*(\pi \times \mathbf{Z}) = L^{\langle -i\rangle}_*(\pi) \oplus L^{\langle -i-1\rangle}_*(\pi) \quad (i \geqslant 0,\ L^{\langle 0\rangle}_* = L^p_*)$$

and exact sequences

$$\cdots \to \widehat{H}^{m+1}\big(\mathbf{Z}_2; \widetilde{K}_{-i}(\mathbf{Z}[\pi])\big) \to L_m^{\langle -i+1\rangle}(\pi) \to L_m^{\langle -i\rangle}(\pi) \\ \to \widehat{H}^{m}\big(\mathbf{Z}_2; \widetilde{K}_{-i}(\mathbf{Z}[\pi])\big) \to \dots .$$

Intermediate L-groups. Cappell [C1] introduced the intermediate L-groups $L_*^U(\pi)$, *which are defined for any* $*$*-invariant subgroup* $U \subseteq Wh(\pi)$ *using quadratic forms and automorphisms (or formations) with torsion in* U, *in connection with the extension of the splitting theorem for* $\pi \times \mathbf{Z}$ *to an exact sequence in the twisted case* $\pi \times_\alpha \mathbf{Z}$

$$\cdots \to L_m(\pi) \to L_m(\pi \times_\alpha \mathbf{Z}) \to L_{m-1}^{Wh(\pi)^\alpha}(\pi) \xrightarrow{1-\alpha} L_{m-1}(\pi) \to \dots$$

(obtained algebraically in Ranicki [R3]). There are also intermediate projective L-groups $L_*^U(\pi)$, *which are defined for any* $*$*-invariant subgroup* $U \subseteq \widetilde{K}_0\big(\mathbf{Z}[\pi]\big)$.

17E. Further Calculations

The reader of this section should be particularly cautioned that the topics discussed are in a state of vigorous development; I have tried to describe the results known as of the end of July 1970, but this may lead to some results being anticipated, and others described in an unsatisfactory interim version.

I start with finiteness theorems for the surgery obstruction groups. First note that since our fundamental groups are all finitely presented, hence countable, one can count the matrices (even $\infty \times \infty$) over them, and it follows that all $L_i(\pi)$ etc. are countable groups. I had originally hoped that all might be finitely generated. However, a result of Bass and Murthy (see [B7, XII, 10.6]) shows that, for example, $Wh(\mathbf{Z} \times \mathbf{Z}_4)$ has a direct summand which is an infinite direct sum of cyclic groups of order a power of 2. The exact sequence relating L^B and L^E (§17D) then shows that at least one of these for $\mathbf{Z} \times \mathbf{Z}_4$ fails to be finitely generated, and if it is L^B, then some L^E for $\mathbf{Z} \times \mathbf{Z} \times \mathbf{Z}_4$ fails to be.

Since $L_k(\pi)$ need not be finitely generated for π finitely generated abelian, the only reasonable class of π to look at is the class of finite groups. For $L_{2k}(\pi)$, we gave the result in (13A.4). Now $L_{2k+1}(\pi)$ is a commutator quotient group of an infinite unitary group. It follows from a stability theorem of Bak [B1], [B3] that a finite unitary group already maps onto this, and from a result of Borel and Harish-Chandra [B13] that these groups are finitely generated. Hence $L_{2k+1}(\pi)$ is finitely generated for π finite. Now by a very recent result of Passman and Petrie [P1], $L^B_{2k+1}(\pi)$ has exponent dividing 4 when π has odd order. Thus for π of finite, odd order, the groups $L_{2k+1}(\pi)$ are finite 2-groups. The same conclusion is probably valid when π has even order; in fact I would also guess that $L_{2k+1}(\pi)$ vanishes when π has odd order*.

This last conjecture is supported by the recent result of Lee [L9] that $L_3(\mathbf{Z}_p) = 0$ for p an odd prime. It is rumoured that this can also be extended to groups with order the product of two distinct odd primes. A comparable, but distinct result is the recent proof by Petrie [P5] that such groups can act freely and smoothly on homotopy spheres. More precisely, let G have a cyclic normal subgroup of odd order, with quotient of odd prime order q. Then Petrie shows, using surgery in the style of [W18, §6], that G acts freely and smoothly on a homotopy sphere of dimension $2q - 1$. This solves a problem of some years standing. The result of [P1] has a rather similar proof.

There has been progress, too, on the conjecture following (12.4). The following

**It was shown by Wall [W32] that the odd L-groups $L_{2*+1}(\pi)$ are finite of exponent 8 for finite groups π, and by Bak [B2] that they vanish for finite π of odd order. See the notes at the end of §13A for a brief account of the general computation scheme for the L-theory of finite groups.*

is due to Lee [L6]. Suppose that

$$\Phi : \quad \begin{array}{ccc} A & \longrightarrow & C \\ \downarrow & & \downarrow \\ B & \longrightarrow & D \end{array}$$

is a pushout diagram of groups, with all maps injective. Lee assumes (i) A is trivial and (ii) B and C contain no element of order 2; and shows that $L_n(\Phi) = 0$ for all n. The argument needs both conditions, though (ii) can be somewhat weakened for no extra work. I would expect that condition (i) can be dispensed with too, but this leads to considerable technical complications. It is interesting to note here that many times in the development of surgery, trouble has arisen with π_2 (or w_2), which can be avoided by such assumptions as (ii) here, but always it seems that subsequent work has avoided the difficulty altogether. Cappell has announced [C1] a proof of the full conjecture*, but I have not had access to any part of the argument. Strictly speaking, he claims that given a finite 2-sided Poincaré embedding $M^m \to V^{m+1}$, inducing a monomorphism of fundamental groups, there is a submanifold $M' \subset V$ realising it as a (non-finite) embedding. Thus the corresponding L-group is again a subquotient of $Wh(M)$ having exponent 2. The result generalises (12.5), and corollaries can be obtained in essentially the same way: see also [L10], [L7]. There is a general feeling that this and a general version of (12.6) should hold for K-theory and L-theory in an abstract algebraic setting. For K-theory, this has been worked out by Waldhausen [W3]. Another attack on splitting theorems, due to Farrell and Wagoner (and communicated to me orally), starting from a geometrical problem on non-compact manifolds, arrived at the following algebraic situation.

Let A be a ring with unit and anti-involution. Define ℓA to be the ring of infinite matrices over A with only a finite number of non-zero entries in each row and in each column; $\mathscr{M}A$ to be the ideal in ℓA of matrices with only finitely many non-zero entries. Then if $\lambda A = \ell A/\mathscr{M}A$, there are isomorphisms $L_n^E(\lambda A) \cong L_{n-1}^B(A)$, $L_n^B(\lambda A) \cong L_{n-1}^A(A)$, where E, B, A have the meanings described in the preceding section. There are also results for abstract K-groups; in particular $K_1(\lambda A) \cong K_0(A)^\dagger$.

Other algebraic ideas on the splitting theorem are contained in Novikov's recent papers [N8].

** The full conjecture is that $L_*(\Phi) = 0$ for any pushout square Φ of groups with injective morphisms. This is false in general, with*

$$L_{n+2}(\Phi) \;=\; LS_n(\Phi) \;=\; \widehat{H}^{n+1} \oplus \mathrm{UNil}_{n+2}(\Phi)$$

the sum of a Tate $\mathbf{Z}_2$-cohomology algebraic K-theory group and the appropriate UNil*-group of Cappell [C2], [C4] – see the notes at the beginning of §12 and at the end of §12A.*

† See Farrell, Taylor and Wagoner [F8].

Added in September, 1970 (after the Congress at Nice)

Hyman Bass, in collaboration with Amit Roy, has obtained further calculations of the groups $L_n(\pi)$ with n odd and π finite abelian ([B8]). Work is still in progress, but here is an interim statement of results.

Their most striking theorem is that for π finite abelian, with Sylow 2-subgroup π_2, in the orientable case, the map $L_3(\pi_2) \to L_3(\pi)$ is an isomorphism. Since $L_3(1) = 0$, $L_3(\pi)$ vanishes for π abelian (or, in particular, cyclic) of odd order. We also have $L_3(\mathbf{Z}_2) \cong \mathbf{Z}_2$. In general, the calculation has been reduced to one over the group ring $\mathbf{Z}_2[\pi]$, so in particular $L_3(\pi)$ is finite: more precisely, it has exponent 2 or 4.

If π is an elementary 2-group (i.e. all elements have order 2), the results are more specific. There is an exact sequence

$$\mathbf{Z}_2[\pi] \to L_3(\pi) \to \mathbf{Z}_2[\pi]^\times \to 0\,,$$

where the invariant of a unitary matrix in the group of units of $\mathbf{Z}[\pi]$ is the spinor norm of the orthogonal matrix obtained by reduction mod 2. For the same groups π, $L_1(\pi)$ is the subgroup of elements of order 2 in the Picard group $\mathrm{Pic}(\mathbf{Z}_2[\pi])$.

This is probably the most appropriate place to mention another idea which helps in such calculations. In the situation of (11.5) and (11.6) we can define new groups, in which we consider surgery both on the large manifold and on the submanifold: denote these obstructions to 'surgery of pairs' by $LP_n(\Phi)$. By forgetting one or other of the two manifolds, we obtain homomorphisms

$$LP_n(\Phi) \to L_n(B)\,,\ LP_n(\Phi) \to L_{n+q}(D)\,,$$

each lying in an an exact sequence: these sequences form the diagram

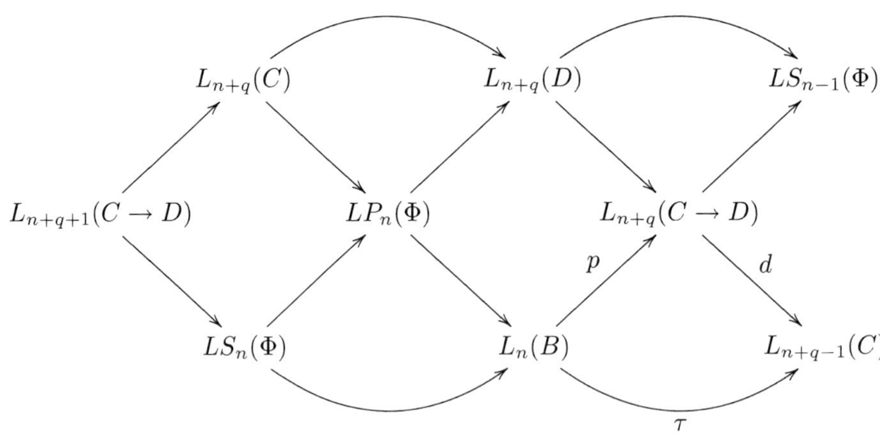

As in (12.3) and (12.4), the most interesting Φ are those of the form $s_2\phi$, i.e. with $A = C$, $B = D$.

A particularly interesting case is when $q = 2$, $A = C = \mathbf{Z}$, $B = D = \mathbf{Z}_N$. Here, τ can be identified with the transfer $L_{2k}(\mathbf{Z}_N) \to L_{2k+1}(\mathbf{Z}) \cong L_{2k}(1)$. If N is odd this, hence also d, is a split surjection. Further, $L_{2k}(\mathbf{Z}) \to L_{2k}(\mathbf{Z}_N)$ is a split injection. We also claim that

$$q : L_{2k-2}(\mathbf{Z}_N) \to L_{2k}(\mathbf{Z} \to \mathbf{Z}_N)$$

is an isomorphism. If this, and Bass's result $L_3(\mathbf{Z}_N) = 0$ are inserted, diagram chasing leads to the isomorphisms

$$\begin{aligned} L_3(\mathbf{Z}_N) &= L_3(\mathbf{Z} \to \mathbf{Z}_N) = LP_3 = LS_3 = LS_0 = 0, \\ L_1(\mathbf{Z}_N) &= L_1(\mathbf{Z} \to \mathbf{Z}_N) = LP_1 = LS_2 = LS_1, \end{aligned}$$

and

$$LP_{2k} = L_{2k+2}(\mathbf{Z}_N).$$

The proof that q is an isomorphism comes by explicit calculation based on (13A.4): we have $L_{2k-2}(\mathbf{Z}_N) \cong L_{2k-2}(1) \oplus \widetilde{L}_{2k-2}(\mathbf{Z}_N)$, the exact sequence shows $L_{2k}(\mathbf{Z} \to \mathbf{Z}_N) \cong L_{2k-2}(1) \oplus \widetilde{L}_{2k}(\mathbf{Z}_N)$, but q does not respect the splitting: roughly, it multiplies the first component by N and the second by $(1+\chi)/(1-\chi)$. By the way, although the proofs of (13A.4) and (13A.5) are not yet fully written up, a first instalment will appear in [W28].

I do not seriously doubt that, for N odd, $L_1(\mathbf{Z}_N) = 0$ and thus $LS_n(\mathbf{Z} \to \mathbf{Z}_N) = 0$ for all n. Observe that this implies that the embedding theorem (11.3) (though not its corollaries) is valid for many codimension 2 embeddings, and fits nicely with the corollary to (14E.7)*.

There have also been several developments concerning the Kervaire-Arf invariant. Browder's work [B22] has been clarified as follows by Ed Brown: see also pp. 9–18 of mimeographed notes on the Conference on Algebraic Topology held at the University of Illinois at Chicago Circle in June, 1968†.

Let $v(M) = \sum_{i \geqslant 0} v_i(M)$ denote the total Wu class of a closed manifold M, characterised by the formula

$$Sq\, x[M] = v(M) x[M]$$

valid for all $x \in H^*(M; \mathbf{Z}_2)$, and also satisfying $Sq\, v = w$, the total Stiefel-Whitney class (see Milnor and Stasheff [M16]; also [B34]). Thus $v_i(M) = 0$ for $\dim M < 2i$; $v_n U = \chi(Sq^n) U$ if U is the Thom class of the normal bundle, and χ the involution of the Steenrod algebra. Also the Adem relations imply that for M orientable, $v_i = 0$ for i odd; and if M is a spin manifold, $v_i = 0$ also for $i \equiv 2 \pmod 4$. If ${w_1}^2 = 0$, then $v_i^2 = 0$ for i odd.

Let $\dim M = 2n$: then $v_{n+1}(M) = 0$. Thus the classifying map for the tangent bundle of M lifts to a map into the universal bundle with this property: call such a lift a v_{n+1}-orientation on M.

* *See Hambleton and Taylor [H9] for a survey of the computations of the L-groups of finite groups.*

† *See also Brown [B31].*

Better, let E_l^n have homotopy groups $\mathbf{Z}_2$ in dimensions l, $l+n$ (and zero elsewhere) and k-invariant $\chi(Sq^{n+1})$. Then for any Poincaré complex X^{2n} with normal fibration ξ^l, the Thom class in $H^l(X^\xi)$ defines a map $U : X^\xi \to K(\mathbf{Z}_2, l)$ which can be lifted to a map $V : X^\xi \to E_l^n$. Now there is a non-split exact sequence

$$0 \to \mathbf{Z}_2 \to \pi_{2n+1}\big(K(\mathbf{Z}_2, n) \wedge E_l^n\big) \to H^{2n}(\mathbf{Z}_2, n; \mathbf{Z}_2) \to 0\,,$$

so we can choose a surjection η of the middle group onto $\mathbf{Z}_4$.

This allows us to define a quadratic map

$$\phi : H^n(X; \mathbf{Z}_2) \to \mathbf{Z}_4$$

with associated bilinear map the cup product. Let $\alpha \in \pi_{2n+1}(X^\xi)$ have degree 1. Define, for $x \in H^n(X; \mathbf{Z}_2)$, $\phi(x)$ to be the image of α under

$$\pi_{2n+1}(X^\xi) \xrightarrow{\Delta_*} \pi_{2n+1}(X^+ \wedge X^\xi) \xrightarrow{(x \wedge V)_*} \pi_{2n+1}\big(K(\mathbf{Z}_2, n) \wedge E_l^n\big) \xrightarrow{\eta} \mathbf{Z}_4\,.$$

The Witt group of nonsingular quadratic maps such as ϕ has order 8: thus we obtain an invariant*

$$K \;=\; K(X, \xi, \alpha, V) \in \mathbf{Z}_8\,.$$

Since (by [W21]) the pair (ξ, α) is unique up to stable isomorphism, this can be regarded as depending only on V.

Any other lift V' differs from V by a map of X^ξ into the fibre $K(\mathbf{Z}_2, l+n)$, corresponding to a class $z \in H^n(X; \mathbf{Z}_2)$. Then $\phi'(x) = \phi(x) + j(x \cdot z)$, where $j : \mathbf{Z}_2 \subset \mathbf{Z}_4$. The case when V is fixed but α changed can be reduced to this: Brown obtains a formula $z = \sum z_i w_{n-2^i+1}(\xi)$ for suitable z_i, leading to some results on independence of choice of α. There is, of course, also the choice of η, but this is universal.

For X a manifold, we can use a v_{n+1}-orientation to fix (ξ, α, V) and hence $K(X)$. If now $Y \xrightarrow{f} X$ is a normal map, and we use the induced v_{n+1}-orientation on Y, we have

$$4c(f) \;=\; K(Y) - K(X)\,(\mathrm{mod}\ 8)\,.$$

There is also some new work on the related characteristic classes. The following was told to me (in different notation) by Brumfiel, and ascribed to Ib Madsen.

Write $k^1 \in H^*(G/PL; \mathbf{Z}_2)$ for Sullivan's class, which is characterised by the identity

$$c(M, f) \;=\; w(M) f^*(k^1)[M]$$

valid for even-dimensional $\mathbf{Z}_2$-manifolds: k^1 only has components in dimensions $4n+2$. Then write k^2 for the class of (13B.5): this satisfies the identity for all even-dimensional manifolds, has components in dimensions $2n$, and is given by

* *See Weiss [W42] for non-simply connected generalisations of this invariant. See Brown [B32] for a survey of the Arf invariant problem.*

$$k^2 = (1 + Sq^2 + Sq^2Sq^2)k^1 \tag{13B.5}$$

Next set $k^3 = (1 + Sq^1)k^2$: I claim this satisfies the identity for all manifolds. For this amounts to proving

$$(m\, w_{m-2i}k_{2i} + w_{m-2i-1}Sq^1k_{2i})[M^m] = 0$$

for any manifold M and even-dimensional class k_{2i}; this follows from the calculation

$$\begin{aligned} w_1w_{m-2i-1}k_{2i}[M] &= Sq^1(w_{m-2i-1}k_{2i})[M] \\ &= (m\, w_{m-2i}k_{2i} + w_{m-2i-1}w_1k_{2i} + w_{m-2i-1}Sq^1k_{2i})[M]. \end{aligned}$$

Now set $k^4 = \chi(Sq)k^3 = (Sq)^{-1}k^3$. Then for any M, f,

$$\begin{aligned} c(M,f) &= w(M)f^*(k^3)[M] \\ &= Sq\, v(M)f^*(Sq\, k^4)[M] \\ &= Sq\big(v(M)f^*k_4\big)[M] = v^2(M)f^*k_4[M], \end{aligned}$$

giving a new sort of formula. If we write $Sq^{r*} = \sum_{i\geqslant 0} Sq^{ri}$, so $Sq^{1*} = Sq$, then $k^2 = (1 + Sq^1)k^3 = (1 + Sq^1)(Sq)k^4 = Sq^{2*}k^4$, and since the Adem relations lead quickly to the identity

$$(1 + Sq^2 + Sq^2Sq^2)Sq^{4*} = Sq^{2*},$$

and we deduce $k^1 = Sq^{4*}k^4$. Hence k^4 (like k^1) only has nonzero components in dimensions $4n+2$. Once can also obtain this directly. Arguing as in (13B.5) using the product formula gives a class k^4 giving $c(M, f)$ for all f. Since $\mathbf{Z}_2$-manifolds detect mod 2 homology, for M a $\mathbf{Z}_2$-manifold $c(M, f) = 0$ except when $\dim M \equiv 2 \pmod 4$, and $v^2(M)$ has nonzero components only in dimensions divisible by 4, it follows that k^4 is concentrated in dimensions $4n+2$.

There seems good reason to regard k^4 as the most natural class to choose from all these. This is supported by the conjecture that if $n+1$ is not a power of 2, k^4_{4n+2} is in the image of $H^{4n+2}(BPL; \mathbf{Z}_2)$. This is true for k^4_{10}. It would follow that these are zero in $H^*(G/O; \mathbf{Z}_2)$, so only the $k^4_{2^n-2}$ $(n \geqslant 2)$ would appear in $c(M, f)$ for $f : M \to G/O$ a smooth normal invariant. Suppose M^{8n+2} a spin manifold, thus $v_{4i+2}(M) = 0$. The conjecture implies in this case that

$$c(M, f) = v^2_{4n}(M)f^*(k_2)[M],$$

a formula which has been proved independently by Brumfiel.

Applying this last formula to the special case

$$f : P_{4n+1}(\mathbf{C}) \to G/O$$

shows at once that $c(M, f) = [k_2] = \mathbf{s}_2 = \mathbf{s}_4 \pmod 2$, in the notation of §14C, and thus can be computed using Pontrjagin classes. For this and further work in this direction see Brumfiel [B41].

17F. Sullivan's Results

As a final new development in techniques of calculation, but one which deserves a special place to itself, I refer to Dennis Sullivan's notes [S23] on localisation, periodicity and Galois symmetry. I would like to describe very briefly the underlying ideas. Given the ring $\mathbf{Z}$ of integers, one may localise at a prime p to obtain the ring of rational numbers with denominator prime to p; one may also (or next) complete this to obtain the ring $\mathbf{Z}_{(p)}$ of p-adic integers. One can also take the profinite completion $\widehat{\mathbf{Z}}$ which gives the product over all p of $\mathbf{Z}_{(p)}$: in modern number-theoretic terminology, this is the ring of integral (finite) adeles. We also have the quotient field $\mathbf{Q}$ and the tensor product $\mathbf{Q}\widehat{\mathbf{Z}}$: the ring of finite adeles. The 'arithmetic square'

$$\begin{array}{ccc} \mathbf{Z} & \longrightarrow & \widehat{\mathbf{Z}} \\ \downarrow & & \downarrow \\ \mathbf{Q} & \longrightarrow & \mathbf{Q}\widehat{\mathbf{Z}} \end{array}$$

is then a pullback.

Sullivan shows how to localise and complete the homotopy type of a CW complex X, obtaining a corresponding pullback square

$$\begin{array}{ccc} X & \longrightarrow & \widehat{X} \\ \downarrow & & \downarrow \\ X_0 & \longrightarrow & X_A \end{array}$$

(X must be 1-connected, or at least simple in some sense); the homology (resp. homotopy) groups of $\widehat{X}$, X_0, X_A are the tensor products of those of X with $\widehat{\mathbf{Z}}$, $\mathbf{Q}$, $\mathbf{Q}\widehat{\mathbf{Z}}$ (provided e.g. X has finite skeletons); and X is determined up to homotopy by the other spaces and maps. X_0 is, of course, familiar to homotopy theorists, but $\widehat{X}$ is obtained using a form of Brown's representability theorem, and (for any K), $[K, \widehat{X}]$ comes equipped with a topology.

Next, Sullivan applies these concepts to spherical fibrations, and obtains the following basic result. There is a natural transformation 'fibrewise profinite completion' which corresponds to a map of classifying spaces $BG_n \to \widehat{B}G_n$. The induced map of fundamental groups is $\{\pm 1\} = \mathbf{Z}^\times \to \widehat{\mathbf{Z}}^\times$ (here $R^\times$, for R a ring, denotes the multiplicative group of units), and the induced map of universal covers is profinite completion. Note that BSG is already profinitely complete (its homotopy groups are finite). Note also the action of $\widehat{\mathbf{Z}}^\times$ on $\widehat{B}SG_n$ and BSG, and that (as one can see by completing one prime at a time) $\widehat{\mathbf{Z}}^\times$ acts via $\mathbf{Z}_{(p)}^\times$ on the p-part of BSG_n. Stabilising gives a homotopy equivalence $K(\widehat{\mathbf{Z}}^\times, 1) \times BSG \to \widehat{B}G$: the action of $\widehat{\mathbf{Z}}^\times$ on BSG is (homotopically) trivial. Since the rational homotopy theory is easy (cf. [Q1]), the arithmetic square now gives a criterion for fitting spherical fibrations 'at different primes'.

The real force of these ideas comes in via using étale homotopy theory to define a functor from normal algebraic varieties (over subfields of $\mathbf{C}$) to profinite homotopy types, which on, say, nonsingular varieties just gives the profinite completion of the space of points defined over $\mathbf{C}$. Naturality of the construction shows that for a variety defined over $\mathbf{Q}$, such as a unitary group or Grassmannian, the Galois group of the algebraic closure $\overline{\mathbf{Q}}$ acts on the space. The induced action on cohomology is not too hard to compute for projective spaces and others related to them: its commutator quotient is identifiable (by class field theory) with $\widehat{\mathbf{Z}}^\times$, and this acts on H^2 of projective spaces by multiplication. Since profinite K-theory is classified by a limit of Grassmannians, the Galois group of $\overline{\mathbf{Q}}/\mathbf{Q}$ acts on each $\widehat{K(X)}$ too; Sullivan shows that this action also reduces to one of $\widehat{\mathbf{Z}}^\times$, and that for k prime to p the action by k on th p-adic part of $\widehat{K(X)}$ is given by the Adams operation ψ^k. The 'Adams conjecture' now follows by a stabilisation argument like the above. Sullivan obtains many other interesting results and problems in the framework of his étale homotopy theory, but it would be out of place to give much discussion here.

In a final century of pages, Sullivan collects his ideas to give a complete determination at odd primes of the homotopy types of the maps in the diagram

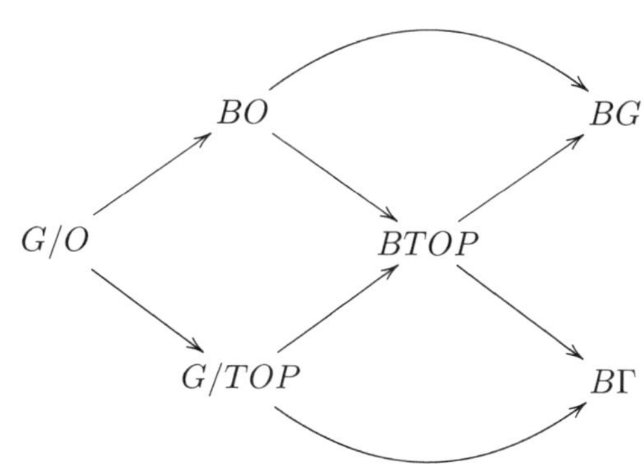

with $\Gamma = TOP/O$. First, using the ideas of [S22] in a more organised fashion, he defines what I call the *Sullivan orientation* $\Delta \in K^*(MTOP) \otimes \mathbf{Z}[\frac{1}{2}]$, where $MTOP$ denotes the Thom space of the universal bundle over $BTOP$ (essentially the same as BPL). The construction of Δ for a topological bundle $\pi : E \to X$

rests on a diagram

$$\begin{array}{ccccc}
\Omega_*(E;\mathbf{Q}) & \longrightarrow & \Omega_*^{TOP}(X;\mathbf{Q}) & \longrightarrow & \mathbf{Q} \\
\downarrow & & \downarrow & & \downarrow \\
\Omega_*(E;\mathbf{Q}/\mathbf{Z}) & \longrightarrow & \Omega_*^{TOP}(X;\mathbf{Q}/\mathbf{Z}) & \longrightarrow & \mathbf{Q}/\mathbf{Z}
\end{array}$$

where the first horizontal map represents the intersection (transverse) with the zero cross-section; the second, taking the signature. Further, Sullivan (see [S23, p. 6.47]) shows that $BTOP$ localised away from 2 classifies the *universal* $K^* \otimes \mathbf{Z}[\frac{1}{2}]$-oriented spherical fibration. This is true even unstably, provided block bundles are used, and the fibre dimension is $\geqslant 3$ (see [R17], [R19] as well as [S23]).

We now have an action of $\widehat{\mathbf{Z}}^\times$ on $[X, \widehat{BTOP}]$ (using this universality) via its action on K-theory, and hence on the K-class Δ; in this terminology, the maps

$$\widehat{BO} \to \widehat{BTOP} \to \widehat{BG} \quad \text{(ignore the 2-primary part)}$$

are $\widehat{\mathbf{Z}}^\times$-equivariant (recall we have the Galois action on BO and the trivial action on $\widehat{BG}$).

Using Δ in the traditional way one obtains a (Bott) characteristic class Θ_E for a topological bundle $\pi : E \to X$. The map

$$\Theta_E : \widehat{\mathbf{Z}}^\times \to \big(\widehat{K}(X)\big)^\times$$

is defined using the action of $\widehat{\mathbf{Z}}^\times$ on $\widehat{K}$ by

$$\Theta_E(\alpha) \cdot \Delta_E \;=\; \Delta_E^\alpha .$$

It has the following properties:

(i) It is a product of functions $\mathbf{Z}_{(p)}^\times \to \big(K(X) \otimes \mathbf{Z}_{(p)}\big)^\times$ over (odd) primes p,

(ii) It is a cocycle: $\Theta_E(\alpha\beta) = \Theta_E(\alpha)^\beta \Theta_E(\beta)$,

(iii) Θ_E is continuous,

(iv) $\Theta_{E\oplus F} = \Theta_E \cdot \Theta_F$,

(v) If k is prime to p, then for an SO_2- (or U_1-) bundle η, the p-component of $\Theta_k(\eta)$ is

$$\frac{1}{k}\left(\frac{\eta^k - \eta^{-k}}{\eta - \eta^{1}}\right)\left(\frac{\eta^k + \eta^{-k}}{\eta + \eta^{-1}}\right).$$

The multiplicative group of functions satisfying (0), (i), (ii) is denoted by $Z_d^1(X) = Z_d^1(\widehat{\mathbf{Z}}^\times; \widehat{K}^\times(X))$. We have a 'coboundary' homomorphism

$$\delta : \widehat{K}^\times(X) \to Z_d^1(X)$$

defined by $\delta u(\alpha) = u^\alpha/u$: its cokernel is written $H_d^1(\widehat{\mathbf{Z}}^\times; \widehat{K}^\times)$. We also have the invariant given for vector bundles (and computable by (iv)),

$$\Theta : \widehat{K}(X) \to Z_d^1(X)$$

and the pullback $\mathscr{K}(X)$ of (δ, Θ). Define $\mathscr{C}(X)$ to be the subgroup of $[X, BTOP]$ corresponding to bundles with trivial invariant Θ – equivalently, for which the Thom isomorphism (given by Δ) is $\widehat{\mathbf{Z}}^\times$-equivariant.

MAIN THEOREM *There is a natural epimorphism of diagrams*

$$\begin{array}{ccc} [X, \widehat{G/O}] & \longrightarrow & [X, \widehat{BO}] \\ \downarrow & & \downarrow \\ [X, \widehat{G/TOP}] & \longrightarrow & [X, \widehat{BTOP}] \end{array} \quad \text{on} \quad \begin{array}{ccc} \mathscr{K}(X) & \longrightarrow & \widehat{K}(X) \\ \downarrow & & \downarrow (\Theta, 0) \\ \widehat{K}^\times(X) & \xrightarrow{(\delta, 0)} & Z_d^1(X) \oplus \mathscr{C}(X) \end{array}$$

which is an isomorphism on all but $[X, \widehat{G/O}]$.

It follows, for example, that a vector bundle E is topologically trivial iff $\Theta_E = 1$, and fibre homotopy trivial iff Θ_E is cohomologous to 1.

Subtler corollaries are obtained by analysing the maps Θ, δ: let us concentrate on an odd prime p. One can identify the torsion subgroup of $\mathbf{Z}_{(p)}^\times$ with the multiplicative group $\mathbf{Z}_p^\times$ of the field with p elements; any torsion free $\mathbf{Z}_{(p)}^\times$-module M is the direct sum of the submodule M_1 fixed by $\mathbf{Z}_p^\times$ and that, M_ξ, annihilated by the sum of its elements. A corresponding decomposition can be made for $[X, BO]$ etc. (which are *not* torsion free) by using a splitting theorem, or decomposing the universal example. Then δ is an isomorphism on the ξ-summand, Θ is (I believe) an isomorphism on the 1-summand. It should be observed that $\mathbf{Z}_{(p)}^\times$ is topologically cyclic: if $k^{p-1} \not\equiv 1 \pmod{p^2}$, then powers of k are dense in $\mathbf{Z}_{(p)}^\times$, so $\Theta_E(k) \in \widehat{K}(X)$ (at p) determines $\Theta_E \in Z_d^1(X)$ (at p). Further calculation shows that Θ fails to be an isomorphism precisely if p is irregular, or if 2 defines an element of odd order in $\mathbf{Z}_p^\times$.

Perhaps I should emphasise that the above holds in the profinitely completed part at *odd* primes. Also, it is difficult to summarise Sullivan's work so briefly: the full philosophical exposition in [S23] should be read.

See also Sullivan [S24].

17G. Reformulations of the Algebra

The algebra in this work, particularly in §6, is complicated, and even so it is not altogether satisfactory: most obviously, in §8. Some work has been done in attempts to improve the situation.

We have made much use of 'simple hermitian' forms, defined by pairs (λ, μ) satisfying (i)–(v) of (5.2) (with $\chi_N = 0$) and G based so that $A\lambda$ is a simple isomorphism. These axioms (i)–(v) are somewhat complicated, and we have given elsewhere [W27] a better formulation, briefly stated as follows.

Let A be a ring, α an anti-automorphism of A, u a unit of A such that $\alpha(u) = u^{-1}$, and $\alpha^2(x) = uxu^{-1}$ for all $x \in A$ (in practice, usually $u = \pm 1$ and α^2 is the identity). For M a right A-module we define the group $S_\alpha(M)$ of sesquilinear forms

$$\phi : M \times M \to A$$

so that e.g. $\phi(ma, n) = \alpha(a)\phi(m, n)$. For such a ϕ, define $T_u(\phi)$ by

$$T_u(\phi)(m, n) \;=\; \alpha\big(\phi(n, m)\big)u\,.$$

Then $T_u : S_\alpha(M) \to S_\alpha(M)$ satisfies $T_u^2 = 1$, and we define the space of *quadratic forms* by

$$Q_{(\alpha,u)}(M) \;=\; \mathrm{Coker}(T_u - 1)\,.$$

It is shown in [W27] how the notion of quadratic form (with $u = \pm 1$) is equivalent to that of a pair (λ, μ) satisfying (i)–(v). Moreover, the generalisation with arbitrary u covers the case of axioms (Q1)–(Q5) of (12.9). There seems no doubt that this is a better formulation of the concept (also, a better terminology).

There is not yet a viable alternative to the presentation in §6, particularly (6.2). It would be more appropriate, perhaps, to replace $RU(\Lambda)$ by the subgroup $EU(\Lambda)$ generated by $UU(\Lambda)$ and its conjugate by Σ: this is the commutator subgroup both of $U(\Lambda)$ and of itself. The details should be worked in the generality above (this is needed for (12.9)).

Pursuing the analogy with the algebraic K-theory, it would be nice to have a definition of L_2 like Milnor's K_2 with a Steinberg group defined by elementary relations between elementary unitary matrices: one would hope to give the relative group L_2 also an abstract algebraic description. Such a description has in fact been given by Sharpe in his thesis (see [S7], [S8]) but the details are more complicated than one would wish. Consider pairs (P, Q) of η-symmetric quadratic forms over Λ on a free module of dimension r. Write $(P, Q) \sim (P', Q')$

if there is an $r \times r$ matrix X such that

$$\begin{aligned} A &= I + \eta Q X \in SL_r(\Lambda)\,, \\ P' - P &= X + \eta X^* + X^* Q X\,, \\ Q &= A Q' A^*\,. \end{aligned}$$

The set $St\,U_r(\Lambda)$ of equivalence classes maps to $RU_r(\Lambda)$ by

$$(P, Q) \mapsto \begin{pmatrix} I & 0 \\ P & I \end{pmatrix} \begin{pmatrix} 0 & I \\ -\eta I & 0 \end{pmatrix} \begin{pmatrix} I & 0 \\ Q & I \end{pmatrix}.$$

Now given a morphism $\psi : \Lambda \to \Lambda'$ of anti-involuted rings, form the pullback diagram

$$\begin{array}{ccc} \widetilde{F}_r & \longrightarrow & St\,U_r(\Lambda') \\ \downarrow & & \downarrow \\ RU_r(\Lambda) & \longrightarrow & RU_r(\Lambda')\,. \end{array}$$

It is necessary to consider a subgroup of $\widetilde{F}_r$ defined by a rather subtle splitting condition on the diagonal, to stabilise under $\oplus$, and to factor out $SL_r(\Lambda)$ – which acts via $SL_r(\Lambda')$ on $St\,U_r(\Lambda')$ as $(P, Q)^\alpha = (\alpha^{*-1} P \alpha^{-1}, \alpha Q \alpha^*)$ and on $RU_r(\Lambda)$ as $A^\alpha = AH(\alpha)$; as well as to contemplate a rather complex algebraic form of surgery; a certain subquotient of $\widetilde{F}_r$ is eventually identified as $L_{2k}(\psi)$ $\big($where $\eta = (-1)^k\big)$.

My general feeling – which has now been justified several times – is that whatever can be done for abstract K-groups can be done (usually with more trouble) for the L_k. Since one of Whitehead's original definitions for his group was in terms of chain complexes, one is naturally led to seek such formulations here also. Such a theory ought to run as follows.

Define the notion of quadratic (reflexive) form on a free (or based, or just projective) finitely generated chain complex; the pairing has some degree, k. A good definition of ~ 0 should generalise (a) a cobordism and (b) existence of a lagrangian. Forming the universal group of forms modulo null-equivalent ones gives $L_k(\Lambda)$; consideration of a form over Λ, and a null-equivalence of the induced form by $\psi : \Lambda \to \Lambda'$ leads to construction of relative groups. The identification with present definitions runs by an algebraic construction imitating surgery until the chain complex has only 1 or 2 terms left. Most of this programme I can do in outline, but one crucial gap at present is that I cannot show with my definition that cobordism is an equivalence relation.

For simplicity, assume the anti-involution is the identity (hence the ring commutative). Given a chain complex C_*, form its dual C^* and consider the interchange (with appropriate signs) $T : C^* \otimes C^* \to C^* \otimes C^*$. Define

$$Q_n H(C) \;=\; \operatorname{Coker}(T_* - 1) \text{ on } H_n(C^* \otimes C^*)\,.$$

An $x \in Q_nH(C)$ is called a *quadratic form of degree* n on C. For $x \in Q_nH(C)$, 'bilinearisation' gives $T_*x + x \in H_n(C^* \otimes C^*)$; slant product with this induces

$$Ax : H_p(C_*) \to H_{n+p}(C^*)\,;$$

call x *nonsingular* if these maps are isomorphisms. There is a similar natural notion of simple form in the based case, using representatives at the chain level, and a chain map $Ax : C_* \to C^*$ of degree n.

A *cobordism* to zero should consist (at least) of a quintuple (C, D, x, y, f) where C, D are complexes as above, $x \in Q_nH(C)$ is nonsingular, $y \in Q_{n+1}H(D)$ is not (in general), $f : C \to D$ is a chain map given up to chain homotopy, and we have an exact triangle (in the derived category)

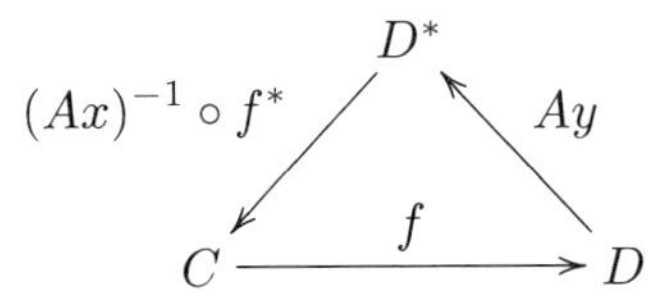

We can then define a cobordism of (C, x) to (C', x') as one of $(C \oplus C', x - x')$ to zero. The trouble with this definition is like that in §8: given two cobordisms (D, y) and (E, z) of (C, x) to zero, it ought to be possible to 'glue' along C to get a nonsingular form on $D \oplus E^*$: but I have only succeeded in getting a symmetric bilinear (not a quadratic) form. (The glueing will not be unique: this does not matter). To justify the picture, think of C as $C_*(\partial M)$, D as $C_*(M)$ – which is dual to $C_*(M, \partial M)$ for a manifold M. Also write the triangle as an exact sequence of chain complexes

$$0 \to D^* \to C \to D \to 0$$

and think of D^* as a lagrangian.

The above theory offers many advantages: quick proofs of (cobordism) exact sequences, including those suggested in §17D relating K- and L-theory as well as those for maps of rings, and a simple and satisfactory algebraic version of the whole setup. I hope it can be made to work.† Note that an independent algebraic treatment gives a payoff in the topology too: topologically motivated results like (12.6), proved in an algebraic setting, can apply more generally, and lead to new results with a different topological application.

The chain complex theory was developed in Ranicki [R4], [R5], [R7], [R9], hopefully providing the 'simple and satisfactory algebraic version of the whole setup' required. The n-dimensional quadratic structures on a chain complex C_* are elements of the $\mathbf{Z}_2$-hyperhomology group

$$Q_n(C) \;=\; H_n(\mathbf{Z}_2; C_* \otimes C_*) \;=\; H_n\big(W \otimes_{\mathbf{Z}[\mathbf{Z}_2]} (C_* \otimes C_*)\big)$$

†It will then be interesting to give a purely algebraic account of the relation between [W18], the argument of §§ 5, 6, and this new approach.

with

$$W \; : \; \cdots \longrightarrow \mathbf{Z}[\mathbf{Z}_2] \xrightarrow{1-T} \mathbf{Z}[\mathbf{Z}_2] \xrightarrow{1+T} \mathbf{Z}[\mathbf{Z}_2] \xrightarrow{1-T} \mathbf{Z}[\mathbf{Z}_2]$$

the standard free $\mathbf{Z}[\mathbf{Z}_2]$*-module resolution of* $\mathbf{Z}$ *with the trivial* $\mathbf{Z}_2$*-action. By definition, an* n*-dimensional quadratic Poincaré complex* (C,ψ) *over a ring with involution* Λ *is an* n*-dimensional f. g. free* Λ*-module chain complex* C *together with an element* $\psi \in Q_n(C)$ *such that there are induced Poincaré duality isomorphisms*

$$(1+T)\psi_0 \; : \; H^{n-*}(C) \;\cong\; H_*(C)\,.$$

The quadratic L-group $L_n(\Lambda)$ *is the cobordism group of* n*-dimensional quadratic Poincaré complexes over* Λ *(with* C *based and* $(1+T)\psi_0 : C^{n-*} \simeq C_*$ *a simple chain equivalence). An* n*-dimensional normal map* $(f,b) : M \to X$ *determines an* n*-dimensional quadratic Poincaré complex* (C,ψ) *with* $C = \mathscr{C}(f^!)$ *the algebraic mapping cone of the Umkehr* $\mathbf{Z}[\pi_1(X)]$*-module chain map*

$$f^! \; : \; C(\widetilde{X}) \simeq C(\widetilde{X})^{n-*} \xrightarrow{\widetilde{f}^*} C(\widetilde{M})^{n-*} \simeq C(\widetilde{M})$$

such that the homology $\mathbf{Z}[\pi_1(X)]$*-modules of* C *are the kernel modules of* §2

$$H_*(C) \;=\; K_*(M) \;=\; \mathrm{Ker}\big(\widetilde{f}_* : H_*(\widetilde{M}) \to H_*(\widetilde{X})\big)\;,$$

with $\widetilde{X}$ *the universal cover of* X *and* $\widetilde{M} = f^*\widetilde{X}$ *the pullback cover of* M*. The surgery obstruction of* (f,b) *is the cobordism class*

$$\theta(f,b) \;=\; (C,\psi) \in L_n\big(\mathbf{Z}[\pi_1(X)]\big)$$

– it is no longer necessary to perform preliminary surgeries below the middle dimension in order to define the surgery obstruction, as was originally done in §§ 5, 6. The element $\psi \in Q_n(C)$ *is obtained from a stable* $\pi_1(X)$*-equivariant map* $F : \Sigma^\infty \widetilde{X}^+ \to \Sigma^\infty \widetilde{M}^+$ *inducing* $f^!$ *by a* $\pi_1(X)$*-equivariant chain level quadratic construction, generalising the homotopy theoretic method used by Browder [B24] to obtain the Kervaire-Milnor [K4] quadratic form in the* $(4k+2)$*-dimensional simply connected case. In particular, for* $n = 2k$ *the surgery obstruction is the Witt class*

$$\theta(f,b) \;=\; (K,\lambda,\mu) \in L_{2k}\big(\mathbf{Z}[\pi_1(X)]\big)$$

of the 'instant surgery obstruction' $(-1)^k$*-hermitian form on the stably f. g. free* $\mathbf{Z}[\pi_1(X)]$*-module*

$$K \;=\; \mathrm{Coker}\left(\begin{pmatrix} d^* & 0 \\ (1+T)\psi_0 & d \end{pmatrix} : C^{k-1} \oplus C_{k+2} \to C^k \oplus C_{k+1}\right)\,,$$

and for $n = 2k+1$ *there is a similar expression for a formation.*

The quadratic chain complex theory is a further development of the symmetric chain complex theory of Mishchenko [M18]. By definition, an n*-dimensional* *symmetric Poincaré complex* (C,ψ) *over a ring with involution* Λ *is an* n*-dimensional f. g. free* Λ*-module chain complex* C *together with an element of the* $\mathbf{Z}_2$*-hypercohomology group*

$$\phi \in Q^n(C) \;=\; H^n(\mathbf{Z}_2; C_* \otimes C_*) \;=\; H_n\big(\mathrm{Hom}_{\mathbf{Z}[\mathbf{Z}_2]}(W, C_* \otimes C_*)\big)$$

such that there are induced Poincaré duality isomorphisms

$$\phi_0 \;:\; H^{n-*}(C) \;\cong\; H_*(C)\,.$$

The symmetric L*-group* $L^n(\Lambda)$ *is the cobordism group of* n*-dimensional symmetric Poincaré complexes over* Λ*. The symmetrisation maps*

$$1+T \;:\; L_n(\Lambda) \to L^n(\Lambda)\;;\; (C,\psi) \mapsto (C,(1+T)\psi)$$

are isomorphisms modulo 8-torsion. The symmetric L*-groups of a group ring* $\mathbf{Z}[\pi]$ *are denoted by* $L^*(\pi)$*, by analogy with* $L_*(\pi)$*. An* n*-dimensional Poincaré complex* X *has a symmetric signature invariant*

$$\sigma^*(X) \;=\; (C(\widetilde{X}),\phi) \in L^n\big(\mathbf{Z}[\pi_1(X)]\big)$$

with

$$\phi_0 \;=\; [X] \cap - \;:\; C(\widetilde{X})^{n-*} \simeq C(\widetilde{X})\,.$$

The surgery obstruction $\theta(f,b) \in L_m\big(\pi_1(X)\big)$ *of a normal map* $(f,b) : M \to X$ *has symmetrisation*

$$(1+T)\theta(f,b) \;=\; \sigma^*(M) - \sigma^*(X) \in L^m\big(\pi_1(X)\big)\,.$$

The standard computation of the simply connected surgery obstruction groups

$$L_n(1) \;=\; \begin{cases} \mathbf{Z}\ \left(\tfrac{1}{8}(\mathit{signature})\right) \\ 0 \\ \mathbf{Z}_2\ (\mathit{Kervaire\text{-}Arf\ invariant}) \\ 0 \end{cases} \quad \mathit{if}\ n \equiv \begin{cases} 0 \\ 1 \\ 2 \\ 3 \end{cases} \pmod 4$$

was extended in [R4] to the simply connected symmetric L*-groups*

$$L^n(1) \;=\; \begin{cases} \mathbf{Z}\ (\mathit{signature}) \\ \mathbf{Z}_2\ (\mathit{de\ Rham\ invariant}) \\ 0 \\ 0 \end{cases} \quad \mathit{if}\ n \equiv \begin{cases} 0 \\ 1 \\ 2 \\ 3 \end{cases} \pmod 4\,.$$

For any rings with involution Λ, Λ' *there are defined products*

$$L_m(\Lambda) \otimes_{\mathbf{Z}} L^n(\Lambda') \;\to\; L_{m+n}(\Lambda \otimes_{\mathbf{Z}} \Lambda')\;;\; (C,\psi) \otimes (D,\phi) \;\to\; (C \otimes D, \psi \otimes \phi)\,.$$

Sullivan's simply connected surgery product formula was generalised in [R5] to the non-simply connected case: the product of an m*-dimensional normal map*

$(f,b) : M \to X$ and an n-dimensional manifold N is an $(m+n)$-dimensional normal map

$$(g,c) \;=\; (f,b)\times 1 \;:\; M\times N \;\to\; X\times N$$

with surgery obstruction the product

$$\begin{aligned}\theta(g,c) &= \theta(f,b)\otimes\sigma^*(N)\\ &\in \mathrm{Im}\Big(L_m(\pi_1(X))\otimes L^n(\pi_1(N)) \to L_{m+n}(\pi_1(X\times N))\Big).\end{aligned}$$

In particular, the 4-periodicity isomorphism of Theorem 9.9 is given algebraically by the product

$$-\otimes\sigma^*\big(P_2(\mathbf{C})\big) \;:\; L_m(\pi) \;\cong\; L_{m+4}(\pi)$$

with the generator $\sigma^*\big(P_2(\mathbf{C})\big) = 1 \in L^4(1) = \mathbf{Z}$.

17H. Rational Surgery

Added in September, 1970 (after the Congress at Nice).

The theory sketched at the end of §17G, with minor modifications, has been successfully done by A. S. Mishchenko, but only in the case when 2 is invertible in Λ; this does, in fact, avoid the main difficulty I had run up against*. This can be found in [M17].

It follows that given any Poincaré complex X, it defines an element in $L_x^B\big(\mathbf{Z}[\frac{1}{2}][\pi_1(X)]\big)$, and that for a normal map $Y \to X$, the image of the surgery obstruction is the difference of the invariants of X and Y. Moreover, this vanishes if the map $Y \to X$ is a homotopy equivalence modulo the class of finite 2-groups. Going further, the image in $L_x^B\big(\mathbf{Q}[\pi_1(X)]\big)$ gives an invariant of the rational homotopy type of X.

I will now give some results which, while weaker than what I sought in the preceding section, seem of some importance, and have more direct geometrical relevance than the above. They were inspired by ideas of Novikov, which I mention at the end.

Denote by Q^8 Milnor's [M10] closed framed 3-connected PL manifold with signature 8, so that the surgery obstruction for $Q^8 \to S^8$ is the generator x of $L_8(1) \cong L_0(1)$. For any manifold M^m with fundamental group π, define the invariant†

$$\psi(M) \in L_m(\pi)$$

to be the surgery obstruction for the product map $M^m \times Q^8 \to M^m \times S^8$.

THEOREM. *For any normal map $f : N \to M$, the surgery obstruction $\theta(f)$ satisfies*

$$8\,\theta(f) \;=\; \psi(N) - \psi(M)\,.$$

Note that the exponent 8 is needed here, since already in the 1-connected case, M may have signature 1, which does not correspond to an element of $L_0(1)$; also, the exponent 8 just suffices to kill the generalised Arf invariant of Brown, also (it now seems very probable) the torsion subgroups of the $L_m(\pi)$, π finite.

** See the notes at the end of §17G.*

† This invariant is the product $\psi(M) = \theta(Q^8 \to S^8) \otimes \sigma^*(M) \in L_{m+8}(\pi) = L_m(\pi)$ of the generator $\theta(Q^8 \to S^8) = 1 \in L_8(1) = \mathbf{Z}$ and the symmetric signature $\sigma^*(M) \in L^m(\pi)$ of Mishchenko [M18] and Ranicki [R5].

For the proof, multiply the normal maps $Q^8 \to S^8$, $f : N^m \to M^m$ to obtain a diagram

$$\begin{array}{ccc} N^m \times Q^8 & \longrightarrow & N^m \times S^8 \\ \downarrow{\scriptstyle f\times 1} & & \downarrow{\scriptstyle f\times 1} \\ M^m \times Q^8 & \longrightarrow & M^m \times S^8 . \end{array}$$

The theorem is now an immediate consequence of the two following assertions.

LEMMA 1.* *The surgery obstruction for* $f \times 1 : N^m \times Q^8 \to M^m \times Q^8$ (*resp. for* $f \times 1 : N^m \times S^8 \to M^m \times S^8$) *is* $8\,\theta(f)$ (*resp.* 0).

LEMMA 2.† *The surgery obstruction for the composite of two normal maps is the sum of their obstructions.*

Proof of Lemma 1. This is essentially the same as (9.9): certainly those methods make the assertion for S^8 clear. We offer a slight variant for Q^8. Let S_i^4 $(1 \leqslant i \leqslant 8)$ be the embedded spheres provided by the construction of Q^8. First suppose $n = 2k$ even; we may suppose we have spheres $S_j^k \subset N$ representing a base of $K_k(N)$. Then as for (9.9), the surgery to kill $K_k(N \times Q^8)$ can be chosen to replace the $S_j^n \times S_i^4$ by spheres. These clearly give a preferred base for $K_{k+4}(N \times Q^8)$: their intersections and self-intersections are obvious for geometrical reasons, so to show our quadratic form is the tensor product of that on $K_k(N)$ with that on $H_4(Q)$ it remains only to show that we have framed immersions in the right regular homotopy classes. But if they were not, only the coefficient of $1 \in \pi$ in the $\mu(\xi)$ would be wrong: but this is correct by Browder's product formula. Note finally that the direct sum of the symmetric bilinear form on $H_4(Q)$ with the matrix (-1) is equivalent to the form whose matrix is diagonal, with 8 $+1$'s and a -1 on the diagonal. So the class in $L_{m+8}(\pi)$ of the tensor product form equals 8 times that of the form on $K_k(N)$.

For n odd, we must argue more or less as in (9.9): however, a slightly weaker result is obtained at once using the natural splitting (§17D)

$$L_{n+1}(\pi \times \mathbf{Z}) \cong L_{n+1}(\pi) \oplus L_n^B(\pi) \ :$$

viz., that the surgery obstruction in $L_n^B(\pi)$, to homotopy equivalence, is as asserted.

Proof of Lemma 2. This is clearly a basic result which should have been treated earlier in this book. Suppose (cf. proof of (9.4)) surgery done so that the normal maps

$$A \xrightarrow{f} B \xrightarrow{g} C$$

* *This is a special case of the surgery product formula of Ranicki [R5], which is stated in the notes at the end of* §*17G.*

† *The surgery obstruction is defined in [R5] for any normal map of Poincaré complexes, and it is proved there that the obstruction for the composite of two normal maps is the sum of their obstructions.*

are already 2-connected. By induction on simplices make g transverse to the 2-skeleton of C, and do surgery on the preimages of the simplices to make them contractible (this is easy using 2-connectivity of g and the condition $\dim B \geqslant 5$). Then do the same for f.

It follows that we can find regular neighbourhoods of these complexes mapped by simple homotopy equivalences: then (cf. §4) so are their boundaries. Deleting these neighbourhoods, we may suppose we have boundary components with fundamental group π mapped by simple homotopy equivalences

$$\partial A \to \partial B \to \partial C\,.$$

But now (by (5.8), (6.5)) f is normally cobordant to a map inducing a simple homotopy equivalence outside a collar neighbourhood $\partial B \times I$, whereas we can suppose g a simple homotopy equivalence on such a neighbourhood. Since the maps now have disjoint supports, additivity follows. We have used this principle several times already: it follows formally from additivity with disjoint unions and naturality for inclusion.

Similar ideas can be used to provide purely geometric proofs of (5.8), (6.5): see [Q2].

If f is a (simple) homotopy equivalence, $\theta(f) = 0$ in $L^B_m(\pi)$ (or $L_m(\pi)$), so $\psi(N) = \psi(M)$. Thus $\psi(M)$ is an invariant of simple homotopy type; its image in $L^B_m(\pi)$ is a homotopy invariant of M.

Next, we can regard $\psi(M)$ as the image of $M \times x$ under the pairing

$$\Omega_*(\pi) \times L_*(1) \to L_*(\pi)\,,$$

and hence ψ as the corresponding map

$$\psi : \Omega_*(\pi) \to L_*(\pi)\,.$$

I assert that ψ is an Ω_*-module map, where Ω_* acts via the signature $\in \mathbf{Z}$ on $L_*(\pi)$. Using the definition, it suffices in fact to check that

$$\Omega_* \times L_*(1) \to L_*(1)$$

defines this pairing. But this follows at once from product formulae for the signature and Arf invariant.

It follows that ψ factors through $\Omega_*(\pi) \otimes_{\Omega_*} \mathbf{Z}$. Now according to Sullivan [S22], we can identify $\Omega_*(\pi) \otimes_{\Omega_*} \mathbf{Z}[\frac{1}{2}] \cong KO_*(K(\pi,1)) \otimes \mathbf{Z}[\frac{1}{2}]$. Thus we have an invariant (one eighth ψ) given by a map

$$\Omega_*(K(\pi,1)) \to KO_*(K(\pi,1)) \otimes \mathbf{Z}[\tfrac{1}{2}] \xrightarrow{\ell'_\pi} L_*(\pi) \otimes \mathbf{Z}[\tfrac{1}{2}]$$

and detecting the image of the surgery obstruction.

If we are prepared to tensor further with $\mathbf{Q}$ (and by the example at the end of §15 this does lose information), we can be even more explicit. For any space X we can identify

$$\Omega_*(X) \otimes_{\Omega_*} \mathbf{Q} \cong H_*(X; \mathbf{Q})\,,$$

where homology is graded by $\mathbf{Z}_4$. Moreover, if $f : M \to X$ represents a bordism class, we can write down the corresponding homology class. If $\ell(M)$ is the Hirzebruch class of M, this is $f_*\big([M] \cap \ell(M)\big)$. For this is clearly a bordism class invariant, and has the top component correct: it thus suffices to check that we have defined an Ω_*-linear map. But for any N, we have $f \circ \pi_1 : M \times N \to X$, and

$$\begin{aligned}(f \circ \pi_1)_*\big([M \times N] \cap \ell(M \times N)\big) \\ &= (f \circ \pi_1)_*\{\big([M] \cap \ell(M)\big) \otimes \big(N \cap \ell(N)\big)\} \\ &= f_*\{\big(M \cap \ell(M)\big) \cdot \sigma(N)\}\,,\end{aligned}$$

since π_{1*} is trivial except on $H_0(N)$.

Hence our map factors through a homomorphism

$$\ell_\pi : H_*(\pi; \mathbf{Q}) \to L_*(\pi) \otimes \mathbf{Q}\,,$$

and $\psi(M) \otimes \mathbf{Q} = 8\,\ell_\pi f_*\big([M] \cap \ell(M)\big)$. Or we can regard ℓ_π as a cohomology class, in $H^*\big(\pi; L_*(\pi) \otimes \mathbf{Q}\big)$; then we have $\psi(M) \otimes \mathbf{Q} = 8\, f^*\ell_\pi \cdot \ell(M)[M]$. We can thus determine (13B.3) rationally: for if $h : N \to M$ is a normal map, corresponding to $g : M \to G/TOP$, then $\ell(N) = h^*\ell(M)h^*g^*\big(1 + \ell(G/TOP)\big)$, so $\theta(M, f, g) \otimes \mathbf{Q} = f^*\ell_\pi g^*(G/TOP)\ell(M)[M]$. In principle, we also get a formula of this type after tensoring with $\mathbf{Z}[\frac{1}{2}]$.

Note that since all our maps, in particular ℓ_π, are geometrically defined, they are natural for finite coverings (i.e. transfer). This gives a more conceptual proof of (15A.2).

I am now ready to discuss the relation with the above Russian results. It seems clear that the image of $\psi(M)$ in $L_m(\mathbf{Z}[\frac{1}{2}][\pi])$ is 8 times Mishchenko's invariant. The most interesting invariant is the *higher signature* σ_π, defined for $f : M \to K(\pi, 1)$ as

$$\sigma_\pi(M) \;=\; f_*\big([M] \cap \ell(M)\big) \in H_*(\pi; \mathbf{Q})\,.$$

Novikov conjectures that this is a homotopy invariant. Since $\ell_\pi(\sigma_\pi)$ is, it would suffice to prove ℓ_π injective: indeed, I think the conjecture is equivalent to ℓ_π being injective. Novikov further conjectures a natural map

$$L_*(\pi) \to H_*(\pi; \mathbf{Q}) \;:$$

one would probably want it left inverse to ℓ_π.

These conjectures seem remote with present techniques: I personally do not expect them to hold. But there are no counterexamples yet: for π finite, it is trivial that ℓ_π is injective, and we can choose a left inverse naturally. For π a poly-$\mathbf{Z}$ group, it follows from §15B that ℓ_π is an isomorphism. More generally, Cappell's splitting theorem shows for an amalgamated free product $A *_C B$ or $A *_C$ that if ℓ_π is an isomorphism for A, B and C then it is for the large group*. But this is not clear for monomorphism.

**See Cappell [C6].*

By working over $\mathbf{Q}$, one can quickly estimate $\mathscr{S}(M)$, and in particular give criteria for it to be finite. For (using $f \mapsto [M] \cap \ell(M) \cdot g^*\ell(G/PL)$) we can identify

$$[M, G/O] \otimes \mathbf{Q} \cong [M, G/TOP] \otimes \mathbf{Q} \cong H_m(M; \mathbf{Q})$$

(recall homology is $\mathbf{Z}_4$-graded here), and $\theta \otimes \mathbf{Q}$ becomes $\ell_\pi \circ f_*$, where $f : M \to K(\pi, 1)$. If, for example, ℓ_π is a rational isomorphism, we can identify

$$\mathscr{S}(M) \otimes \mathbf{Q} \cong H_{m+1}(f; \mathbf{Q}) .$$

For the recent Russian results see particularly [G1], [N8] and references in the latter.

The assembly map, the Novikov conjecture and the Borel conjecture. The assembly map of Quinn [Q4] and Ranicki [R9]

$$A \ : \ H_*(X; \mathbf{L}_\bullet) \to L_*(\pi_1(X))$$

is defined for any space X, with an algebraic surgery exact sequence

$$\cdots \to L_{m+1}(\pi_1(X)) \to \mathscr{S}_{m+1}(X) \to H_m(X; \mathbf{L}_\bullet) \xrightarrow{A} L_m(\pi_1(X)) .$$

See the notes at the end of §10 for the identification of the geometric surgery sequence for an m-dimensional topological manifold X

$$\cdots \to L_{m+1}(\pi_1(X)) \to \mathscr{S}^{TOP}(X) \to \mathscr{T}^{TOP}(X) \xrightarrow{\theta} L_m(\pi_1(X))$$

with the algebraic surgery exact sequence, including

$$\begin{aligned} \mathscr{S}^{TOP}(X) \ &= \ \mathscr{S}_{m+1}(X) \ , \\ \theta \ = \ A \ : \ \mathscr{T}^{TOP}(X) \ &= \ [X, G/TOP] \ = \ H_m(X; \mathbf{L}_\bullet) \to L_m(\pi_1(X)) \ . \end{aligned}$$

Write the assembly map for an Eilenberg-MacLane space $X = K(\pi, 1)$ (not necessarily a manifold) as

$$A_\pi \ = \ A \ : \ H_*(K(\pi, 1); \mathbf{L}_\bullet) \to L_*(\pi) \ .$$

The morphisms ℓ_π, ℓ'_π defined above are determined by A_π, with

$$\begin{aligned} \ell_\pi \ &: \ H_*(\pi; \mathbf{Q}) \ = \ H_*(K(\pi, 1); \mathbf{L}_\bullet) \otimes \mathbf{Q} \xrightarrow{A_\pi \otimes 1} L_*(\pi) \otimes \mathbf{Q} \ , \\ \ell'_\pi \ &: \ KO_*(K(\pi, 1)) \otimes \mathbf{Z}[\tfrac{1}{2}] \ = \ H_*(K(\pi, 1); \mathbf{L}_\bullet) \otimes \mathbf{Z}[\tfrac{1}{2}] \xrightarrow{A_\pi \otimes 1} L_*(\pi) \otimes \mathbf{Z}[\tfrac{1}{2}] \ . \end{aligned}$$

The conjecture of Novikov [N8] on the homotopy invariance of the higher signatures $\sigma_\pi(M) \in H_*(\pi; \mathbf{Q})$ is equivalent to the rational injectivity of A_π ([R9, 24.5]). The conjecture of Borel (originally formulated only for arithmetic groups and isometries, on the basis of the Mostow rigidity theorem) is that an aspherical Poincaré complex $K(\pi, 1)$ is homotopy equivalent to a compact topological manifold M, and that any homotopy equivalence $M \to M'$ of such manifolds is *homotopic to a homeomorphism; this is equivalent to* A_π *being an isomorphism. The main algebraic result of §15C is that this is indeed the case for polycyclic* π*, in particular free abelian* $\pi = n\mathbf{Z}^+$.

The collection of papers edited by Ferry, Ranicki and Rosenberg [F10] gives the 1995 status of the Novikov and Borel conjectures, which have been verified for a large class of geometrically significant infinite groups, using a wide variety of methods. On the other hand, it should be noted that Gromov [G4] proposes a search for counterexamples among pathological groups.

References

[A1] Akiba, T. On the homotopy type of PL_2. *J. Fac. Sci. Tokyo, sec.* 1, **14** (1967), 197–204.

[A2] Anderson, D. W., Brown, E. H. and Peterson, F. P. Spin cobordism. *Bull. Amer. Math. Soc.* **72** (1966), 256–260.

[A3] Armstrong, M., Cooke, G.E. and Rourke, C. P. "The Princeton notes on the Hauptvermutung" (1968). *In* "The Hauptvermutung book" [R11], 105–192.

[A4] Armstrong, M. and Rourke, C. P. "Mersions of topological manifolds." Preprint (as Appendix to [R19]), Warwick University, 1969.

[A5] Armstrong, M. and Zeeman, E. C. Piecewise linear transversality. *Bull. Amer. Math. Soc.* **73** (1967), 184–188.

[A6] Atiyah, M. F. Bordism and cobordism. *Proc. Camb. Phil. Soc.* **57** (1961), 200–208.

[A7] Atiyah, M. F. and Bott, R. A Lefschetz fixed point formula for elliptic complexes: II. Applications. *Ann. of Math.* **88** (1968), 451–491.

[A8] Atiyah, M. F. and Singer, I. M. The index of elliptic operators III. *Ann. of Math.* **87** (1968), 546–604.

[A9] Auslander, L. and Schenkman, E. Free groups, Hirsch-Plotkin radicals and applications to geometry. *Proc. Amer. Math. Soc.* **16** (1965), 784–788.

[B1] Bak, A. On modules with quadratic forms. *In* "Algebraic K-theory and its geometric applications." Lecture Notes in Mathematics **108**, Springer, 1969, 55–66.

[B2] Bak, A. Odd dimension surgery groups of odd torsion groups vanish. *Topology* **14** (1975), 367–374.

[B3] Bak, A. "K-theory of forms" Ann. Math. Stud. **98**, Princeton, 1981.

[B4] Barden, D. "h-cobordisms between 4-manifolds." Notes, Cambridge University, 1964.

[B5] Bass, H. K-theory and stable algebra. *Publ. Math. I.H.E.S.* **22** (1964), 5–60.

[B6] Bass, H. "Topics in algebraic K-theory." Tata Institute, Bombay, 1967.

[B7] Bass, H. "Algebraic K-theory." W. A. Benjamin Inc., 1968.

[B8] Bass, H. L_3 of finite abelian groups. *Ann. of Math.* **99** (1974), 118–153.

[B9] Bass, H., Heller, A. and Swan, R. G. The Whitehead group of a polynomial extension. *Publ. Math. I.H.E.S.* **22** (1964), 61–79.

[B10] Berstein, I. "A proof of the vanishing of the simply connected surgery obstruction in the odd-dimensional case." Preprint, Cornell University, 1969.

[B11] Berstein, I. and Livesay, G. R. "Non-unique desuspension of involutions." Preprint, Cornell University, 1968.

[B12] Boardman, J. M. and Vogt, R. M. Homotopy everything H-spaces. *Bull. Amer. Math. Soc.* **74** (1968), 1117–1122.

[B13] Borel, A. and Harish-Chandra. Arithmetic subgroups of algebraic groups. *Ann. of Math.* **75** (1962), 485–535.

[B14] Bowers, J. F. On composition series of polycyclic groups. *J. London Math. Soc.* **35** (1960), 433–444.

[B15] Browder, W. Homotopy type of differentiable manifolds. *In* "Colloq. on Alg. Top." Notes, Aarhus, 1962, 42–46, and *in* "Proc. 1993 Oberwolfach Conference on the Novikov conjectures, Rigidity and Index Theorems, Vol. 1." London Math. Soc. Lecture Notes **226**, Cambridge, 1995, 97–100.

[B16] Browder, W. "Cap products and Poincaré duality." Notes, Cambridge University, 1964.

[B17] Browder, W. Structures on $M \times \mathbf{R}$. *Proc. Camb. Phil. Soc.* **61** (1965), 337–345.

[B18] Browder, W. Embedding 1-connected manifolds. *Bull. Amer. Math. Soc.* **72** (1966), 225–231 and 736.

[B19] Browder, W. Manifolds with $\pi_1 = \mathbf{Z}$. *Bull. Amer. Math. Soc.* **72** (1966), 238–244.

[B20] Browder, W. Embedding smooth manifolds. *In* "Proc. I.C.M. (Moscow, 1966)." Mir, 1968, 712–719.

[B21] Browder, W. Surgery and the theory of differentiable transformation groups. *In* "Proc. Conference on Transformation Groups (New Orleans, 1967)." Springer, 1968, 1–46.

[B22] Browder, W. The Kervaire invariant of framed manifolds and its generalisation. *Ann. of Math.* **90** (1969), 157–186.

[B23] Browder, W. Free $\mathbf{Z}_p$-actions on homotopy spheres. *In* "Topology of Manifolds. Proc. 1969 Georgia Conference." Markham Press, 1970, 217–226.

[B24] Browder, W. "Surgery on simply connected manifolds." Ergebnisse der Mathematik und ihrer Grenzgebiete **65**, Springer, 1972.

[B25] Browder, W. Poincaré spaces, their normal fibrations, and surgery. *Invent. Math.* **17** (1972), 191–202.

[B26] Browder, W. and Hirsch, M. W. Surgery on piecewise linear manifolds and applications. *Bull. Amer. Math. Soc.* **72** (1966), 959–964.

[B27] Browder, W. and Levine, J. Fibering manifolds over a circle. *Comm. Math. Helv.* **40** (1966), 153–160.

[B28] Browder, W., Liulevicius, A. and Peterson, F. P. Cobordism theories. *Ann. of Math.* **84** (1966), 91–101.

[B29] Browder, W. and Livesay, G. R. Fixed point free involutions on homotopy spheres. *Bull. Amer. Math. Soc.* **73** (1967), 242–245. Also *Tohôku J. Math.* **25** (1973), 69–88.

[B30] Browder, W., Petrie, T. and Wall, C. T. C. The classification of free actions of cyclic groups of odd order on homotopy spheres. *Bull. Amer. Math. Soc.* **77** (1971), 455–459.

[B31] Brown, E. H. Generalizations of the Kervaire invariant. *Ann. of Math.* **95** (1972), 368–383.

[B32] Brown, E. H. The Arf invariant problem. *In* [C7].

[B33] Brown, E. H., Anderson, D. W. and Peterson, F. P. The structure of the spin cobordism ring. *Ann. of Math.* **86** (1967), 271–298.

[B34] Brown, E. H. and Peterson, F. P. Relations among characteristic classes I. *Topology* **3** supplement 1 (1964), 39–52.

[B35] Brown, E. H. and Peterson, F. P. The Kervaire invariant of $(8k+2)$-manifolds. *Bull. Amer. Math. Soc.* **71** (1965), 190–193 and *Amer. J. Math.* **88** (1966), 815–826.

[B36] Brown, E. H. and Peterson, F. P. SU-cobordism, KO-characteristic numbers, and the Kervaire invariant. *Ann. of Math.* **83** (1966), 54–67.

[B37] Brown, M. Locally flat embeddings of topological manifolds. *Ann. of Math.* **75** (1962), 331–341.

[B38] Brown, R. Groupoids and van Kampen's theorem. *Proc. London Math. Soc.* **17** (1967), 385–401.

[B39] Brown, R. "Algebraic Topology." McGraw-Hill, 1968.

[B40] Brumfiel, G. "Differentiable S^1-actions on homotopy spheres." Preprint, Princeton University, 1969.

[B41] Brumfiel, G. Homotopy equivalences of almost smooth manifolds. *In* "Proc. Sympos. in Pure Math. 22. (Algebraic topology, Madison, Wisconsin, 1970)." Amer. Math. Soc., 1971, 73–79.

[C1] Cappell, S. A splitting theorem for manifolds and surgery groups. *Bull. Amer. Math. Soc.* **77** (1971), 281–286.

[C2] Cappell, S. Splitting obstructions for hermitian forms and manifolds with $\mathbf{Z}_2 \subset \pi_1$. *Bull. Amer. Math. Soc.* **79** (1973), 909–913.

[C3] Cappell, S. On connected sums of manifolds. *Topology* **13** (1974), 395–400.

[C4] Cappell, S. Unitary nilpotent groups and hermitian K-theory. *Bull. Amer. Math. Soc.* **80** (1974), 1117–1122.

[C5] Cappell, S. A splitting theorem for manifolds. *Invent. Math.* **33** (1976), 69–170.

[C6] Cappell, S. On homotopy invariance of higher signatures. *Invent. Math.* **33** (1976), 171–179.

[C7] Cappell, S., Ranicki, A. A. and Rosenberg, J. (eds.) "Surveys on surgery theory." Ann. Math. Stud., Princeton (to appear).

[C8] Cappell, S. and Shaneson, J. On four dimensional surgery and applications. *Comm. Math. Helv.* **46** (1971), 500–528.

[C9] Cappell, S. and Shaneson, J. The codimension two placement problem, and homology equivalent manifolds. *Ann. of Math.* **99** (1974), 277–348.

[C10] Casson, A. J. Fibrations over spheres. *Topology* **6** (1967), 489–499.

[C11] Casson, A. J. Generalisations and applications of block bundles, Trinity College, Cambridge fellowship dissertation, 1967. *In* "Thc Hauptvermutung book" [R11], 33–68.

[C12] CERF, J. Topologie de certains espaces de plongements. *Bull. Soc. Math. France* **89** (1961), 227–380.

[C13] CERF, J. La nullité de $\pi_0(\text{Diff } S^3)$. *Sém. H. Cartan*, 1962/63, Nos. 9, 10, 20, 21. Also, "Sur les difféomorphismes de la sphère de dimension trois ($\Gamma_4 = 0$)." Lecture Notes in Mathematics **53**, Springer, 1968.

[C14] DE CHRIST, S. DE NEYMET. Some relations in Whitehead torsion. (In Spanish). *Bol. Soc. Mat. Mexicana* **12** (1967), 55–70.

[C15] CONNER, P. E. A bordism theory for actions of an abelian group. *Bull. Amer. Math. Soc.* **69** (1963), 244–247.

[C16] CONNER, P. E. "Seminar on Periodic Maps." Lecture Notes in Mathematics **46**, Springer, 1967.

[C17] CONNER, P. E. and FLOYD, E. E. "Differentiable Periodic Maps." Springer, 1964.

[C18] CONNER, P. E. and FLOYD, E. E. Maps of odd period. *Ann. of Math.* **84** (1966), 132–156.

[D1] DAVIS, M. Poincaré duality groups. *In* [C7].

[D2] DONALDSON, S.K. and KRONHEIMER, P.B. "The geometry of four-manifolds." Oxford, 1990.

[D3] DOUADY, A. Variétés a bord anguleux et voisinages tubulaires. *Sém. H. Cartan* 1961/62, No. 1.

[D4] DOVERMANN, K.-H. and SCHULTZ, R. "Equivariant surgery theories and their periodicity properties." Lecture Notes in Mathematics **1443**, Springer, 1990.

[D5] DRESS, A. Induction and structure theorems for orthogonal representations of finite groups. *Ann. of Math.* **102** (1975), 291–325.

[E1] EDWARDS, R. H. and KIRBY, R. C. Deformations of spaces of embeddings. *Ann. of Math.* **93** (1971), 63–88.

[F1] FADELL, E. Generalized normal bundles for locally flat embeddings. *Trans. Amer. Math. Soc.* **114** (1965), 488–513.

[F2] FARRELL, F. T. "The Obstruction to Fibering a Manifold Over a Circle." Ph.D. thesis, Yale University, 1967. Also, *Bull. Amer. Math. Soc.* **73** (1967), 737–740, and *Indiana Univ. J.* **21** (1971), 315–346.

[F3] FARRELL, F. T. and HSIANG, W.-C. A formula for $K_1 R_\alpha[T]$. *In* "Proc. Symp. in Pure Math. 17 (Categorical algebra)." Amer. Math. Soc., 1970, 192–218.

[F4] FARRELL, F. T. and HSIANG, W.-C. Manifolds with $\pi_1 = G \times_\alpha T$. *Amer. J. Math.* **95** (1973), 813–845.

[F5] FARRELL, F. T. and HSIANG, W.-C. Rational L-groups of Bieberbach groups. *Comm. Math. Helv.* **52** (1977), 89–109.

[F6] FARRELL, F. T. and JONES, L. E. The surgery L-groups of poly-(finite or cyclic) groups. *Invent. Math.* **91** (1988), 559–586.

[F7] FARRELL, F. T. and JONES, L. E. "Classical aspherical manifolds." C. B. M. S. Regional Conference in Mathematics **75**, Amer. Math. Soc., 1990.

[F8] FARRELL, F. T., TAYLOR, L. and WAGONER, J. B. Infinite matrices in algebraic K-theory and topology. *Comm. Math. Helv.* **47** (1972), 474–501.

[F9] FERRY, S. and PEDERSEN, E.K. Epsilon surgery theory. *In* "Proc. 1993 Oberwolfach Conference on the Novikov conjectures, Rigidity and Index Theorems, Vol. 2." London Math. Soc. Lecture Notes **227**, Cambridge, 1995, 167–226.

[F10] FERRY, S., RANICKI, A. A. and ROSENBERG, J. (eds.) "Proc. 1993 Oberwolfach Conference on the Novikov conjectures, Rigidity and Index Theorems, Vols. 1,2." London Math. Soc. Lecture Notes **226**, **227**, Cambridge, 1995.

[F11] FREEDMAN, M. and QUINN, F. S., "The topology of 4-manifolds." Princeton, 1990.

[F12] FRÖHLICH, A. and MCEVETT, A. M. Forms over rings with involution. *J. Algebra* **12** (1969), 79–104.

[G1] GELFAND, I. M. and MISHCHENKO, A. S. Quadratic forms over commutative group rings and K-theory. (In Russian). *Functional Analysis* **3** (1969), 28–33.

[G2] GIFFEN, C. H. Smooth homotopy projective spaces. *Bull. Amer. Math. Soc.* **75** (1969), 509–513.

[G3] GOLO, V. L. Realization of Whitehead torsion and discriminants of bilinear forms. *Sov. Math. Doklady* **9** (1968), 1532–1534.

[G4] GROMOV, M. Geometric reflections on the Novikov conjecture. *In* "Proc. 1993 Oberwolfach Conference on the Novikov conjectures, Rigidity and Index Theorems, Vol. 1." London Math. Soc. Lecture Notes **226**, Cambridge, 1995, 164–173.

[H1] HAEFLIGER, A. Differentiable embeddings. *Bull. Amer. Math. Soc.* **67** (1961), 109–112.

[H2] HAEFLIGER, A. Plongements différentiables de variétés dans variétés. *Comm. Math. Helv.* **36** (1961), 47–82.

[H3] HAEFLIGER, A. and POENARU, V. La classification des immersions combinatoires. *Publ. Math. I.H.E.S.* **23** (1964), 75–91.

[H4] HAEFLIGER, A. Differentiable embeddings of S^n in S^{n+q} for $q > 2$. *Ann. of Math.* **83** (1966), 402–436.

[H5] HAEFLIGER, A. Knotted spheres and related geometric problems. *In* "Proc. I.C.M. (Moscow, 1966)." Mir, 1968, 437–444.

[H6] HAEFLIGER, A. Lissage des immersions I. *Topology* **6** (1967), 221–239.

[H7] HAMBLETON, I. Projective surgery obstructions on closed manifolds. *In* "Algebraic K-theory. Proc. Oberwolfach 1980. Part II." Lecture Notes in Mathematics **967**, Springer, 1982, 101–131.

[H8] HAMBLETON, I., MILGRAM, J., TAYLOR, L. and WILLIAMS, B. Surgery with finite fundamental group. *Proc. Lond. Math. Soc.* **56** (1988), 349–379.

[H9] HAMBLETON, I. and TAYLOR, L. A guide to the calculation of the surgery obstruction groups for finite groups. *In* [C7].

[H10] HAMBLETON, I., TAYLOR, L. and WILLIAMS, B., An introduction to maps between surgery obstruction groups. *In* "Proc. 1982 Arhus Algebraic Topology Conference". Lecture Notes in Mathematics **1051**, Springer, 1984, 49–127.

[H11] HAUSMANN, J.-C. and VOGEL, P. "Geometry on Poincaré spaces." Mathematical Notes **41**, Princeton, 1993.

[H12] HELLER, A. Some exact sequences in algebraic K-theory. *Topology* **3** (1965), 389–408.

[H13] HIGGINS, P. J. Presentations of groupoids, with applications to groups. *Proc. Camb. Phil. Soc.* **60** (1964), 7–20.

[H14] HIGMAN, G. The units of group rings. *Proc. Lond. Math. Soc.* **46** (1940), 231–248.

[H15] HIRSCH, M. W. Immersions of manifolds. *Trans. Amer. Math. Soc* **93** (1959), 242–276.

[H16] HIRSCH, M. W. On non-linear cell bundles. *Ann. of Math.* **84** (1966), 373–385.

[H17] HIRSCH, M. W. and MILNOR, J. W. Some curious involutions of spheres. *Bull. Amer. Math. Soc.* **70** (1964), 372–377.

[H18] HIRZEBRUCH, F. "Topological methods in algebraic geometry" (3rd edition). Springer, 1966.

[H19] HIRZEBRUCH, F. Involutionen auf Mannigfaltigkeiten, *In* "Proc. Conference on Transformation Groups (New Orleans, 1967)." Springer, 1968, 148–166.

[H20] HIRZEBRUCH, F. and JÄNICH, K. Involutions and singularities. *In* "Proc. Int. Colloq. on Alg. Geom (Bombay 1968)." Oxford, 1969, 219–240.

[H21] HSIANG, W.-C. A note on free differentiable actions of S^1 and S^3 on homotopy spheres. *Ann. of Math.* **83** (1966), 266–272.

[H22] HSIANG, W.-C. A splitting theorem and the Künneth formula in algebraic K-theory. *In* "Algebraic K-theory and its Geometric Applications." Lecture Notes in Mathematics **108**, Springer, 1969, 72–77.

[H23] HSIANG, W.-C. and SHANESON, J. Fake tori, the annulus conjecture and the conjectures of Kirby. *Proc. Nat. Acad. Sci.* **62** (1969), 687–691.

[H24] HSIANG, W.-C. and SHANESON, J. Fake tori. *In* "Topology of Manifolds. Proc. 1969 Georgia Conference." Markham Press, 1970, 18–51.

[H25] HSIANG, W.-C. and SHARPE, R. W. Parametrized surgery and isotopy. *Pac. J. Maths.* **67** (1976), 401–459.

[H26] HSIANG, W.-C. and WALL, C. T. C. On homotopy tori II. *Bull. London Math. Soc.* **1** (1969), 341–342.

[H27] HUDSON, J. F. P. Concordance and isotopy of PL embeddings. *Bull. Amer. Math. Soc.* **72** (1966), 534–535. Also, *Ann. of Math.* **91** (1970), 425–448.

[H28] HUDSON, J. F. P. Embedding bounded manifolds. *Proc. Camb. Phil. Soc.* **72** (1972), 11–20.

[H29] HUDSON, J. F. P. and ZEEMAN, E. C. On regular neighbourhoods. *Proc. London Math. Soc.* **14** (1964), 719–745.

[I1] IRWIN, M. C. Embeddings of piecewise linear manifolds. *Ann. of Math.* **82** (1965), 1–14. (See also *Bull. Amer. Math. Soc.* **68** (1962), 25–27).

[J1] JONES, L. E. Patch spaces: a geometric representation for Poincaré spaces. *Ann. of Math.* **97** (1973), 276–306; corrigendum *ibid.* **102** (1975), 183–185.

[K1] KERVAIRE, M. A. "La méthode de Pontrjagin pour la classification des applications sur une sphère." Notes, CIME, Rome University, 1962.

[K2] KERVAIRE, M. A. Les noeuds de dimension supérieure. *Bull. Soc. Math. France* **93** (1965), 225–271.

[K3] KERVAIRE, M. A. La théorème de Barden-Mazur-Stallings. *Comm. Math. Helv.* **40** (1965), 31–42.

[K4] KERVAIRE, M. A. and MILNOR, J. W. On 2-spheres in 4-manifolds. *Proc. Nat. Acad. Sci.* **47** (1961), 1651–1657.

[K5] KERVAIRE, M. A. and MILNOR, J. W. Groups of homotopy spheres I. *Ann. of Math.* **77** (1963), 504–537.

[K6] KIRBY, R. C. "Lectures on triangulation of manifolds." Notes, U.C.L.A., 1969.

[K7] KIRBY, R. C. Stable homeomorphisms and the annulus conjecture. *Ann. of Math.* **89** (1969), 575–582.

[K8] KIRBY, R. C. Locally flat codimension 2 submanifolds have normal bundles. *In* "Topology of Manifolds. Proc. 1969 Georgia Conference." Markham Press, 1970, 416–423.

[K9] KIRBY, R. C. and SIEBENMANN, L. C. On the triangulation of manifolds and the Hauptvermutung. *Bull. Amer. Math. Soc.* **75** (1969), 742–749.

[K10] KIRBY, R. C. and SIEBENMANN, L. C. Foundations of topology. *Notices Amer. Math. Soc.* **16** (1969), 848.

[K11] KIRBY, R. C. and SIEBENMANN, L. C. "Foundational essays on topological manifolds, smoothings, and triangulations." Ann. Math. Stud. **88**, Princeton University Press (1977)

[K12] KIRBY, R. C. and SIEBENMANN, L. C. Deformation of smooth and piecewise linear manifold structures. Essay I. of [K11].

[K13] KIRBY, R. C. and TAYLOR, L. A survey of 4-manifolds through the eyes of surgery. *In* [C7].

[K14] KLEIN, J. Poincaré duality spaces. *In* [C7].

[K15] KNESER, M. Hasse principle for H^1 of simply connected groups. *In* "Proc. Symp. in Pure Math. 9 (Algebraic Groups and Discontinuous Sub-groups)." Amer. Math. Soc., 1966, 187–196.

[K16] KUIPER, N. H. and LASHOF, R. K. Microbundles and bundles I: Elementary theory. *Invent. Math.* **1** (1966), 1–17.

[K17] KWUN, K. W. and SZCZARBA, R. Product and sum theorems for Whitehead torsion. *Ann. of Math.* **82** (1965), 183–190.

[L1] LANDHERR, W. Äquivalenz Hermitescher Formen über einen beliebigen algebraischen Zahlkörper. *Abh. Math. Sem. Hamb.* **11** (1936), 245–248.

[L2] LASHOF, R. K. Poincaré duality and cobordism. *Trans. Amer. Math. Soc.* **109** (1963), 257–277.

[L3] LASHOF, R. K. and ROTHENBERG, M. Microbundles and smoothing. *Topology* **3** (1965), 357–388.

[L4] LASHOF, R. K. and ROTHENBERG, M. On the Hauptvermutung, triangulation of manifolds, and h-cobordism. *Bull. Amer. Math. Soc.* **72** (1966), 1040–1043.

[L5] LEE, R. "Unlinking spheres in codimension 2 and their applications." Lecture notes, Institute for Advanced Study, 1968.

[L6] LEE, R. "Splitting a manifold into two parts." Preprint, Institute for Advanced Study, 1968.

[L7] LEE, R. "Bott periodicity of the Wall obstruction groups and the unlinking problem." Preprint, Institute for Advanced Study, 1969.

[L8] LEE, R. Piecewise linear classification of some free $\mathbf{Z}_p$-actions on S^{4k+3}. *Mich. Math. J.* **17** (1970), 149–160.

[L9] LEE, R. Computation of Wall groups. *Topology* **10** (1971), 149–176.

[L10] LEE, R. and ORLIK, P. "On a codimension 1 embedding problem." Preprint, Institute for Advanced Study, 1969.

[L11] LEES, J. A. Immersions and surgeries of topological manifolds. *Bull. Amer. Math. Soc.* **75** (1969), 529–534.

[L12] LEFSCHETZ, S. "Introduction to Topology." Princeton, 1949.

[L13] LEVINE, J. On differentiable embeddings of simply connected manifolds. *Bull. Amer. Math. Soc.* **69** (1963), 806–809.

[L14] LEVINE, J. A classification of differentiable knots. *Ann. of Math.* **82** (1965), 15–50.

[L15] LEVINE, J. and ORR, K. A survey of applications of surgery to knot and link theory. *In* [C7].

[L16] LEVITT, N. On the structure of Poincaré duality spaces. *Topology* **7** (1968), 369–388.

[L17] LEVITT, N. Generalized Thom spectra and transversality for spherical fibrations. *Bull. Amer. Math. Soc.* **76** (1970), 727–731.

[L18] LEVITT, N. Poincaré duality cobordism. *Ann. of Math.* **96** (1972), 211–244.

[L19] LEWIS, D. W. Forms over real algebras and the multisignature of a manifold. *Adv. Math.* **23** (1977), 272–284.

[L20] LÓPEZ DE MEDRANO, S. "Involutions." Ph.D. thesis, Princeton University, 1968.

[L21] LÓPEZ DE MEDRANO, S. "Involutions on manifolds." Ergebnisse der Mathematik und ihrer Grenzgebiete **59**, Springer, 1971.

[L22] LÓPEZ DE MEDRANO, S. Invariant knots and surgery in codimension 2. *In* "Proc. I.C.M. (Nice, 1970), Vol. 2." Gauthier–Villars, 1971, 99–112.

[L23] LÜCK, W. and RANICKI, A. A. Surgery transfer. *In* "Algebraic topology, Göttingen 1987." Lecture Notes in Mathematics **1361**, Springer, 1988, 167–246.

[M1] MADSEN, I. and MILGRAM, J. "The classifying spaces for surgery and cobordism of manifolds." Ann. Math. Stud. **92**, Princeton, 1979.

[M2] MADSEN, I., THOMAS, C. B. and WALL, C. T. C. The topological spherical space-form problem II: Existence of free actions. *Topology* **15** (1976), 375–382.

[M3] MARIN, A. La transversalite topologique. *Ann. of Math.* **106** (1977), 269–293.

[M4] MAZUR, B. Simple neighbourhoods. *Bull. Amer. Math. Soc.* **68** (1962), 87–92.

[M5] MAZUR, B. Differential topology from the point of view of simple homotopy theory. *Publ. Math. I.H.E.S.* **15** (1963), 5–93.

[M6] MAZUR, B. Relative neighbourhoods and the theorems of Smale. *Ann. of Math.* **77** (1963), 232–249.

[M7] MILNOR, J. W. Construction of universal bundles *Ann. of Math.* **63** (1956), I. 272–284, II. 430–436.

[M8] MILNOR, J. W. On simply connected 4-manifolds. *In* "Symp. Internat. de Top. Alg. (Mexico, 1956)." Mexico, 1958, 122–128.

[M9] MILNOR, J. W. Groups which act on S^n without fixed points. *Amer. J. Math.* **79** (1957), 623–630.

[M10] MILNOR, J. W. "Differentiable Manifolds which are Homotopy Spheres." Notes, Princeton University, 1959.

[M11] MILNOR, J. W. "Microbundles and Differentiable Structures." Notes, Princeton University, 1961.

[M12] MILNOR, J. W. A procedure for killing the homotopy groups of differentiable manifolds. *In* "Proc. Symp. in Pure Math. 3 (Differential Geometry)." Amer. Math. Soc., 1961, 39–55.

[M13] MILNOR, J. W. "Lectures on the h-cobordism theorem" (notes by L. Siebenmann and J. Sondow). Mathematical Notes **1**, Princeton, 1965.

[M14] MILNOR, J. W. Whitehead torsion. *Bull. Amer. Math. Soc.* **72** (1966), 358–426.

[M15] MILNOR, J. W. "Introduction to algebraic K-theory." Ann. Math. Stud. **72**, Princeton, 1971.

[M16] MILNOR, J. W. and STASHEFF, J. "Characteristic classes." Ann. Math. Stud. **76**, Princeton, 1974.

[M17] MISHCHENKO, A. S. Homotopy invariants of nonsimply connected manifolds I. Rational invariants. (In Russian). *Izv. Akad. Nauk S.S.S.R. ser. mat.* **34** (1970), 501–514.

[M18] MISHCHENKO, A. S. Homotopy invariants of nonsimply connected manifolds III. Higher signatures. (In Russian). *Izv. Akad. Nauk S.S.S.R. ser. mat.* **35** (1971), 1316–1355.

[M19] MONTGOMERY, D. and YANG, C. T. Free differentiable actions on homotopy spheres. *In* "Proc. Conference on Transformation Groups (New Orleans, 1967)." Springer, 1968, 175–192.

[M20] MORGAN, J. W. and SULLIVAN, D. P. The transversality characteristic class and linking cycles in surgery theory. *Ann. of Math.* **99** (1974), 463–554.

[M21] MORLET, C. Les voisinages tubulaires des variétés semi-linéaires. *Comptes Rendus* **262** (1966), 740–743.

[M22] MUNKRES, J. R. "Elementary differential topology." Princeton, 1963.

[M23] MURANOV, Y. V. The splitting problem. (In Russian). *Tr. Mat. Inst. Steklova* **212** (1996), 123–146.

[N1] NAMIOKA, I. Maps of pairs in homotopy theory. *Proc. London Math. Soc.* **12** (1962), 725–738.

[N2] NEUWIRTH, L. P. The algebraic determination of the genus of a knot. *Amer. J. Math.* **82** (1960), 791–798.

[N3] NEUWIRTH, L. P. A topological classification of certain 3-manifolds. *Bull. Amer. Math. Soc.* **69** (1963), 372–375.

[N4] NOVIKOV, S. P. Diffeomorphisms of simply connected manifolds. *Doklady Akad. Nauk. S.S.S.R.* **143** (1962), 1046–1049 – *Soviet Math. Doklady* **3**, 540–543.

[N5] NOVIKOV, S. P. Homotopy equivalent smooth manifolds I. (In Russian). *Izv. Akad. Nauk S.S.S.R. ser. mat.* **28** (2) (1964), 365–474. Also, Translations Amer. Math. Soc. **48** (1965), 271–396.

[N6] NOVIKOV, S. P. On manifolds with free abelian fundamental group and applications (Pontrjagin classes, smoothings, high–dimensional knots). *Izv. Akad. Nauk SSSR, ser. mat.* **30** (1966), 208–246.

[N7] NOVIKOV, S. P. Pontrjagin classes, the fundamental group, and some problems of stable algebra. *In* "Essays on Topology and Related Topics. Mémoires dédiés à Georges de Rham." Springer, 1970, 147–155. Also Translations Amer. Math. Soc. **70** (1968), 172–179.

[N8] NOVIKOV, S. P. The algebraic construction and properties of hermitian analogues of K-theory for rings with involution from the point of view of Hamiltonian formalism. Some applications to differential topology and the theory of characteristic classes. *Izv. Akad. Nauk. SSSR ser. mat.* **34** (1970), I. 253–238. II. 475–500.

[N9] NOVIKOV, S. P. "Topology I." Encyclopaedia of Mathematical Sciences, Vol. **12**, Springer, 1996.

[O1] ORLIK, P. "Seminar notes on simply connected surgery." Institute for Advanced Study, 1968.

[O2] ORLIK, P. Smooth homotopy lens spaces. *Mich. Math. J.* **16** (1969), 245–255.

[P1] PASSMAN, D. S. and PETRIE, T. Surgery with coefficients in a field. *Ann. of Math.* **95** (1972), 385–405.

[P2] PEDERSEN, E. K. and RANICKI, A. A. Projective surgery theory, *Topology* **19** (1980), 239–254.

[P3] PETRIE, T. The Atiyah-Singer invariant, the Wall groups $L_n(\pi, 1)$ and the function $te^x + 1/te^x - 1$. *Ann. of Math.* **92** (1970), 174–187.

[P4] PETRIE, T. Representation theory, surgery and free actions of finite groups on varieties and homotopy spheres. *In* "The Steenrod Algebra and its Applications (Proc. Conf. to celebrate N. E. Steenrod's sixtieth birthday, Columbus, Ohio, 1970)." Lecture Notes in Mathematics **168**, Springer, 1970, 250–266.

[P5] PETRIE, T. Free metacyclic group actions on homotopy spheres. *Ann. of Math.* **94** (1971), 108–124.

[P6] PRESSMAN, I. S. Functors whose domain is a category of morphisms. *Acta Math.* **118** (1967), 223–249.

[Q1] QUILLEN, D. G. Rational homotopy theory. *Ann. of Math.* **90** (1969), 205–295.

[Q2] QUINN, F. S. "A geometric formulation of surgery." Ph.D. thesis, Princeton University, 1969.

[Q3] QUINN, F. S. "A geometric formulation of surgery." *In* "Topology of Manifolds. Proc. 1969 Georgia Conference." Markham Press, 1970, 500–511.

[Q4] QUINN, F. S. $\widetilde{BTOP}_n$ and the surgery obstruction. *Bull. Amer. Math. Soc.* **77** (1971), 596–600.

[Q5] QUINN, F. S. Surgery on Poincaré and normal spaces. *Bull. Amer. Math. Soc.* **78** (1972), 262–267.

[Q6] QUINN, F. S. Resolutions of homology manifolds, and the topological characterization of manifolds. *Invent. Math.* **72** (1983), 264–284; corrigendum *ibid.* **85** (1986) 653.

[Q7] QUINN, F. S. Topological transversality holds in all dimensions. *Bull. Amer. Math. Soc.* **18** (1988), 145–148.

[Q8] QUINN, F. S. Problems in 4-dimensional topology. *In* [C7].

[R1] RANICKI, A. A. Algebraic L-theory I. Foundations, *Proc. London Math. Soc.* **27** (1973), 101–125.

[R2] RANICKI, A. A. Algebraic L-theory II. Laurent extensions. *Proc. London Math. Soc.* **27** (1973), 126–158.

[R3] RANICKI, A. A. Algebraic L-theory III. Twisted Laurent extensions. *In* "Algebraic K-theory III. Hermitian K-theory and geometric applications. Proc. 1972 Seattle Battelle Inst. Conference." Lecture Notes in Mathematics **343**, Springer (1973), 412–463.

[R4] RANICKI, A. A. The algebraic theory of surgery I. Foundations. *Proc. Lond. Math. Soc.* **40** (1980), 87–192.

[R5] RANICKI, A. A. The algebraic theory of surgery II. Applications to topology. *Proc. Lond. Math. Soc.* **40** (1980), 193–283.

[R6] RANICKI, A. A. The L-theory of twisted quadratic extensions. *Canad. J. Math.* **39** (1987), 345–364.

[R7] RANICKI, A. A. "Exact sequences in the algebraic theory of surgery." Mathematical Notes **26**, Princeton, 1981.
Available from http://www.maths.ed.ac.uk/~aar

[R8] RANICKI, A. A. "Lower K- and L-theory." London Math. Soc. Lecture Notes **178**, Cambridge, 1992.

[R9] RANICKI, A. A. "Algebraic L-theory and topological manifolds." Tracts in Mathematics **102**, Cambridge, 1992.

[R10] RANICKI, A. A. On the Novikov conjecture. *In* "Proc. 1993 Oberwolfach Conference on the Novikov Conjectures, Rigidity and Index Theorems, Vol. 1." London Math. Soc. Lecture Notes **226**, 272–337, Cambridge, 1995.

[R11] RANICKI, A. A. (ed.) "The Hauptvermutung book." Collection of papers by Casson, Sullivan, Armstrong, Cooke, Rourke and Ranicki, K-Monographs in Mathematics **1**, Kluwer, 1996.

[R12] RANICKI, A. A. On the Hauptvermutung. *In* [R11], 3–31.

[R13] RANICKI, A. A. "High-dimensional knot theory." Mathematical Monograph, Springer, 1998.

[R14] DE RHAM, G. "Torsion et type simple d'homotopie." Lecture Notes in Mathematics **48**, Springer, 1967.

[R15] Rochlin, V. A. New examples of 4-dimensional manifolds. (In Russian). *Doklady Akad. Nauk S.S.S.R.* **162** (1965), 273–276.

[R16] Rourke, C. P. "The Hauptvermutung according to Sullivan, I, II." Lecture notes, Institute for Advanced Study, 1967. *In* "The Hauptvermutung book" [R11, A3].

[R17] Rourke, C. P. and Sanderson, B. J. Block bundles. *Bull. Amer. Math. Soc.* **72** (1966), 1036–1039. *Ann. of Math.* **87** (1968), I. 1–28, II. 256–278, III. 431–483.

[R18] Rourke, C. P. and Sanderson, B. J. An embedding without a normal microbundle. *Invent. Math.* **3** (1967), 293–299.

[R19] Rourke, C. P. and Sanderson, B. J. On topological neighbourhoods. *Compositio Math.* **22** (1970), 387–424.

[R20] Rourke, C. P. and Sullivan, D. P. On the Kervaire obstruction. *Ann. of Math.* **94** (1971), 397–413.

[S1] Scott, G. P. A note on the homotopy of PL_2. *Proc. Camb. Phil. Soc.* **69** (1971), 257–258.

[S2] Scott, W. R. "Group theory." Prentice Hall, 1964.

[S3] Serre, J.-P. Groupes d'homotopie et classes de groupes abéliens. *Ann. of Math.* **58** (1953), 258–294.

[S4] Shaneson, J. Wall's surgery obstruction groups for $\mathbf{Z} \times G$, for suitable groups G. *Bull. Amer. Math. Soc.* **74** (1968), 467–471.

[S5] Shaneson, J. Embeddings with codimension two of spheres in spheres and h-cobordisms of $S^1 \times S^3$. *Bull. Amer. Math. Soc.* **74** (1968), 972–974.

[S6] Shaneson, J. Wall's surgery obstruction groups for $G \times \mathbf{Z}$. *Ann. of Math.* **90** (1969), 296–334.

[S7] Sharpe, R. W. On the structure of the unitary Steinberg group. *Ann. of Math.* **96** (1972), 444–479.

[S8] Sharpe, R. W. Surgery of compact manifolds: the bounded even-dimensional case. *Ann. of Math.* **98** (1973), 187–209.

[S9] Shimura, G. Arithmetic of unitary groups. *Ann. of Math.* **79** (1964), 369–409.

[S10] Smale, S. Generalized Poincaré's conjecture in dimensions greater than 4. *Ann. of Math.* **74** (1961), 391–406.

[S11] Smale, S. Differentiable and combinatorial structures on manifolds. *Ann. of Math.* **74** (1961), 498–502.

[S12] Smale, S. On the structure of manifolds. *Amer. J. Math.* **84** (1962), 387–399.

[S13] Spanier, E. H. "Algebraic Topology." McGraw-Hill, 1966.

[S14] Spivak, M. Spaces satisfying Poincaré duality. *Topology* **6** (1967), 77–102.

[S15] Stallings, J. R. On fibering certain 3-manifolds. *In* "Topology of 3-manifolds and related topics." (Ed. M. K. Fort, jr.) Prentice Hall, 1962, 95–99.

[S16] Stallings, J. R. On infinite processes leading to differentiability in the complement of a point. *In* "Differential and Combinatorial Topology." (Ed. S. S. Cairns). Princeton, 1965, 245–254.

[S17] STALLINGS, J. R. "On polyhedral topology." Tata Institute, Bombay, 1968.

[S18] STARK, C. W. Surgery theory and infinite fundamental groups. *In* [C7].

[S19] STASHEFF, J. A classification theorem for fibre spaces. *Topology* **2** (1963), 239–246.

[S20] SULLIVAN, D. P. "Triangulating homotopy equivalences." Notes, Warwick University, 1966.

[S21] SULLIVAN, D. P. "Smoothing homotopy equivalences." Notes, Warwick University, 1966.

[S22] SULLIVAN, D. P. "Triangulating and smoothing homotopy equivalences and homeomorphisms." Geometric Topology seminar notes, Princeton University, 1967. *In* "The Hauptvermutung book" [R11], 69–104.

[S23] SULLIVAN, D. P. "Geometric Topology, Part I. Localization, Periodicity, and Galois Symmetry." Notes, M.I.T., 1970.

[S24] SULLIVAN, D. P. Geometric periodicity and the invariants of manifolds. *In* "Proc. 1970 Amsterdam Conference on Manifolds." Lecture Notes in Mathematics **197**, Springer, 1971, 44–75.

[S25] SWAN, R. G. Projective modules over group rings and maximal orders. *Ann. of Math.* **76** (1962), 55–61.

[T1] TAYLOR, L. and WILLIAMS, B. Surgery spaces: formulae and structure. *In* "Proc. 1978 Waterloo Algebraic Topology Conference." Lecture Notes in Mathematics **741**, Springer, 1979, 170–195.

[T2] THOM, R. Quelques propriétés globales des variétés différentiables. *Comm. Math. Helv.* **28** (1954), 17–86.

[T3] THOMAS, C. B. Frobenius reciprocity of hermitian forms. *J. Algebra* **18** (1971), 237–244.

[T4] THOMAS, C. B. and WALL, C. T. C. The topological spherical space-form problem I. *Compositio Math.* **23** (1971), 101–114.

[V1] VASERSTEIN, L. Stability of the unitary and orthogonal groups over rings with involution. *Math. Sb.* **81** (1970), 328–351.

[W1] WAGONER, J. B. Smooth and piecewise linear surgery. *Bull. Amer. Math. Soc.* **73** (1967), 72–77.

[W2] WAGONER, J. B. Producing PL homeomorphisms by surgery. *Bull. Amer. Math. Soc.* **73** (1967), 78–83.

[W3] WALDHAUSEN, F. Algebraic K-theory of generalized free products. *Ann. of Math.* **108** (1978), 135–256.

[W4] WALL, C. T. C. Determination of the cobordism ring. *Ann. of Math.* **72** (1960), 292–311.

[W5] WALL, C. T. C. Cobordism of pairs. *Comm. Math. Helv.* **35** (1961), 136–145.

[W6] WALL, C. T. C. Killing the middle homotopy groups of odd dimensional manifolds. *Trans. Amer. Math. Soc.* **103** (1962), 421–433.

[W7] WALL, C. T. C. Classification problems in differential topology. I. Classification of handlebodies. *Topology* **2** (1963), 253–261.

[W8] WALL, C. T. C. Classification problems in differential topology. II. Diffeomorphisms of handlebodies. *Topology* **2** (1963), 263–272.

[W9] WALL, C. T. C. Quadratic forms on finite groups and related topics. *Topology* **2** (1963), 281–298.
[W10] WALL, C. T. C. Diffeomorphisms of 4-manifolds. *J. London Math. Soc.* **39** (1964), 131–140.
[W11] WALL, C. T. C. On simply connected 4-manifolds. *J. London Math. Soc.* **39** (1964), 141–149.
[W12] WALL, C. T. C. Unknotting tori in codimension one and spheres in codimension two. *Proc. Camb. Phil. Soc.* **61** (1965), 659–664.
[W13] WALL, C. T. C. "Differential Topology." Notes, Liverpool University, 1965, IV Theory of handle decompositions, VA Cobordism: Geometric Theory.
[W14] WALL, C. T. C. Finiteness conditions for *CW* complexes. I. *Ann. of Math.* **81** (1965), 56–69.
[W15] WALL, C. T. C. Finiteness conditions for *CW* complexes. II. *Proc. Roy. Soc. A* **295** (1966), 129–139.
[W16] WALL, C. T. C. On the exactness of interlocking sequences. *l'Enseignement Math.* **12** (1966), 95–100.
[W17] WALL, C. T. C. An extension of results of Novikov and Browder. *Amer. J. Math.* **88** (1966), 20–32.
[W18] WALL, C. T. C. Surgery of non-simply connected manifolds. *Ann. of Math.* **84** (1966), 217–276.
[W19] WALL, C. T. C. Locally flat *PL* submanifolds with codimension two. *Proc. Camb. Phil. Soc.* **63** (1967), 5–8.
[W20] WALL, C. T. C. On bundles over a sphere with fibre Euclidean space. *Fund. Math.* **61** (1967), 57–72.
[W21] WALL, C. T. C. Poincaré complexes I. *Ann. of Math.* **86** (1967), 213–245.
[W22] WALL, C. T. C. Graded algebras, anti-involutions, simple groups and symmetric spaces. *Bull. Amer. Math. Soc.* **74** (1968), 198–202.
[W23] WALL, C. T. C. Free piecewise linear involutions on spheres. *Bull. Amer. Math. Soc.* **74** (1968), 554–558.
[W24] WALL, C. T. C. Non-additivity of the signature. *Invent. Math.* **7** (1969), 269–274.
[W25] WALL, C. T. C. On homotopy tori and the annulus theorem. *Bull. London Math. Soc.* **1** (1969), 95–97.
[W26] WALL, C. T. C. The topological space-form problem. *In* "Topology of Manifolds. Proc. 1969 Georgia Conference." Markham Press, 1970, 319–351.
[W27] WALL, C. T. C. On the axiomatic foundations of the theory of hermitian forms. *Proc. Camb. Phil. Soc.* **67** (1970), 243–250.
[W28] WALL, C. T. C. On the classification of hermitian forms. I. Rings of algebraic integers. *Compositio Math.* **22** (1970), 425–451.
[W29] WALL, C. T. C. On the classification of hermitian forms. II. Semisimple rings. *Invent. Math.* **18** (1972) 119–141.
[W30] WALL, C. T. C. On the classification of hermitian forms. III. Complete semilocal rings. *Invent. Math.* **19** (1973), 59–71.

[W31] WALL, C. T. C. On the classification of hermitian forms. IV. Adèle rings. *Invent. Math.* **23** (1974), 241–260.

[W32] WALL, C. T. C. On the classification of hermitian forms. V. Global rings. *Invent. Math.* **23** (1974), 261–288.

[W33] WALL, C. T. C. On the classification of hermitian forms. VI. Group rings. *Ann. of Math.* **103** (1976), 1–80.

[W34] WALL, C. T. C. Formulae for surgery obstructions. *Topology* **15** (1976), 189–210; corrigendum *ibid.* **16** (1977), 495–496.

[W35] WALL, C. T. C. Free actions of finite groups on spheres. *In* "Proc. Symp. in Pure Math. 32, Vol. 1, (Algebraic and geometric topology)." Amer. Math. Soc., 1978, 115–124.

[W36] WALLACE, A. H. Modifications and cobounding manifolds I. *Canad. J. Math.* **12** (1960), 503–528.

[W37] WALLACE, A. H. Modifications and cobounding manifolds II. *J. Math. Mech.* **10** (1961), 773–809.

[W38] WALLACE, A. H. Modifications and cobounding manifolds III. *J. Math. Mech.* **11** (1961), 971–990.

[W39] WANG, H.-C. Discrete subgroups of solvable Lie groups I. *Ann. of Math.* **64** (1956), 1–19.

[W40] WEIL, A. Algebras with involutions and the classical groups. *J. Indian Math. Soc.* **24** (1961), 589–623.

[W41] WEINBERGER, S. "The topological classification of stratified spaces." Chicago, 1994.

[W42] WEISS, M. Surgery and the generalized Kervaire invariant. *Proc. London Math. Soc.* **51** (1985), I. 146–192, II. 193–230.

[W43] WEISS, M. and WILLIAMS, B. Automorphisms of manifolds and K-theory II. *J. Pure Appl. Algebra* **62** (1989), 47–107.

[W44] WHITEHEAD, J. H. C. Combinatorial homotopy I. *Bull. Amer. Math. Soc.* **55** (1949), 213–245.

[W45] WILLIAMSON, R. E. Cobordism of combinatorial manifolds. *Ann. of Math.* **83** (1966), 1–33.

[W46] WILLIAMSON, R. E. Surgery in $M \times N$ with $\pi_1 M \neq 1$. *Bull. Amer. Math. Soc.* **75** (1969), 582–585.

[W47] WOLF, J. A. "Spaces of constant curvature." McGraw-Hill, 1967.

[W48] WOLF, J. A. Growth of finitely generated solvable groups and curvature of Riemannian manifolds. *J. Diff. Geom.* **2** (1968), 421–446.

[W49] WU, W.-T. Les i-carrés dans une variété grassmannienne. *C.R. Acad. Sci. Paris*, **230** (1950), 918.

[Y1] YANG, C. T. On involutions of the five sphere. *Topology* **5** (1966), 17–20.

[Z1] ZEEMAN, E. C. Unknotting combinatorial balls. *Ann. of Math.* **78** (1963), 501–526.

Index

Notations **Z**, **Q**, **R**, **C**, **H** denote (as usual) the rings of integers, rational, real or complex numbers, and of quaternions. $\mathbf{R}^n$ is real number space, with its usual Euclidean structure.

$$\begin{aligned} D^n &= \{x \in \mathbf{R}^n : \|x\| \leqslant 1\}, \\ S^{n-1} &= \{x \in \mathbf{R}^n : \|x\| = 1\}, \\ D_+^{n-1} &= \{x \in S^{n-1} : x_n \geqslant 0\}, \\ D_-^{n-1} &= \{x \in S^{n-1} : x_n \leqslant 0\}. \end{aligned}$$

We use standard notation for manifolds. τ_M denotes the tangent bundle of M; ν is usually a normal bundle. If ν is a bundle over X, X^ν denotes its Thom space. Our notation for Lie groups (e.g. O, Spin) and their classifying spaces (e.g. BO) is also standard.

Simple homotopy theory

The terms simple homotopy equivalence, simple homotopy type, stably free, s-base, simple equivalence, simple isomorphism, based short exact sequence, $Wh(\pi)$, Whitehead torsion appear first in this book on pp. 22–27; readers unfamiliar with them are referred to Milnor's survey article [M14] for a clear and readable account.

Selected Titles in This Series

(Continued from the front of this publication)

40.2 **Daniel Gorenstein, Richard Lyons, and Ronald Solomon,** The classification of the finite simple groups, number 2, 1995

40.1 **Daniel Gorenstein, Richard Lyons, and Ronald Solomon,** The classification of the finite simple groups, number 1, 1994

39 **Sigurdur Helgason,** Geometric analysis on symmetric spaces, 1994

38 **Guy David and Stephen Semmes,** Analysis of and on uniformly rectifiable sets, 1993

37 **Leonard Lewin, Editor,** Structural properties of polylogarithms, 1991

36 **John B. Conway,** The theory of subnormal operators, 1991

35 **Shreeram S. Abhyankar,** Algebraic geometry for scientists and engineers, 1990

34 **Victor Isakov,** Inverse source problems, 1990

33 **Vladimir G. Berkovich,** Spectral theory and analytic geometry over non-Archimedean fields, 1990

32 **Howard Jacobowitz,** An introduction to CR structures, 1990

31 **Paul J. Sally, Jr. and David A. Vogan, Jr., Editors,** Representation theory and harmonic analysis on semisimple Lie groups, 1989

30 **Thomas W. Cusick and Mary E. Flahive,** The Markoff and Lagrange spectra, 1989

29 **Alan L. T. Paterson,** Amenability, 1988

28 **Richard Beals, Percy Deift, and Carlos Tomei,** Direct and inverse scattering on the line, 1988

27 **Nathan J. Fine,** Basic hypergeometric series and applications, 1988

26 **Hari Bercovici,** Operator theory and arithmetic in H^∞, 1988

25 **Jack K. Hale,** Asymptotic behavior of dissipative systems, 1988

24 **Lance W. Small, Editor,** Noetherian rings and their applications, 1987

23 **E. H. Rothe,** Introduction to various aspects of degree theory in Banach spaces, 1986

22 **Michael E. Taylor,** Noncommutative harmonic analysis, 1986

21 **Albert Baernstein, David Drasin, Peter Duren, and Albert Marden, Editors,** The Bieberbach conjecture: Proceedings of the symposium on the occasion of the proof, 1986

20 **Kenneth R. Goodearl,** Partially ordered abelian groups with interpolation, 1986

19 **Gregory V. Chudnovsky,** Contributions to the theory of transcendental numbers, 1984

18 **Frank B. Knight,** Essentials of Brownian motion and diffusion, 1981

17 **Le Baron O. Ferguson,** Approximation by polynomials with integral coefficients, 1980

16 **O. Timothy O'Meara,** Symplectic groups, 1978

15 **J. Diestel and J. J. Uhl, Jr.,** Vector measures, 1977

14 **V. Guillemin and S. Sternberg,** Geometric asymptotics, 1977

13 **C. Pearcy, Editor,** Topics in operator theory, 1974

12 **J. R. Isbell,** Uniform spaces, 1964

11 **J. Cronin,** Fixed points and topological degree in nonlinear analysis, 1964

10 **R. Ayoub,** An introduction to the analytic theory of numbers, 1963

9 **Arthur Sard,** Linear approximation, 1963

8 **J. Lehner,** Discontinuous groups and automorphic functions, 1964

7.2 **A. H. Clifford and G. B. Preston,** The algebraic theory of semigroups, Volume II, 1961

7.1 **A. H. Clifford and G. B. Preston,** The algebraic theory of semigroups, Volume I, 1961

6 **C. C. Chevalley,** Introduction to the theory of algebraic functions of one variable, 1951

5 **S. Bergman,** The kernel function and conformal mapping, 1950

4 **O. F. G. Schilling,** The theory of valuations, 1950

3 **M. Marden,** Geometry of polynomials, 1949

(See the AMS catalog for earlier titles)